Two Manuals Combined into One Book

Military First Aid

and

Medical Platoon Leader's Handbook

Department of Defense

Pentagon
Publishing

*FIELD MANUAL
NO. 4-25.11
NAVY TACTICAL
REFERENCE
PUBLICATION
NO. 4-02.1
AIR FORCE MANUAL
NO. 44-163(I)

HEADQUARTERS
DEPARTMENT OF THE ARMY,
THE NAVY, AND THE AIR FORCE
Washington, DC, 23 December 2002

FIRST AID

TABLE OF CONTENTS

Page

*This publication supersedes FM 21-11, 27 October 1988

PREFACE

This manual meets the first aid training needs of individual service members. Because medical personnel will not always be readily available, the nonmedical service members must rely heavily on their own skills and knowledge of life-sustaining methods to survive on the integrated battlefield. This publication outlines both self-aid and aid to other service members (buddy aid). More importantly, it emphasizes prompt and effective action in sustaining life and preventing or minimizing further suffering and disability. First aid is the emergency care given to the sick, injured, or wounded before being treated by medical personnel. The term *first aid* can be defined as "urgent and immediate lifesaving and other measures, which can be performed for casualties by nonmedical personnel when medical personnel are not immediately available." Nonmedical service members have received basic first aid training and should remain skilled in the correct procedures for giving first aid. This manual is directed to *all* service members. The procedures discussed apply to all types of casualties and the measures described are for use by both male and female service members.

This publication is in consonance with the following North Atlantic Treaty Organization (NATO) International Standardization Agreements (STANAGs) and American, British. Canadian, and Australian Quadripartite Standardization Agreements (QSTAGs).

TITLE	STANAG	QSTAG
Medical Training in First Aid, Basic Hygiene and Emergency Care	2122	535
First Aid Kits and Emergency Medical Care Kits	2126	
Medical First Aid and Hygiene Training in NBC Operations	2358	
First Aid Material for Chemical Injuries	2871	

These agreements are available on request, using Department of Defense (DD) Form 1425 from the Standardization Documents Order Desk, 700 Robins Avenue, Building 4, Section D, Philadelphia, Pennsylvania 19111-5094.

Unless this publication states otherwise, masculine nouns and pronouns do not refer exclusively to men.

Use of trade or brand names in this publication is for illustrative purposes only and does not imply endorsement by the Department of Defense (DOD).

The proponent for this publication is the US Army Medical Department Center and School. Submit comments and recommendations for the improvement of this publication directly to the **Commander, US Army Medical Department Center and School, ATTN: MCCS-FCD-L, 1400 East Grayson Street, Fort Sam Houston, Texas 78234-5052**.

CHAPTER 1

FUNDAMENTAL CRITERIA FOR FIRST AID

"The fate of the wounded rests in the hands of the ones who apply the first dressing."

Nicholas Senn (1898) (49th President of the American Medical Association)

1-1. General

When a nonmedical service member comes upon an unconscious or injured service member, he must accurately evaluate the casualty to determine the first aid measures needed to prevent further injury or death. He should seek medical assistance as soon as possible, but he should not interrupt the performance of first aid measures. To interrupt the first aid measures may cause more harm than good to the casualty. Remember that in a chemical environment, the service member should not evaluate the casualty until the casualty has been masked. After performing first aid, the service member must proceed with the evaluation and continue to monitor the casualty for development of conditions which may require the performance of necessary basic lifesaving measures, such as clearing the airway, rescue breathing, preventing shock, and controlling bleeding. He should continue to monitor the casualty until relieved by medical personnel.

Service members may have to depend upon their first aid knowledge and skills to save themselves (self-aid) or other service members (buddy aid/ combat lifesaver). They may be able to save a life, prevent permanent disability, or reduce long periods of hospitalization by knowing **WHAT** to do, **WHAT NOT** to do, and **WHEN** to seek medical assistance.

NOTE

The prevalence of various body armor systems currently fielded to US service members, and those in development for future fielding, may present a temporary obstacle to effective evaluation of an injured service member. You may have to *carefully remove* the body armor from the injured service member to complete the evaluation or administer first aid. Begin by removing the outer-most hard or soft body armor components (open, unfasten or cut the closures, fasteners, or straps), then remove any successive layers in the same manner. Be sure to follow other notes, cautions and warnings regarding procedures in contaminated situations and when a broken back or neck is suspected. Continue to evaluate.

1-2. Terminology

To enhance the understanding of the material contained in this publication, the following terms are used—

- *Combat lifesaver.* This is a US Army program governed by Army Regulation (AR) 350-41. The combat lifesaver is a member of a nonmedical unit selected by the unit commander for additional training beyond basic first aid procedures (referred to as *enhanced first aid*). A minimum of one individual per squad, crew, team, or equivalent-sized unit should be trained. The primary duty of this individual does not change. The additional duty of combat lifesaver is to provide enhanced first aid for injuries based on his training before the trauma specialist (military occupational specialty [MOS] 91W) arrives. The combat lifesaver's training is normally provided by medical personnel assigned, attached, or in direct support (DS) of the unit. The senior medical person designated by the commander manages the training program.

- *Trauma Specialist (US Army) or Hospital Corpsman (HM).* A medical specialist trained in emergency medical treatment (EMT) procedures and assigned or attached in support of a combat or combat support unit or marine forces.

- *Casualty evacuation.* Casualty evacuation (CASEVAC) is a term used by nonmedical units to refer to the movement of casualties aboard nonmedical vehicles or aircraft. See also the term *transported* below. Refer to FM 8-10-6 for additional information.

> **CAUTION**
>
> Casualties transported in this manner do not receive en route medical care.

- *Enhanced first aid (US Army).* Enhanced first aid is administered by the combat lifesaver. It includes measures, which require an additional level of training above self-aid and buddy aid, such as the initiation of intravenous (IV) fluids.

- *Medical* evacuation*.* Medical evacuation is the timely, efficient movement of the wounded, injured, or ill service members from the battlefield and other locations to medical treatment facilities (MTFs). Medical personnel provide en route medical care during the evacuation. Once the casualty has entered the medical stream (trauma specialist, hospital corpsman, evacuation crew, or MTF), the role of first aid in the care of the casualty ceases and the casualty becomes the responsibility of the health service support (HSS) chain. Once he has entered the HSS chain he is referred to as a *patient.*

- *First aid measures.* Urgent and immediate lifesaving and other measures, which can be performed for casualties (or performed by the casualty himself) by nonmedical personnel when medical personnel are not immediately available.

- *Medical treatment.* Medical treatment is the care and management of wounded, injured, or ill service members by medically trained (MOS-trained) HM, and area of concentration (AOC) personnel. It may include EMT, advanced trauma management (ATM), and resuscitative and surgical intervention.

- *Medical treatment facility.* Any facility established for the purpose of providing medical treatment. This includes battalion aid stations, Level II facilities, dispensaries, clinics, and hospitals.

- *Self-aid/buddy aid.* Each individual service member is trained to be proficient in a variety of specific first aid procedures. This training enables the service member or a buddy to apply immediate first aid measures to alleviate a life-threatening situation.

- *Transported.* A casualty is moved to an MTF in a nonmedical vehicle without en route care provided by a medically-trained service member (such as a Trauma Specialist or HM). First aid measures should be continually performed while the casualty is being transported. If the casualty is acquired by a dedicated medical vehicle with a medically-trained crew, the role of first aid ceases and the casualty becomes the responsibility of the HSS chain, and is then referred to as a *patient.* This method of transporting a casualty is also referred to as *CASEVAC.*

1-3. Understanding Vital Body Functions for First Aid

In order for the service member to learn to perform first aid procedures, he must have a basic understanding of what the vital body functions are and what the result will be if they are damaged or not functioning.

a. Breathing Process. All humans must have oxygen to live. Through the breathing process, the lungs draw oxygen from the air and put it into the blood. The heart pumps the blood through the body to be used by the cells that require a constant supply of oxygen. Some cells are more dependent on a constant supply of oxygen than others. For example, cells of the brain may die within 4 to 6 minutes without oxygen. Once these cells die, they are lost forever since they do not regenerate. This could result in permanent brain damage, paralysis, or death.

b. *Respiration.* Respiration occurs when a person inhales (oxygen is taken into the body) and then exhales (carbon dioxide [CO_2] is expelled from the body). Respiration involves the—

- *Airway*. The airway consists of the nose, mouth, throat, voice box, and windpipe. It is the canal through which air passes to and from the lungs.

- *Lungs*. The lungs are two elastic organs made up of thousands of tiny air spaces and covered by an airtight membrane. The *bronchial tree* is a part of the lungs.

- *Rib cage*. The rib cage is formed by the muscle-connected ribs, which join the spine in back, and the breastbone in front. The top part of the rib cage is closed by the structure of the neck, and the bottom part is separated from the abdominal cavity by a large dome-shaped muscle called the *diaphragm* (Figure 1-1). The diaphragm and rib muscles, which are under the control of the respiratory center in the brain, automatically *contract* and *relax*. *Contraction* increases and *relaxation* decreases the size of the rib cage. When the rib cage increases and then decreases, the air pressure in the lungs is first less and then more than the atmospheric pressure, thus causing the air to rush into and out of the lungs to equalize the pressure. This cycle of inhaling and exhaling is repeated about 12 to 18 times per minute.

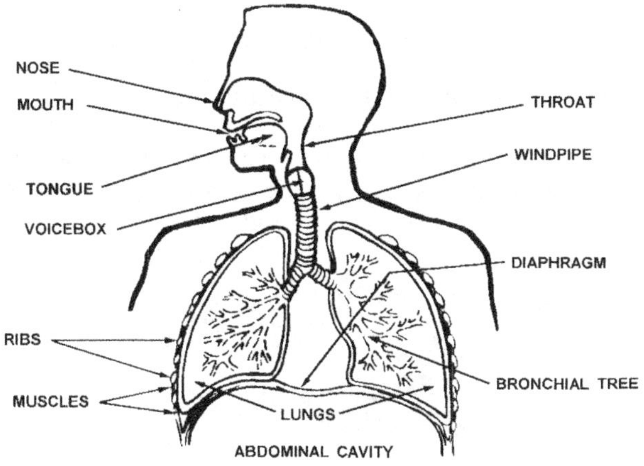

Figure 1-1. Airway, lungs, and rib cage.

c. *Blood Circulation.* The heart and the blood vessels (arteries, veins, and capillaries) circulate blood through the body tissues. The heart is divided into two separate halves, each acting as a pump. The left side pumps oxygenated blood (bright red) through the arteries into the capillaries; nutrients and oxygen pass from the blood through the walls of the capillaries into the cells. At the same time waste products and CO_2 enter the capillaries. From the capillaries the oxygen poor blood is carried through the veins to the right side of the heart and then into the lungs where it expels the CO_2 and picks up oxygen. Blood in the veins is dark red because of its low oxygen content. Blood does not flow through the veins in spurts as it does through the arteries. The entire system of the heart, blood vessels, and lymphatics is called the *circulatory system.*

(1) *Heartbeat.* The heart functions as a pump to circulate the blood continuously through the blood vessels to all parts of the body. It contracts, forcing the blood from its chambers; then it relaxes, permitting its chambers to refill with blood. The rhythmical cycle of contraction and relaxation is called the *heartbeat.* The normal heartbeat is from 60 to 80 beats per minute.

(2) *Pulse.* The heartbeat causes a rhythmical expansion and contraction of the arteries as it forces blood through them. This cycle of expansion and contraction can be felt (monitored) at various points in the body and is called the *pulse.* The common points for checking the pulse are at the—

- Side of the neck (*carotid*).
- Groin (*femoral*).
- Wrist (*radial*).
- Ankle (*posterior tibial*).

(*a*) *Carotid pulse.* To check the carotid pulse, feel for a pulse on the side of the casualty's neck closest to you. This is done by placing the tips of your first two fingers beside his Adam's apple (Figure 1-2).

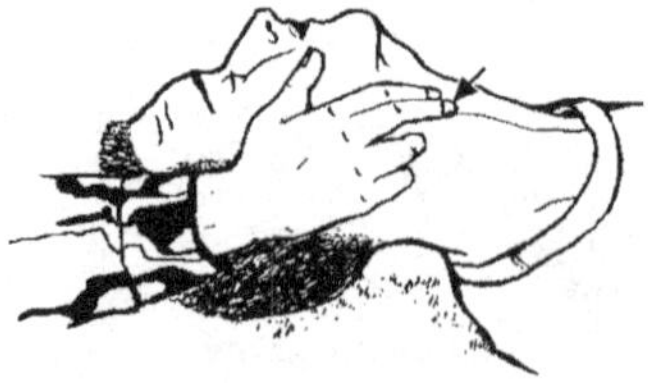

Figure 1-2. Carotid pulse.

(*b*) *Femoral pulse*. To check the femoral pulse, press the tips of your first two fingers into the middle of the groin (Figure 1-3).

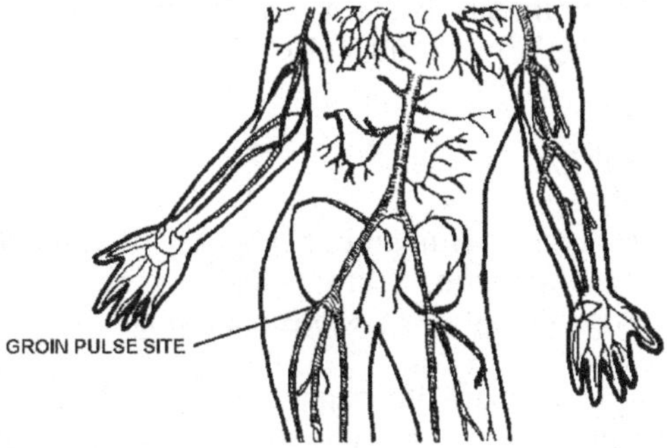

Figure 1-3. Femoral pulse.

(*c*) *Radial pulse*. To check the radial pulse, place your first two fingers on the thumb side of the casualty's wrist (Figure 1-4).

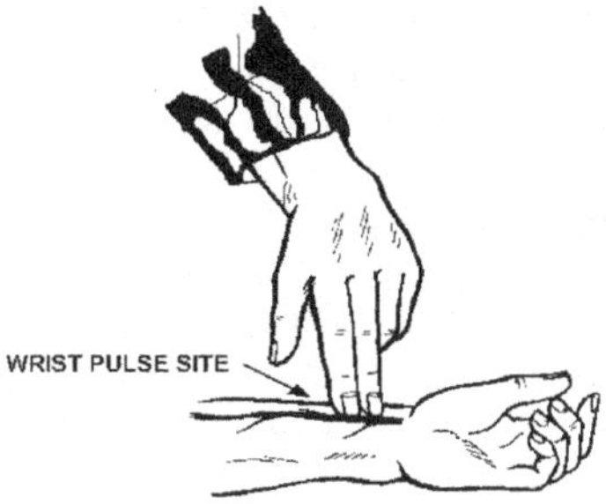

Figure 1-4. Radial pulse.

(*d*) *Posterior tibial pulse*. To check the posterior tibial pulse, place your first two fingers on the inside of the ankle (Figure 1-5).

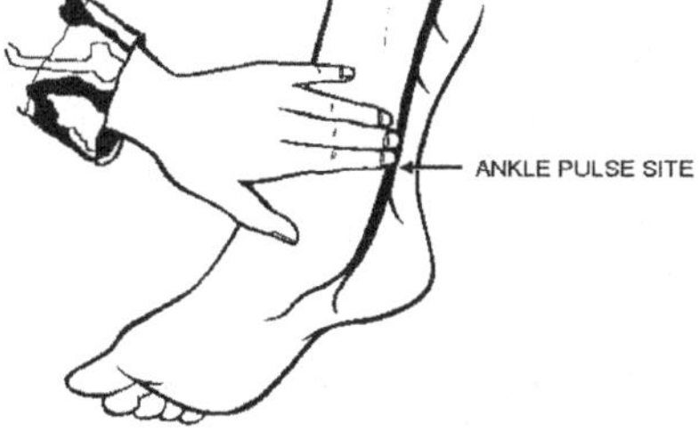

Figure 1-5. Posterior tibial pulse.

NOTE

DO NOT use your thumb to check a casualty's pulse because you may confuse the beat of your pulse with that of the casualty.

1-4. Adverse Conditions

a. Lack of Oxygen. Human life cannot exist without a continuous intake of oxygen. Lack of oxygen rapidly leads to death. First aid involves knowing how to open the airway and restore breathing.

b. Bleeding. Human life cannot continue without an adequate volume of blood circulating through the body to carry oxygen to the tissues. An important first aid measure is to stop the bleeding to prevent the loss of blood.

c. Shock. Shock means there is an inadequate blood flow to the vital tissues and organs. Shock that remains uncorrected may result in death even though the injury or condition causing the shock would not otherwise be fatal. Shock can result from many causes, such as loss of blood, loss of fluid from deep burns, pain, and reaction to the sight of a wound or blood. First aid includes preventing shock, since the casualty's chances of survival are much greater if he does not develop shock. Refer to paragraphs 2-21 through 2-24 for a further discussion of shock.

d. Infection. Recovery from a severe injury or a wound depends largely upon how well the injury or wound was initially protected. Infections result from the multiplication and growth (spread) of harmful microscopic organisms (sometimes referred to as germs). These harmful microscopic organisms are in the air, water, and soil, and on the skin and clothing. Some of these organisms will immediately invade (contaminate) a break in the skin or an open wound. The objective is to keep wounds clean and free of these organisms. A good working knowledge of basic first aid measures also includes knowing how to dress a wound to avoid infection or additional contamination.

1-5. Basics of First Aid

Most injured or ill service members are able to return to their units to fight or support primarily because they are given appropriate and timely first aid followed by the best medical care possible. Therefore, all service members must remember the basics.

- Check for **BREATHING**: Lack of oxygen intake (through a compromised airway or inadequate breathing) can lead to brain damage or death in very few minutes.

- Check for **BLEEDING**: Life cannot continue without an adequate volume of blood to carry oxygen to tissues.

- Check for **SHOCK**: Unless shock is prevented, first aid performed, and medical treatment provided, death may result even though the injury would not otherwise be fatal.

1-6. Evaluating a Casualty

a. The time may come when you must instantly apply your knowledge of first aid measures. This could occur during combat operations, in training situations, or while in a nonduty status. Any service member observing an unconscious and/or ill, injured, or wounded person must carefully and skillfully evaluate him to determine the first aid measures required to prevent further injury or death. He should seek help from medical personnel as soon as possible, but must not interrupt his evaluation of the casualty or fail to administer first aid measures. A second service member may be sent to find medical help. One of the cardinal principles for assisting a casualty is that you (the initial rescuer) must continue the evaluation and first aid measures, as the tactical situation permits, until another individual relieves you. If, during any part of the evaluation, the casualty exhibits the conditions (such as shock) for which the service member is checking, the service member must stop the evaluation and immediately administer first aid. In a chemical environment, the service member should not evaluate the casualty until both the individual and the casualty have been masked. If it is suspected that a nerve agent was used, administer the casualty's own nerve agent antidote autoinjector. After providing first aid, the service member must proceed with the evaluation and continue to monitor the casualty for further complications until relieved by medical personnel.

WARNING

Do not use your own nerve agent antidote autoinjector on the casualty.

NOTE

Remember, when evaluating and/or administering first aid to a casualty, you should seek medical aid as soon as possible. **DO NOT** stop first aid measures, but if the situation allows, send another service member to find medical aid.

b. To evaluate a casualty, perform the following steps:

(1) *Check the casualty for responsiveness.* This is done by gently shaking or tapping him while calmly asking, "Are you OK?" Watch for a response. If the casualty does not respond, go to step (2). If the casualty responds, continue with the evaluation.

(*a*) If the casualty is conscious, ask him where he feels different than usual or where it hurts. Ask him to identify the location of pain if he can, or to identify the area in which there is no feeling.

(*b*) If the casualty is conscious but is choking and cannot talk, stop the evaluation and begin first aid measures. Refer to paragraphs 2-10 and 2-11 for specific information on opening the airway.

WARNING

If a broken back or neck is suspected, do not move the casualty unless his life is in immediate danger (such as close to a burning vehicle). Movement may cause permanent paralysis or death.

(2) *Check for breathing.* (Refer to paragraph 2-6 for this procedure.)

(*a*) If the casualty is breathing, proceed to step (3).

(*b*) If the casualty is not breathing, stop the evaluation and begin first aid measures to attempt to ventilate the casualty. Attempt to open the airway, if an airway obstruction is apparent, clear the airway obstruction, then ventilate (see paragraphs 2-10 and 2-11).

(*c*) After successfully ventilating the casualty, proceed to step (3).

(3) *Check for pulse.* (Refer to paragraph 1-3*c*(2) for specific methods.) If a pulse is present and the casualty is breathing, proceed to step (4).

(*a*) If a pulse is present, but the casualty is still not breathing, start rescue breathing.

(*b*) If a pulse is not present, seek medical personnel for help.

(4) *Check for bleeding.* Look for spurts of blood or blood-soaked clothes. Also check for *both* entry and exit wounds. If the casualty is bleeding from an open wound, stop the evaluation and begin first aid procedures as follows for a—

(*a*) Wound of the arm or leg (refer to paragraphs 2-16 through 2-18 for information on putting on a field or pressure dressing).

(*b*) Partial or complete amputation, apply dressing (refer to paragraph 2-16 to 2-18) and then apply tourniquet if bleeding is not stopped (refer to paragraph 2-20 for information on putting on a tourniquet).

(*c*) Open head wound (refer to paragraph 3-10 for information on applying a dressing to an open head wound).

(*d*) Open chest wound (refer to paragraph 3-5 for information on applying a dressing to an open chest wound).

(*e*) Open abdominal wound (refer to paragraph 3-7 for information on applying a dressing to an open abdominal wound).

WARNING

In a chemically contaminated area, do not expose the wounds. Apply field dressing and then pressure dressing over wound area as needed.

(5) *Check for shock.* (Refer to paragraph 2-24 for first aid measures for shock.) If the signs and symptoms of shock are present, stop the evaluation, and begin first aid measures immediately. The following are the nine signs and symptoms of shock.

(*a*) Sweaty but cool skin (clammy skin).

(*b*) Paleness of skin. (In dark-skinned service members look for a grayish cast to the skin.)

(*c*) Restlessness or nervousness.

(*d*) Thirst.

(*e*) Loss of blood (bleeding).

(*f*) Confusion (does not seem aware of surroundings).

(*g*) Faster than normal breathing rate.

(*h*) Blotchy or bluish skin, especially around the mouth.

(*i*) Nausea or vomiting.

WARNING

Leg fractures must be splinted before elevating the legs as a first aid measure for shock.

(6) *Check for fractures.*

(*a*) Check for the following signs and symptoms of a back or neck injury and perform first aid procedures as necessary.

- Pain or tenderness of the back or neck area.
- Cuts or bruises on the back or neck area.
- Inability of a casualty to move or decreased sensation to extremities (paralysis or numbness).
  - Ask about ability to move (paralysis).
  - Touch the casualty's arms and legs and ask whether he can feel your hand (numbness).
- Unusual body or limb position.

(*b*) Immobilize any casualty suspected of having a back or neck injury by doing the following:

- Tell the casualty not to move.
- If a back injury is suspected, place padding (rolled or folded to conform to the shape of the arch) under the natural arch of the casualty's back. (For example, a blanket/poncho may be used as padding.)

> **WARNING**
>
> **Do not move casualty to place padding.**

- If a neck injury is suspected, immediately immobilize (manually) the head and neck. Place a roll of cloth under the casualty's neck, and put weighted boots (filled with dirt or sand) or rocks on both sides of his head.

(*c*) Check the casualty's arms and legs for open or closed fractures.

- Check for *open* fractures by looking for—
  - Bleeding.
  - Bones sticking through the skin.
  - Check for pulse.
- Check for *closed* fractures by looking for—
  - Swelling.
  - Discoloration.
  - Deformity.
  - Unusual body position.
  - Check for pulse.

(*d*) Stop the evaluation and begin first aid measures if a fracture to an arm or leg is suspected. Refer to Chapter 4 for information on splinting a suspected fracture.

(*e*) Check for signs/symptoms of fractures of other body areas (for example, shoulder or hip) and provide first aid as necessary.

(7) *Check for burns.* Look carefully for reddened, blistered, or charred skin; also check for singed clothing. If burns are found, stop the evaluation and begin first aid procedures. Refer to paragraph 3-9 for information on giving first aid for burns.

NOTE

Burns to the upper torso and face may cause respiratory complications. When evaluating the casualty, look for singed nose hair, soot around the nostrils, and listen for abnormal breath sounds or difficulty breathing.

(8) *Check for possible head injury.*

(*a*) Look for the following signs and symptoms:

- Unequal pupils.
- Fluid from the ear(s), nose, mouth, or injury site.
- Slurred speech.
- Confusion.
- Sleepiness.
- Loss of memory or consciousness.
- Staggering in walking.
- Headache.
- Dizziness.
- Nausea or vomiting.
- Paralysis.
- Convulsions or twitches.
- Bruising around the eyes and behind the ears.

(*b*) If a head injury is suspected, continue to watch for signs which would require performance of rescue breathing, first aid measures for shock, or control of bleeding; seek medical aid. Refer to paragraph 3-10 for information on first aid measures for head injuries.

CHAPTER 2

BASIC MEASURES FOR FIRST AID

2-1. General

Several conditions that require immediate attention are an inadequate airway, lack of breathing, and excessive loss of blood (circulation). A casualty without a clear airway or who is not breathing may die from lack of oxygen. Excessive loss of blood may lead to shock, and shock can lead to death; therefore, you must act immediately to control the loss of blood. All wounds are considered to be contaminated, since infection-producing organisms (germs) are always present on the skin and clothing, and in the soil, water, and air. Any missile or instrument (such as a bullet, shrapnel, knife, or bayonet) causing a wound pushes or carries the germs into that wound. Infection results as these organisms multiply. That a wound is contaminated does not lessen the importance of protecting it from further contamination. You must dress and bandage a wound as soon as possible to prevent further contamination.

NOTE

It is also important that you attend to any airway, breathing, or bleeding problems **IMMEDIATELY** because these problems, if left unattended, may become life threatening.

Section I. OPEN THE AIRWAY AND RESTORE BREATHING

2-2. Breathing Process

All humans must have oxygen to live. Through the breathing process, the lungs draw oxygen from the air and put it into the blood. The heart pumps the blood through the body to be used by the cells that require a constant supply of oxygen. Some cells are more dependent on a constant supply of oxygen than others. For example, cells of the brain may die within 4 to 6 minutes without oxygen. Once these cells die, they are lost forever since they do not regenerate. This could result in permanent brain damage, paralysis, or death.

2-3. Assessment of and Positioning the Casualty

a. **CHECK** for responsiveness (Figure 2-1A)—establish whether the casualty is conscious by gently shaking him and asking, "Are you OK?"

b. **CALL** for help (Figure 2-1B).

c. **POSITION** the unconscious casualty so that he is lying on his back and on a firm surface (Figure 2-1C).

> **WARNING**
>
> **If the casualty is lying on his chest (prone position), cautiously roll the casualty as a unit so that his body does not twist (which may further complicate a back, neck, or spinal injury).**

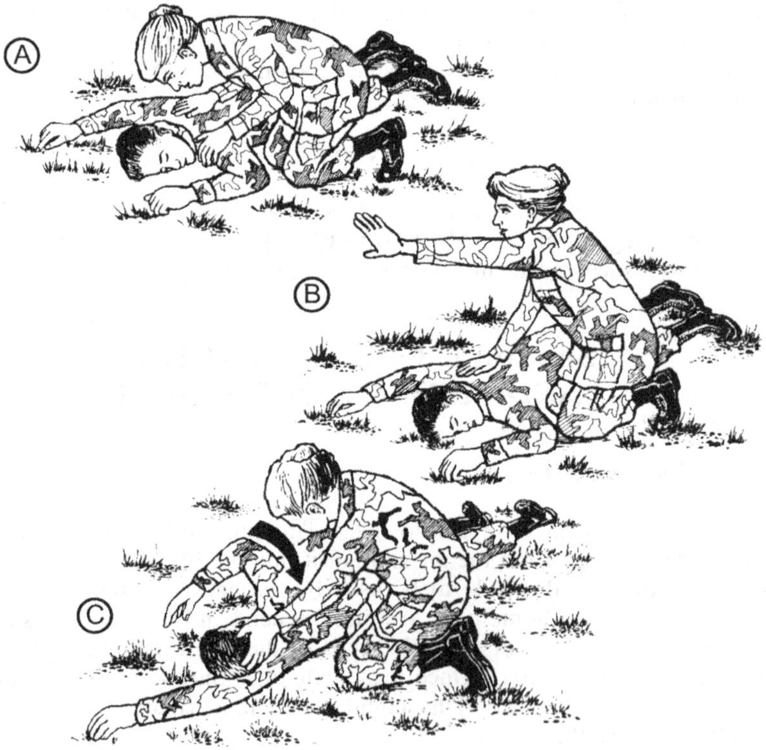

Figure 2-1. Assessment (Illustrated A—C).

(1) Straighten the casualty's legs. Take the casualty's arm that is nearest to you and move it so that it is straight and above his head. Repeat the procedure for the other arm.

(2) Kneel beside the casualty with your knees near his shoulders (leave space to roll his body) (Figure 2-1B). Place one hand behind his head and neck for support. With your other hand, grasp the casualty under his far arm (Figure 2-1C).

(3) Roll the casualty towards you using a steady, even pull. His head and neck should stay in line with his back.

(4) Return the casualty's arms to his side. Straighten his legs. Reposition yourself so that you are now kneeling at the level of the casualty's shoulders. However, if a neck injury is suspected and the jaw-thrust technique will be used, kneel at the casualty's head, looking towards his feet.

2-4. Opening the Airway of an Unconscious or Not Breathing Casualty

The tongue is the single most common cause of an airway obstruction (Figure 2-2). In most cases, simply using the head-tilt/chin-lift technique can clear the airway. This action pulls the tongue away from the air passage in the throat (Figure 2-3).

Figure 2-2. Airway blocked by tongue.

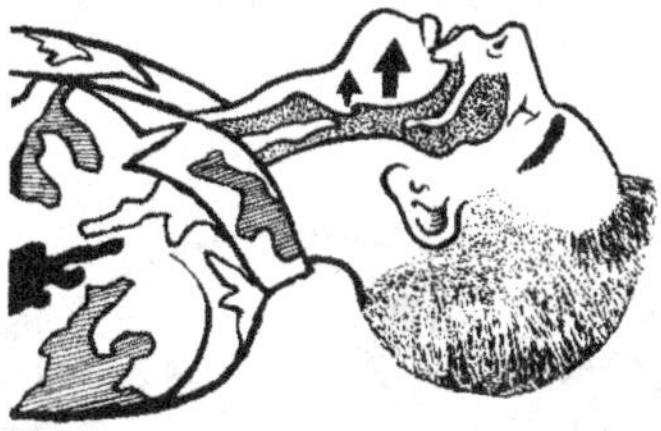

Figure 2-3. Airway opened by extending neck.

a. Call for help and then position the casualty. Move (roll) the casualty onto his back (Figure 2-1C). (Refer to paragraph 2-3*c* for information on positioning the casualty.)

NOTE

Perform finger sweep. If foreign material or vomitus is visible in the mouth, it should be removed, but do not spend an excessive amount of time doing so.

b. Open the airway using the jaw-thrust or head-tilt/chin-lift technique.

CAUTION

The head-tilt/chin-lift technique is an important procedure in opening the airway; however, use extreme care because excess force in performing this maneuver may cause further spinal injury. In a casualty with a suspected neck injury or severe head trauma, the safest approach to opening the airway is the jaw-thrust technique because in most cases it can be accomplished without extending the neck.

(1) *Perform the jaw-thrust technique.* The jaw-thrust may be accomplished by the rescuer grasping the angles of the casualty's lower jaw and lifting with both hands, one on each side, displacing the jaw forward and up (Figure 2-4). The rescuer's elbows should rest on the surface on which the casualty is lying. If the lips close, the lower lip can be retracted with the thumb. If mouth-to-mouth breathing is necessary, close the nostrils by placing your cheek tightly against them. The head should be carefully supported without tilting it backwards or turning it from side to side. If this is unsuccessful, the head should be tilted back very slightly. The jaw-thrust is the safest first approach to opening the airway of a casualty who has a suspected neck injury because in most cases it can be accomplished without extending the neck.

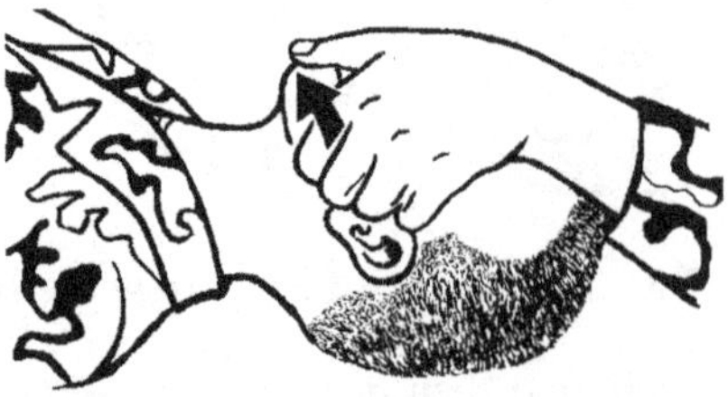

Figure 2-4. Jaw-thrust technique of opening airway.

(2) *Perform the head-tilt/chin-lift technique*. Place one hand on the casualty's forehead and apply firm, backward pressure with the palm to tilt the head back. Place the fingertips of the other hand under the bony part of the lower jaw and lift, bringing the chin forward. The thumb should not be used to lift the chin (Figure 2-5).

NOTE

The fingers should not press deeply into the soft tissue under the chin because the airway may be obstructed.

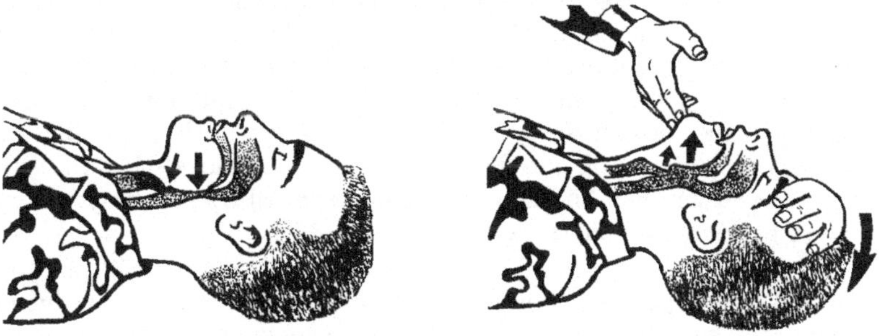

Figure 2-5. Head-tilt/chin-lift technique of opening airway.

(3) *Check for breathing (while maintaining an airway)*. After establishing an open airway, it is important to maintain that airway in an open position. Often the act of just opening and maintaining the airway will allow the casualty to breathe properly. Once the rescuer uses one of the techniques to open the airway (jaw thrust or head tilt/chin-lift), he should maintain that head position to keep the airway open. Failure to maintain the open airway will prevent the casualty from receiving an adequate supply of oxygen. Therefore, while maintaining an open airway the rescuer should check for breathing by observing the casualty's chest and performing the following actions within 3 to 5 seconds:

(*a*) **LOOK** for the chest to rise and fall.

(*b*) **LISTEN** for air escaping during exhalation by placing your ear near the casualty's mouth.

(*c*) **FEEL** for the flow of air on your cheek (see Figure 2-6).

(*d*) **PERFORM** rescue breathing if the casualty does not resume breathing spontaneously.

NOTE

If the casualty resumes breathing, monitor and maintain the open airway. He should be transported to an MTF, as soon as practical.

2-5. Rescue Breathing (Artificial Respiration)

a. If the casualty does not promptly resume adequate spontaneous breathing after the airway is open, rescue breathing (artificial respiration) must be started. Be calm! Think and act quickly! The sooner you begin rescue breathing, the more likely you are to restore the casualty's breathing. If you are in doubt whether the casualty is breathing, give artificial respiration, since it can do no harm to a person who is breathing. If the casualty is breathing, you can feel and see his chest move. If the casualty is breathing, you can feel and hear air being expelled by putting your hand or ear close to his mouth and nose.

b. There are several methods of administering rescue breathing. The mouth-to-mouth method is preferred; however, it cannot be used in all situations. If the casualty has a severe jaw fracture or mouth wound or his jaws are tightly closed by spasms, use the mouth-to-nose method.

2-6. Preliminary Steps—All Rescue Breathing Methods

a. Establish unresponsiveness. Call for help. Turn or position the casualty.

b. Open the airway.

c. Check for breathing by placing your ear over the casualty's mouth and nose, and looking toward his chest.

(1) **LOOK** for rise and fall of the casualty's chest (Figure 2-6).

(2) **LISTEN** for sounds of breathing.

(3) **FEEL** for breath on the side of your face. If the chest does not rise and fall and no air is exhaled, then the casualty is not breathing.

(4) **PERFORM** rescue breathing if the casualty is not breathing.

NOTE

Although the rescuer may notice that the casualty is making respiratory efforts, the airway may still be obstructed and opening the airway may be all that is needed. If the casualty resumes breathing, the rescuer should continue to maintain an open airway.

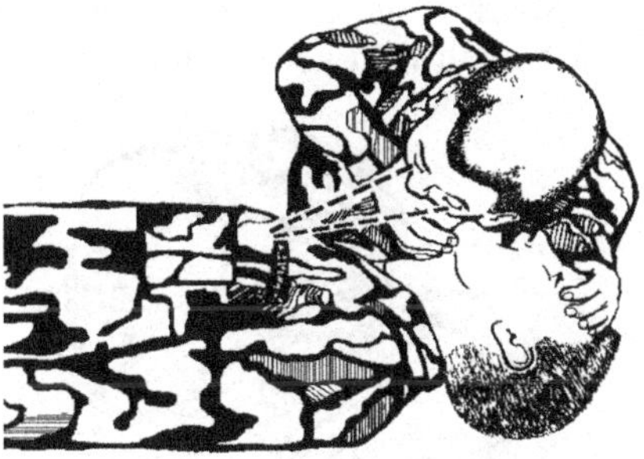

Figure 2-6. Check for breathing.

2-7. Mouth-to-Mouth Method

In this method of rescue breathing, you inflate the casualty's lungs with air from your lungs. This can be accomplished by blowing air into the person's mouth. The mouth-to-mouth rescue breathing method is performed as follows:

a. If the casualty is not breathing, place your hand on his forehead, and pinch his nostrils together with the thumb and index finger of this hand. Let this same hand exert pressure on his forehead to maintain the backward head tilt and maintain an open airway. With your other hand, keep your fingertips on the bony part of the lower jaw near the chin and lift (Figure 2-7).

Figure 2-7. Head tilt/chin lift.

NOTE

If you suspect the casualty has a neck injury and you are using the jaw-thrust technique, close the nostrils by placing your cheek tightly against them.

b. Take a deep breath and place your mouth (in an airtight seal) around the casualty's mouth (Figure 2-8). (If the injured person is small, cover both his nose and mouth with your mouth, sealing your lips against the skin of his face.)

Figure 2-8. Rescue breathing.

c. Blow two full breaths into the casualty's mouth (1 to 1 1/2 seconds per breath), taking a breath of fresh air each time before you blow. Watch out of the corner of your eye for the casualty's chest to rise. If the chest rises, sufficient air is getting into the casualty's lungs. Therefore, proceed as described in step (1). If the chest does not rise, do the following (*a*, *b*, and *c* below) and then attempt to ventilate again.

(1) Take corrective action immediately by reestablishing the airway. Make sure that air is not leaking from around your mouth or out of the casualty's pinched nose.

(2) Reattempt to ventilate.

(3) If the chest still does not rise, take the necessary action to open an obstructed airway (paragraph 2-10).

NOTE

If the initial attempt to ventilate the casualty is unsuccessful, reposition the casualty's head and repeat rescue breathing. Improper chin and head positioning is the most common cause of difficulty with ventilation. If the casualty cannot be ventilated after repositioning the head, proceed with foreign-body airway obstruction maneuvers (see paragraph 2-10).

(4) After giving two slow breaths, which cause the chest to rise, attempt to locate a pulse on the casualty. Feel for a pulse on the side of the casualty's neck closest to you by placing the first two fingers (index and middle fingers) of your hand on the groove beside the casualty's Adam's apple (carotid pulse) (Figure 2-9). (Your thumb should not be used for pulse taking because you may confuse your pulse beat with that of the casualty.) Maintain the airway by keeping your other hand on the casualty's forehead. Allow 5 to 10 seconds to determine if there is a pulse.

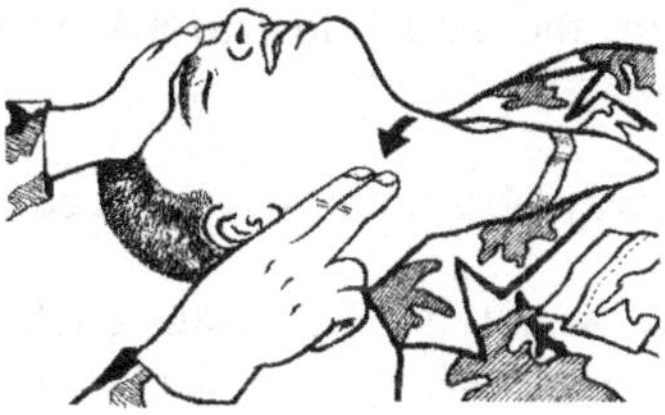

Figure 2-9. Placement of fingers to detect pulse.

(*a*) If signs of circulation are present and a pulse is found and the casualty is breathing—**STOP**; allow the casualty to breathe on his own. If possible, keep him warm and comfortable.

(*b*) If a pulse is found and the casualty is not breathing, continue rescue breathing.

(*c*) If a pulse is not found, seek medically trained personnel for help as soon as possible.

2-8. Mouth-to-Nose Method

Use this method if you cannot perform mouth-to-mouth rescue breathing because the casualty has a severe jaw fracture or mouth wound or his jaws are tightly closed by spasms. The mouth-to-nose method is performed in the same way as the mouth-to-mouth method except that you blow into the nose while you hold the lips closed with one hand at the chin. You then remove your mouth to allow the casualty to exhale passively. It may be necessary to separate the casualty's lips to allow the air to escape during exhalation.

2-9. Heartbeat

If a casualty's heart stops beating, you must immediately seek medical help. **SECONDS COUNT!** Stoppage of the heart is soon followed by cessation of respiration unless it has occurred first. Be calm! Think and act! When a casualty's heart has stopped, there is no pulse at all; the person is unconscious and limp, and the pupils of his eyes are open wide. When evaluating a casualty or when performing the preliminary steps of rescue breathing, feel for a pulse. If you DO NOT detect a pulse, seek medical help.

2-10. Airway Obstructions

In order for oxygen from the air to flow to and from the lungs, the upper airway must be unobstructed.

a. Upper airway obstructions often occur because—

(1) The casualty's tongue falls back into his throat while he is unconscious. The tongue *falls back* and *obstructs* the airway, it is not swallowed by the casualty.

NOTE

Ensure the correct positioning and maintenance of the open airway for an injured or unconscious casualty.

(2) Foreign bodies become lodged in the throat. These obstructions usually occur while eating. Choking on food (usually meat) is associated with—

- Attempting to swallow large pieces of poorly chewed food.
- Drinking alcohol.
- Slipping dentures.

(3) The contents of the stomach are regurgitated and may block the airway.

(4) Blood clots may form as a result of head and facial injuries.

b. Upper airway obstruction may cause either partial or complete airway blockage.

(1) *Partial airway obstruction.* The casualty may still have an air exchange. A good air exchange means that the casualty can cough

forcefully, though he may be wheezing between coughs. You, the rescuer, should not interfere, and should encourage the casualty to cough up the object obstructing his airway on his own. A poor air exchange may be indicated by weak coughing with a high pitched noise between coughs. Further, the casualty may show signs of shock (paragraph 1-6*b*[5]) indicating a need for oxygen. You should assist the casualty and treat him as though he had a complete obstruction.

(2) *Complete airway obstruction.* A complete obstruction (no air exchange) is indicated if the casualty cannot speak, breathe, or cough at all. He may be clutching his neck and moving erratically. In an unconscious casualty, a complete obstruction is also indicated if after opening his airway you cannot ventilate him.

2-11. Opening the Obstructed Airway—Conscious Casualty

Clearing a conscious casualty's airway obstruction can be performed with the casualty either standing or sitting and by following a relatively simple procedure.

WARNING

Once an obstructed airway occurs, the brain will develop an oxygen deficiency resulting in unconsciousness. Death will follow rapidly if breathing is not promptly restored.

a. Ask the casualty if he can speak or if he is choking. Check for the universal choking sign (Figure 2-10).

Figure 2-10. Universal sign of choking.

b. If the casualty can speak, encourage him to attempt to cough; the casualty still has a good air exchange. If he is able to speak or cough effectively, DO NOT interfere with his attempts to expel the obstruction.

c. Listen for high pitched sounds when the casualty breathes or coughs (poor air exchange). If there is poor air exchange or no breathing, CALL FOR HELP and immediately deliver manual thrusts (either an abdominal or chest thrust).

NOTE

The manual thrust with the hands centered between the waist and the rib cage is called an abdominal thrust (or Heimlich maneuver). The chest thrust (the hands are centered in the middle of the breastbone) is used only for an individual in the advanced stages of pregnancy, in the markedly obese casualty, or if there is a significant abdominal wound.

(1) Apply abdominal thrusts. This can be accomplished by using the following procedures:

(*a*) Stand behind the casualty and wrap your arms around his waist.

(*b*) Make a fist with one hand and grasp it with the other. The thumb side of your fist should be against the casualty's abdomen, in the midline and slightly above the casualty's navel, but well below the tip of the breastbone (Figure 2-11).

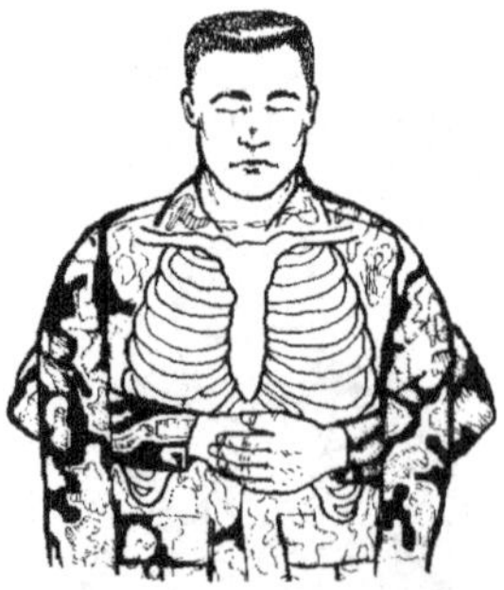

Figure 2-11. Anatomical view of abdominal thrust procedure.

(*c*) Press the fists into the abdomen with a quick backward and upward thrust (Figure 2-12).

Figure 2-12. Profile view of abdominal thrust.

(*d*) Each thrust should be a separate and distinct movement.

NOTE

Continue performing abdominal thrusts until the obstruction is expelled or the casualty becomes unresponsive.

(*e*) If the casualty becomes unresponsive, call for help as you proceed with steps to open the airway, and perform rescue breathing. (Refer to paragraph 2-7 for information on how to perform mouth-to-mouth resuscitation.)

(2) Apply chest thrusts. An alternate technique to the abdominal thrust is the chest thrust. This technique is useful when the casualty has an abdominal wound, when the casualty is pregnant, or when the casualty is so large that you cannot wrap your arms around the abdomen. To apply chest thrusts with casualty sitting or standing:

(*a*) Stand behind the casualty and wrap your arms around his chest with your arms under his armpits.

(*b*) Make a fist with one hand and place the thumb side of the fist in the middle of the breastbone (take care to avoid the tip of the breastbone and the margins of the ribs).

(*c*) Grasp the fist with the other hand and exert thrusts (Figure 2-13).

Figure 2-13. Profile view of chest thrust.

(*d*) Each thrust should be delivered slowly, distinctly, and with the intent of relieving the obstruction.

(*e*) Perform chest thrusts until the obstruction is expelled or the casualty becomes unresponsive.

(*f*) If the casualty becomes unresponsive, call for help as you proceed with steps to open the airway and perform rescue breathing.

2-12. Opening the Obstructed Airway—Casualty Lying Down or Unresponsive

The following procedures are used to expel an airway obstruction in a casualty who is lying down, who becomes unconscious, or who is found unconscious (the cause unknown):

- If a conscious casualty who is choking becomes unresponsive, call for help, open the airway, perform a finger sweep, and attempt rescue breathing (paragraphs 2-4 through 2-8). If you still cannot administer rescue breathing due to an airway blockage, then remove the airway obstruction using the procedures as in *b* below.

- If a casualty is unresponsive when you find him (the cause unknown), assess or evaluate the situation, call for help, position the casualty on his back, open the airway, establish breathlessness, and attempt to perform rescue breathing (paragraphs 2-4 through 2-8).

a. Open the airway and attempt rescue breathing (refer to paragraph 2-7 for information on how to perform mouth-to-mouth resuscitation).

b. If still unable to ventilate the casualty, perform 6 to 10 manual (abdominal or chest) thrusts.

(1) To perform the abdominal thrusts:

(*a*) Kneel astride the casualty's thighs (Figure 2-14).

Figure 2-14. Abdominal thrust on unresponsive casualty.

(*b*) Place the heel of one hand against the casualty's abdomen (in the midline slightly above the navel but well below the tip of the breastbone). Place your other hand on top of the first one. Point your fingers toward the casualty's head.

(*c*) Press into the casualty's abdomen with a quick, forward and upward thrust. You can use your body weight to perform the maneuver. Deliver each thrust quickly and distinctly.

(*d*) Repeat the sequence of abdominal thrusts, finger sweep, and rescue breathing (attempt to ventilate) as long as necessary to remove the object from the obstructed airway.

(*e*) If the casualty's chest rises, proceed to feeling for pulse.

(2) To perform chest thrusts:

(*a*) Place the unresponsive casualty on his back, face up, and open his mouth. Kneel close to the side of the casualty's body.

1. Locate the lower edge of the casualty's ribs with your fingers. Run the fingers up along the rib cage to the notch (Figure 2-15A).

2. Place the middle finger on the notch and the index finger next to the middle finger on the lower edge of the breastbone. Place the heel of the other hand on the lower half of the breastbone next to the two fingers (Figure 2-15B).

3. Remove the fingers from the notch and place that hand on top of the positioned hand on the breastbone, extending or interlocking the fingers (Figure 2-15C).

4. Straighten and lock your elbows with your shoulders directly above your hands without bending the elbows, rocking, or allowing the shoulders to sag. Apply enough pressure to depress the breastbone 1 1/2 to 2 inches, then release the pressure completely (Figure 2-15D). Do this 6 to 10 times. Each thrust should be delivered quickly and distinctly. See Figure 2-16 for another view of the breastbone being depressed.

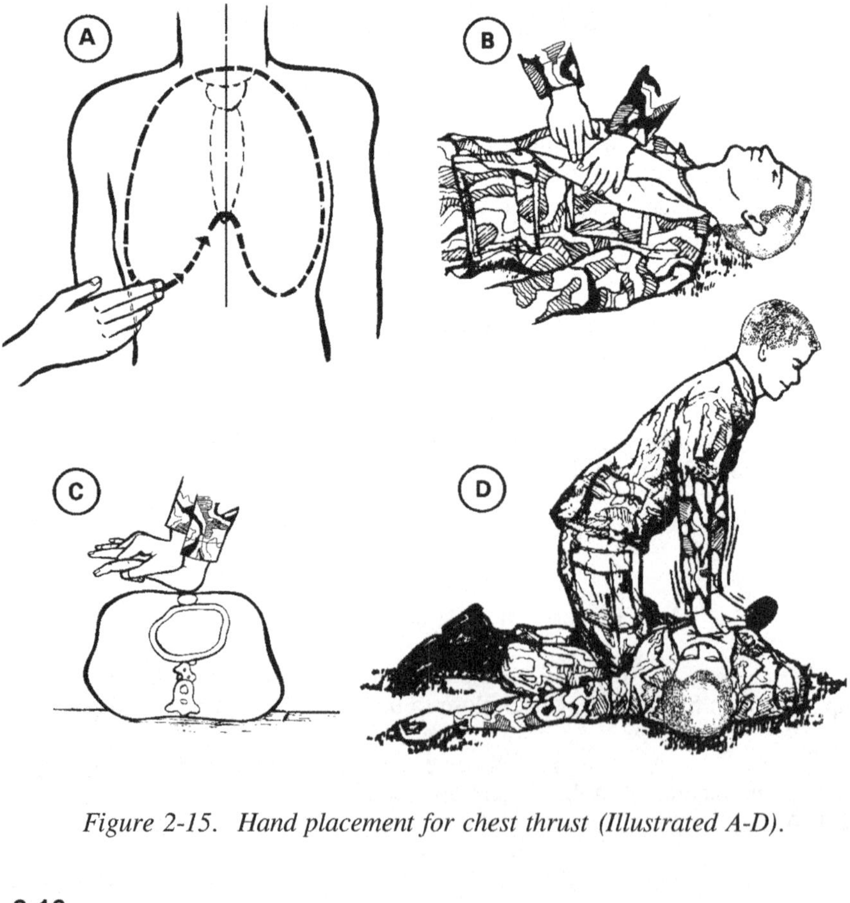

Figure 2-15. Hand placement for chest thrust (Illustrated A-D).

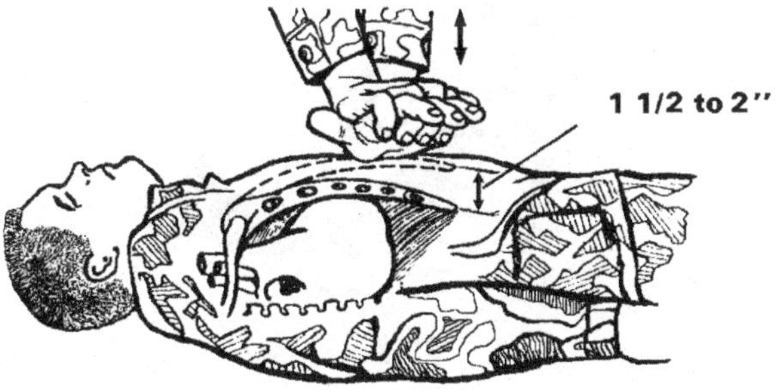

Figure 2-16. Breastbone depressed 1 1/2 to 2 inches.

(*b*) Repeat the sequence of chest thrust, finger sweep, and rescue breathing as long as necessary to clear the object from the obstructed airway. See paragraph (3) below.

(*c*) If the casualty's chest rises, proceed to feeling for his pulse.

(3) If you still cannot administer rescue breathing due to an airway obstruction, then remove the airway obstruction using the procedures in steps (*a*) and (*b*) below.

(*a*) Place the casualty on his back, face up, turn the unresponsive casualty as a unit, and call out for help.

(*b*) Perform finger sweep, keep casualty face up, use tongue-jaw lift to open mouth.

1. Open the casualty's mouth by grasping both his tongue and lower jaw between your thumb and fingers and lifting (tongue-jaw lift) (Figure 2-17). If you are unable to open his mouth, cross your fingers and thumb (crossed-finger method) and push his teeth apart (Figure 2-18) by pressing your thumb against his upper teeth and pressing your finger against his lower teeth.

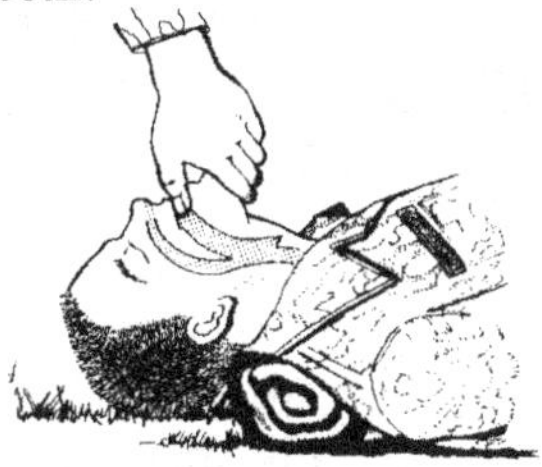

Figure 2-17. Opening casualty's mouth (tongue-jaw lift).

Figure 2-18. Opening casualty's mouth (crossed-finger method).

2. Insert the index finger of the other hand down along the inside of his cheek to the base of the tongue. Use a hooking motion from the side of the mouth toward the center to dislodge the foreign body (Figure 2-19).

Figure 2-19. Using finger to dislodge a foreign body.

WARNING

Take care not to force the object deeper into the airway by pushing it with the finger.

Section II. STOP THE BLEEDING AND PROTECT THE WOUND

2-13. General

The longer a service member bleeds from a major wound, the less likely he will be able to survive his injuries. It is, therefore, important that the first aid provider promptly stop the external bleeding.

2-14. Clothing

In evaluating the casualty for location, type, and size of the wound or injury, cut or tear his clothing and carefully expose the entire area of the wound. This procedure is necessary to properly visualize injury and avoid further contamination. Clothing stuck to the wound should be left in place to avoid further injury. DO NOT touch the wound; keep it as clean as possible.

WARNING

DO NOT REMOVE protective clothing in a chemical environment. Apply dressings over the protective clothing.

2-15. Entrance and Exit Wounds

Before applying the dressing, carefully examine the casualty to determine if there is more than one wound. A missile may have entered at one point and exited at another point. The *EXIT* wound is usually *LARGER* than the entrance wound.

WARNING

The casualty should be continually monitored for development of conditions which may require the performance of necessary basic lifesaving measures, such as clearing the airway and mouth-to-mouth resuscitation. All open (or penetrating) wounds should be checked for a point of entry and exit and first aid measures applied accordingly.

WARNING

If the missile lodges in the body (fails to exit), DO NOT attempt to remove it or probe the wound. Apply a dressing. If there is an object extending from (impaled in) the wound, DO NOT remove the object. Apply a dressing around the object and use additional improvised bulky materials/dressings (use the cleanest material available) to build up the area around the object to stabilize the object and prevent further injury. Apply a supporting bandage over the bulky materials to hold them in place.

2-16. Field Dressing

a. Use the casualty's field dressing; remove it from the wrapper and grasp the tails of the dressing with both hands (Figure 2-20).

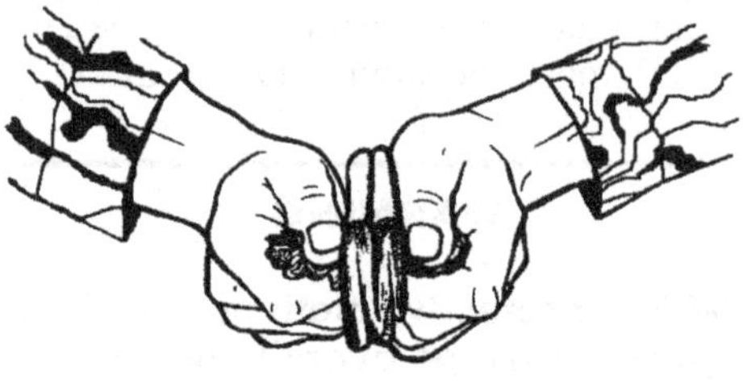

Figure 2-20. Grasping tails of dressing with both hands.

WARNING

DO NOT touch the white (sterile) side of the dressing, and DO NOT allow it to come in contact with any surface other than the wound.

b. Hold the dressing directly over the wound with the white side down. Pull the dressing open (Figure 2-21) and place it directly over the wound (Figure 2-22).

Figure 2-21. Pulling dressing open.

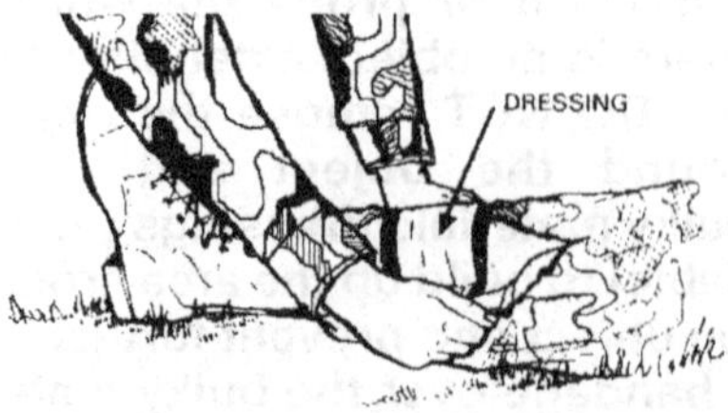

Figure 2-22. Placing dressing directly on wound.

c. Hold the dressing in place with one hand. Use the other hand to wrap one of the tails around the injured part, covering about one-half of the dressing (Figure 2-23). Leave enough of the tail for a knot. If the casualty is able, he may assist by holding the dressing in place.

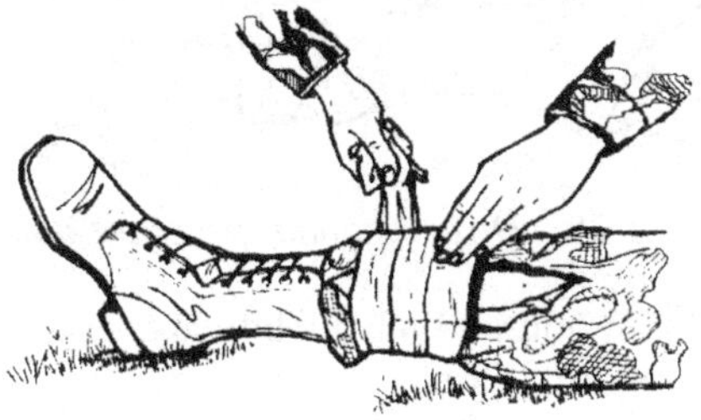

Figure 2-23. Wrapping tail of dressing around injured part.

d. Wrap the other tail in the opposite direction until the remainder of the dressing is covered. The tails should seal the sides of the dressing to keep foreign material from getting under it.

e. Tie the tails into a nonslip knot over the outer edge of the dressing (Figure 2-24). **DO NOT TIE THE KNOT OVER THE WOUND**. In order to allow blood to flow to the rest of an injured limb, tie the dressing firmly enough to prevent it from slipping but without causing a tourniquet-like effect; that is, the skin beyond the injury should not becomes cool, blue, or numb.

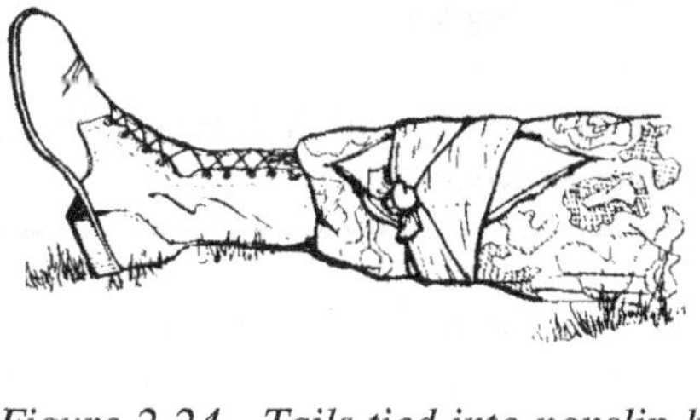

Figure 2-24. Tails tied into nonslip knot.

2-17. Manual Pressure

a. If bleeding continues after applying the sterile field dressing, direct manual pressure may be used to help control bleeding. Apply such pressure by placing a hand on the dressing and exerting firm pressure for 5 to 10 minutes (Figure 2-25). The casualty may be asked to do this himself if he is conscious and can follow instructions.

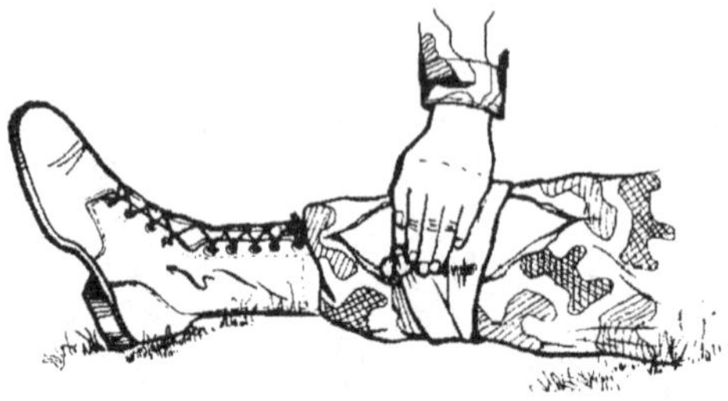

Figure 2-25. Direct manual pressure applied.

b. Elevate an injured limb slightly above the level of the heart to reduce the bleeding (Figure 2-26).

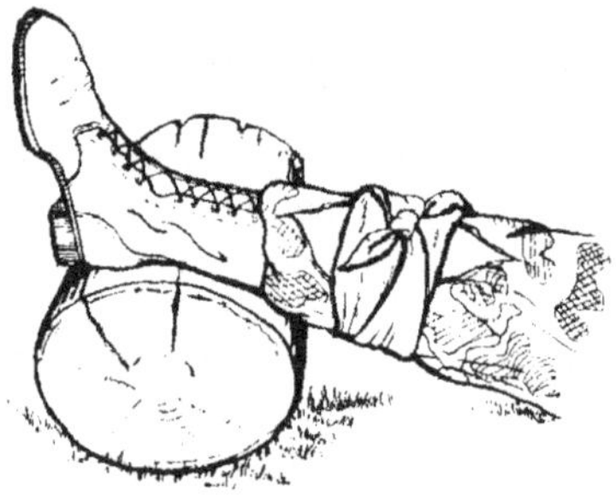

Figure 2-26. Injured limb elevated.

WARNING

DO NOT elevate a suspected fractured limb unless it has been properly splinted.

c. If the bleeding stops, check shock; administer first aid for shock as necessary. If the bleeding continues, apply a pressure dressing.

2-18. Pressure Dressing

Pressure dressings aid in blood clotting and compress the open blood vessel. If bleeding continues after the application of a field dressing, manual pressure, and elevation, then a pressure dressing must be applied as follows:

a. Place a wad of padding on top of the field dressing, directly over the wound (Figure 2-27). Keep the injured extremity elevated.

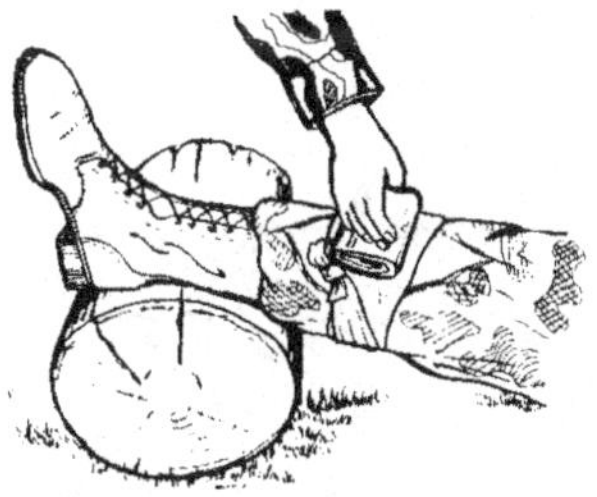

Figure 2-27. Wad of padding on top of field dressing.

NOTE

Improvised bandages may be made from strips of cloth. These strips may be made from T-shirts, socks, or other garments.

b. Place an improvised dressing (or cravat, if available) over the wad of padding (Figure 2-28). Wrap the ends tightly around the injured limb, covering the previously placed field dressing (Figure 2-29).

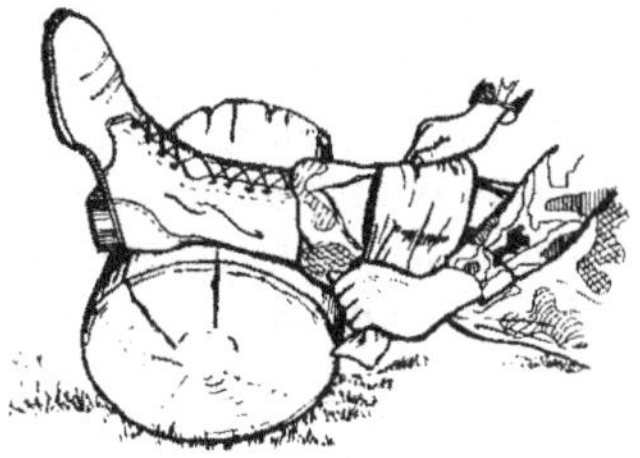

Figure 2-28. Improvised dressing over wad of padding

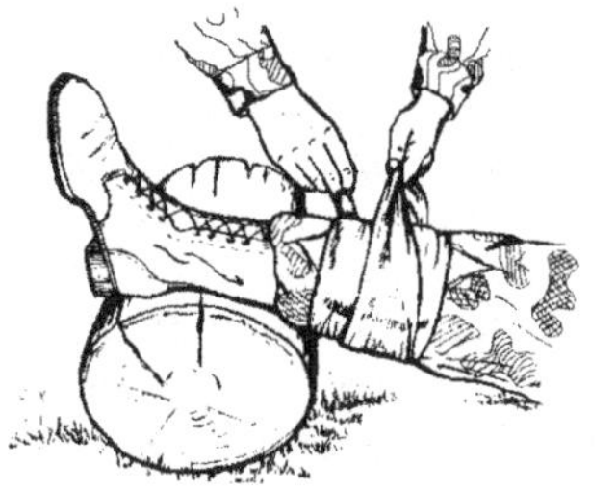

Figure 2-29. Ends of improvised dressing wrapped tightly around limb.

c. Tie the ends together in a nonslip knot, directly over the wound site (Figure 2-30). DO NOT tie so tightly that it has a tourniquet-like effect. If bleeding continues and all other measures have failed, or if the limb is severed, then apply a tourniquet. Use the tourniquet as a **LAST RESORT**. When the bleeding stops, check for shock; administer first aid for shock as necessary.

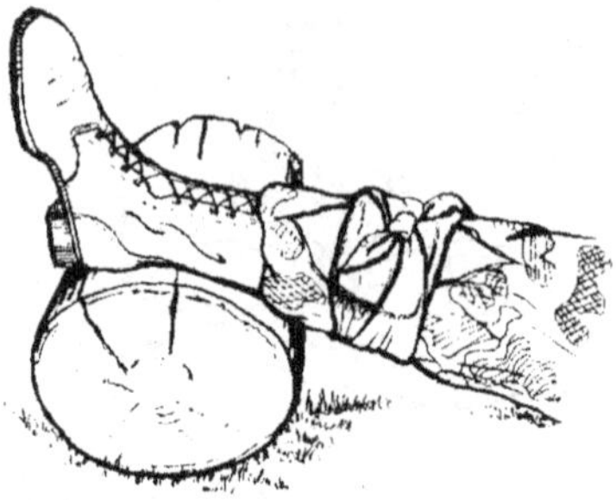

Figure 2-30. Ends of improvised dressing tied together in nonslip knot.

NOTE

Distal end of wounded extremities (fingers and toes) should be checked periodically for adequate circulation. The dressing must be loosened if the extremity becomes cool, blue, or numb.

NOTE

If bleeding continues and all other measures have failed (dressings and covering wound, applying direct manual pressure, elevating the limb above the heart level, and applying a pressure dressing while maintaining limb elevation) *then apply digital pressure* (see paragraph 2-19).

2-19. Digital Pressure

Digital pressure (often called "pressure points") is an alternative method to control bleeding. This method uses pressure from the fingers, thumbs, or hands to press at the site or point where a main artery supplying the wounded area lies near the skin surface or over bone (Figure 2-31). This pressure may help shut off or slow down the flow of blood from the heart to the wound and is used in combination with direct pressure and elevation. It may help in instances where bleeding is not easily controlled, where a pressure dressing has not yet been applied, or where pressure dressings are not readily available.

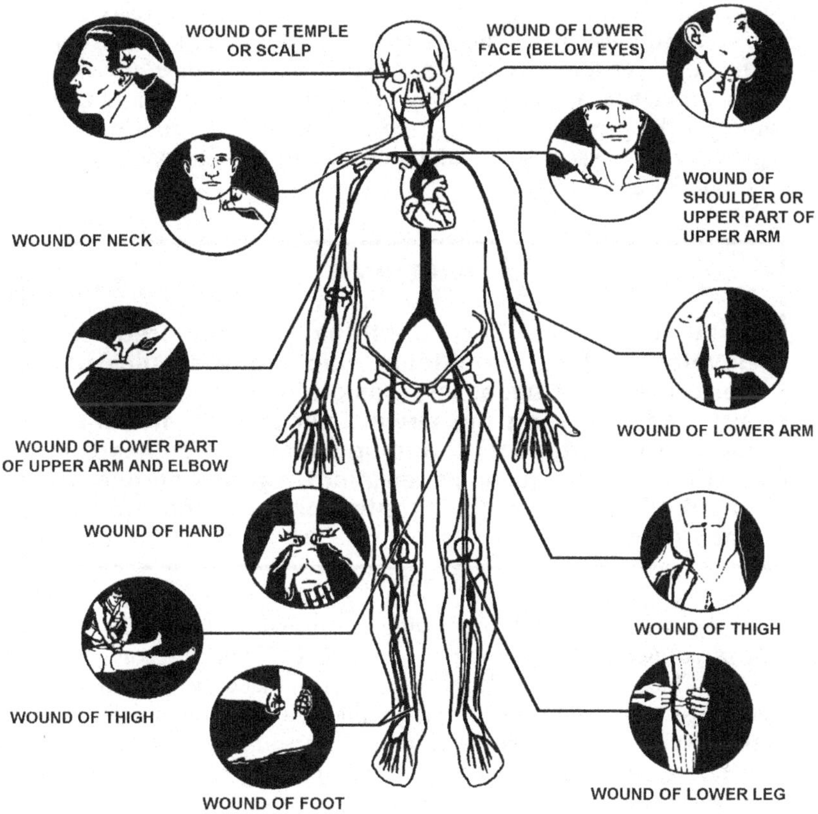

Figure 2-31. Digital pressure (pressure with fingers, thumbs or hands).

2-20. Tourniquet

DANGER

A tourniquet is only used on an arm or leg where there is a danger of the casualty losing his life (bleeding to death).

A tourniquet is a constricting band placed around an arm or leg to control bleeding. A service member whose arm or leg has been completely amputated may not be bleeding when first discovered, but a tourniquet should be applied anyway. This absence of bleeding is due to the body's normal defenses (contraction or clotting of blood vessels) as a result of the amputation, but after a period of time bleeding will start as the blood vessels relax or the clot may be knocked loose by moving the casualty. Bleeding from a major artery of the thigh, lower leg, or arm and bleeding from multiple arteries (which occurs in a traumatic amputation) may prove to be beyond control by manual pressure. If the pressure dressing (see paragraph 2-18, above) under firm hand pressure becomes soaked with blood and the wound continues to bleed, apply a tourniquet.

WARNING

Casualty should be continually monitored for development of conditions which may require the performance of necessary basic lifesaving measures, such as: clearing the airway, performing mouth-to-mouth resuscitation, preventing shock, and/or bleeding control. All open (or penetrating) wounds should be checked for a point of entry or exit and treated accordingly.

The tourniquet should not be used unless a pressure dressing has failed to stop the bleeding or an arm or leg has been cut off. On occasion, tourniquets have injured blood vessels and nerves. If left in place too long, a tourniquet can cause loss of an arm or leg. Once applied, it must stay in place, and the casualty must be taken to the nearest MTF as soon as possible. *DO NOT loosen or release a tourniquet after it has been applied as release could precipitate bleeding and potentially lead to shock.*

a. Improvising a Tourniquet. In the absence of a specially designed tourniquet, a tourniquet may be made from a strong, pliable material, such as gauze or muslin bandages, clothing, or cravats. An improvised tourniquet is used with a rigid stick-like object. To minimize skin damage, ensure that the improvised tourniquet is at least 2 inches wide.

WARNING

The tourniquet must be easily identified or easily seen.

WARNING

DO NOT use wire or shoestring for a tourniquet band.

b. Placing the Improvised Tourniquet.

(1) Place the tourniquet around the limb, between the wound and the body trunk (or between the wound and the heart). Never place it directly over a wound, a fracture, or joint. Tourniquets, for maximum effectiveness, should be placed on the upper arm or above the knee on the thigh (Figure 2-32).

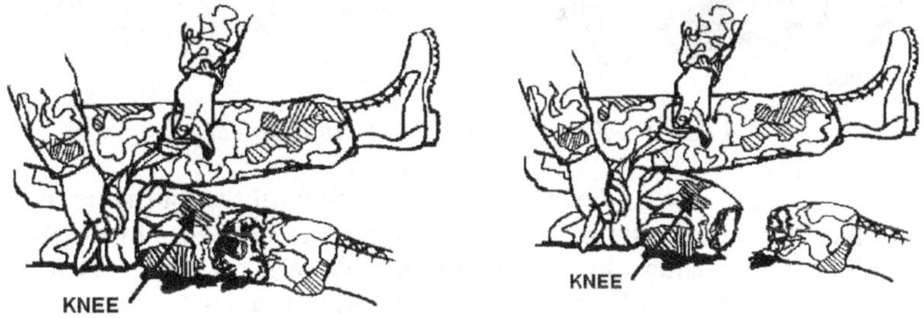

Figure 2-32. Tourniquet above knee.

(2) The tourniquet should be well-padded. If possible, place the tourniquet over the smoothed sleeve or trouser leg to prevent the skin from being pinched or twisted. If the tourniquet is long enough, wrap it around the limb several times, keeping the material as flat as possible. Damaging the skin may deprive the surgeon of skin required to cover an amputation. Protection of the skin also reduces pain.

c. Applying the Tourniquet.

(1) Tie a half-knot. (A half-knot is the same as the first part of tying a shoe lace.)

(2) Place a stick (or similar rigid object) on top of the half-knot (Figure 2-33).

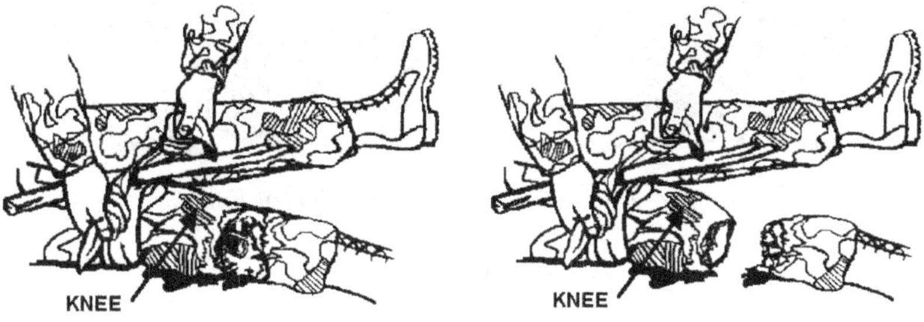

Figure 2-33. Rigid object on top of half-knot.

(3) Tie a full knot over the stick (Figure 2-34).

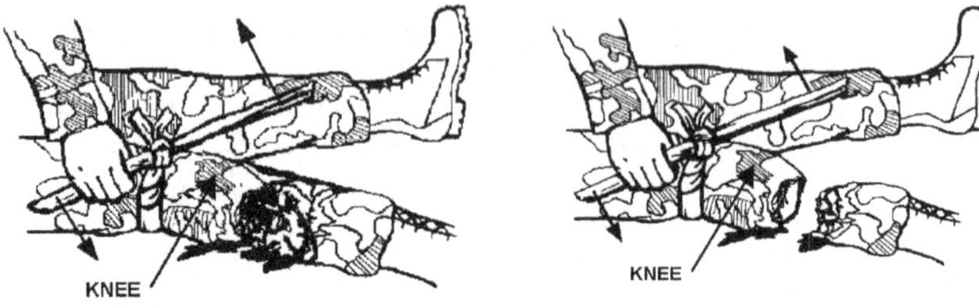

Figure 2-34. Full knot over rigid object.

(4) Twist the stick (Figure 2-35) until the tourniquet is tight around the limb and/or the bright red bleeding has stopped. In the case of amputation, dark oozing blood may continue for a short time. This is the blood trapped in the area between the wound and tourniquet.

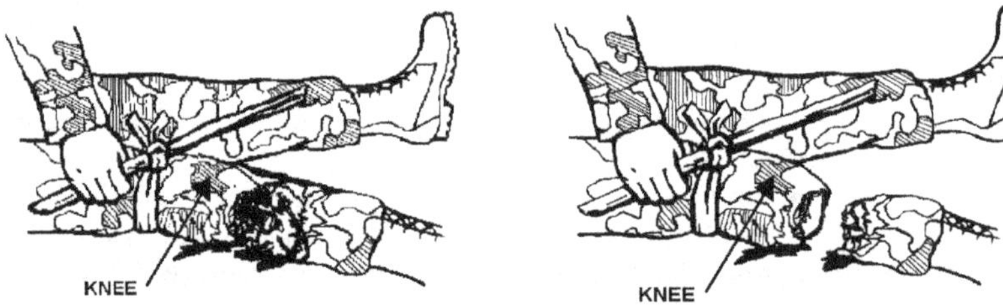

Figure 2-35. Stick twisted.

(5) Fasten the tourniquet to the limb by looping the free ends of the tourniquet over the ends of the stick. Then bring the ends around the limb to prevent the stick from loosening. Tie them together on the side of the limb (Figure 2-36).

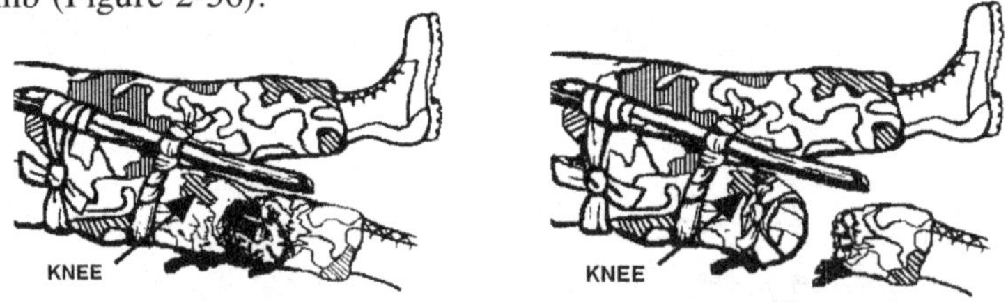

Figure 2-36. Tie free ends on side of limb.

NOTE

Other methods of securing the stick may be used as long as the stick does not unwind and no further injury results.

NOTE

If possible, save and transport any severed (amputated) limbs or body parts with (but out of sight of) the casualty.

(6) DO NOT cover the tourniquet—you should leave it in full view. If the limb is missing (total amputation), apply a dressing to the stump. All wounds should have a dressing to protect the wound from contamination.

(7) Mark the casualty's forehead with a "T" and the time to indicate a tourniquet has been applied. If necessary, use the casualty's blood to make this mark.

(8) Check and treat for shock.

(9) Seek medical aid.

CAUTION

Only appropriately skilled medical personnel may adjust or otherwise remove/release the tourniquet in the appropriate setting.

Section III. CHECK FOR SHOCK AND ADMINISTER FIRST AID MEASURES

2-21. General

The term *shock* has a variety of meanings. In medicine, it refers to a collapse of the body's cardiovascular system which includes an inadequate supply of blood to the body's tissues. Shock stuns and weakens the body. When the normal blood flow in the body is upset, death can result. Early recognition and proper first aid may save the casualty's life.

2-22. Causes and Effects

a. There are three basic mechanisms associated with shock. These are—

- The heart is damaged and fails to work as a pump.

- Blood loss (heavy bleeding) causes the volume of fluid within the vascular system to be insufficient.

- The blood vessels dilate (open wider) so that the blood within the system (even though it is a normal volume [the casualty is not bleeding or dehydrated]) is insufficient to provide adequate circulation within the body.

b. Shock may be the result of a number of conditions. These include—

- Dehydration.

- Allergic reaction to foods, drugs, insect stings, and snakebites.

- Significant loss of blood.

- Reaction to the sight of a wound, blood, or other traumatic scene.

- Traumatic injuries, such as—

  - Burns.

  - Gunshot or shrapnel wounds.

  - Crush injuries.

  - Blows to the body (which can cause broken bones or damage to internal organs).

  - Head injuries.

  - Penetrating wounds (such as from a knife, bayonet, or missile).

2-23. Signs and Symptoms of Shock

Examine the casualty to see if he has any of the following signs and symptoms:

- Sweaty but cool skin (clammy skin).

- Weak and rapid pulse.
- Paleness of skin (in dark-skinned individuals they may have a grayish look to their skin).
- Restlessness, nervousness.
- Thirst.
- Loss of blood (bleeding).
- Confusion (or loss of awareness).
- Faster-than-normal breathing rate.
- Blotchy or bluish skin (especially around the mouth and lips).
- Nausea and/or vomiting.

2-24. First Aid Measures for Shock

In the field, the first aid procedures administered for shock are identical to procedures that would be performed *to prevent shock*. When treating a casualty, assume that shock is present or will occur shortly. By waiting until actual signs and symptoms of shock are noticeable, the rescuer may jeopardize the casualty's life.

a. Position the Casualty. (DO NOT move the casualty or his limbs if suspected fractures have not been splinted. See Chapter 4 for details.)

(1) Move the casualty to cover, if cover is available and the situation permits.

(2) Lay the casualty on his back.

NOTE

A casualty in shock from a chest wound or one who is experiencing breathing difficulty, may breathe easier in a sitting position. If this is the case, allow him to sit upright, but monitor carefully in case his condition worsens.

(3) Elevate the casualty's feet higher than the level of his heart. Use a stable object (field pack or rolled up clothing) so that his feet will not slip off (Figure 2-37).

WARNING

DO NOT elevate legs if the casualty has an unsplinted broken leg, head injury, or abdominal injury.

Figure 2-37. Clothing loosened and feet elevated.

WARNING

Check casualty for leg fracture(s) and splint, if necessary, before elevating his feet. For a casualty with an abdominal wound, place his knees in an upright (flexed) position.

(4) Loosen clothing at the neck, waist, or wherever it may be binding.

CAUTION

DO NOT loosen or remove protective clothing in a chemical environment.

(5) Prevent chilling or overheating. The key is to maintain body temperature. In cold weather, place a blanket or other like item over him to keep him warm and under him to prevent chilling (Figure 2-38). However, if a tourniquet has been applied, leave it exposed (if possible). In hot weather, place the casualty in the shade and protect him from becoming chilled; however, avoid the excessive use of blankets or other coverings.

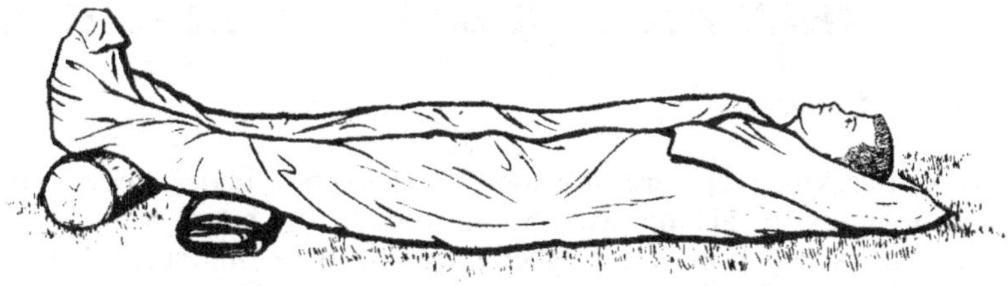

Figure 2-38. Body temperature maintained.

(6) Calm the casualty. Throughout the entire procedure of providing first aid for a casualty, the rescuer should reassure the casualty and keep him calm. This can be done by being authoritative (taking charge) and by showing self-confidence. Assure the casualty that you are there to help him.

(7) Seek medical aid.

b. Food and/or Drink. When providing first aid for shock, DO NOT give the casualty any food or drink. If you must leave the casualty or if he is unconscious, turn his head to the side to prevent him from choking if he vomits (Figure 2-39).

Figure 2-39. Casualty's head turned to side.

c. Evaluate Casualty. Continue to evaluate the casualty until medical personnel arrives or the casualty is transported to an MTF.

CHAPTER 3

FIRST AID FOR SPECIFIC INJURIES

3-1. General

Basic lifesaving steps are discussed in Chapters 1 and 2; they apply to first aid measures for all injuries. Some wounds and burns will require special precautions and procedures when applying these measures. This chapter discusses specific first aid procedures for wounds of the head, face, and neck; chest and stomach wounds; and burns. It also discusses the techniques for applying dressings and bandages to specific parts of the body.

3-2. Head, Neck, and Facial Injuries

a. Head Injuries.

(1) Head injuries range from minor abrasions or cuts on the scalp to severe brain injuries that may result in unconsciousness and sometimes death. Head injuries are classified as open or closed wounds. An open wound is one that is visible, has a break in the skin, and usually has evidence of bleeding. A closed wound may be visible (such as a depression in the skull) or the first aid provider may not be able to see any apparent injury (such as internal bleeding). Some head injuries result in unconsciousness; however, a service member may have a serious head wound and still be conscious. Casualties with head and neck injuries should be treated as though they also have a spinal injury. The casualty should not be moved until the head and neck is stabilized unless he is in immediate danger (such as close to a burning vehicle).

(2) Prompt first aid measures should be initiated for casualties with suspected head and neck injuries. The conscious casualty may be able to provide information on the extent of his injuries. However, as a result of the head injury, he may be confused and unable to provide accurate information. The signs and symptoms a first aid provider might observe are—

- Nausea and vomiting.
- Convulsions or twitches.
- Slurred speech.
- Confusion and loss of memory. (Does he know who he is? Does he know where he is? Does he know what day it is?)
- Recent unconsciousness.

- Dizziness.
- Drowsiness.
- Blurred vision, unequal pupils, or bruising (black eyes).
- Paralysis (partial or full).
- Complaint of headache.
- Bleeding or other fluid discharge from the scalp, nose, or ears.
- Deformity of the head (depression or swelling).
- Staggering while walking.

b. Neck Injuries. Neck injuries may result in heavy bleeding. Apply pressure above and below the injury, *but do not interfere with the breathing process*, and attempt to control the bleeding. Apply a dressing. Always evaluate the casualty for a possible neck fracture/spinal cord injury; if suspected, seek medical treatment immediately.

NOTE

Establish and maintain the airway in cases of facial or neck injuries. If a neck fracture or spinal cord injury is suspected, immobilize the injury and, if necessary, perform basic life support measures.

c. Facial Injuries. Soft tissue injuries of the face and scalp are common. Abrasions (scrapes) of the skin cause no serious problems. Contusions (injury without a break in the skin) usually cause swelling. A contusion of the scalp looks and feels like a lump. Laceration (cut) and avulsion (torn away tissue) injuries are also common. Avulsions are frequently caused when a sharp blow separates the scalp from the skull beneath it. Because the face and scalp are richly supplied with blood vessels (arteries and veins), wounds of these areas usually bleed heavily.

3-3. General First Aid Measures

a. General Considerations. The casualty with a head injury (or suspected head injury) should be continually monitored for the development of conditions that *may require* basic lifesaving measures. After initiating first aid measures, request medical assistance and evacuation. If dedicated medical evacuation assets are not available, transport the casualty to an MTF as soon as the situation permits. The first aid provider should not attempt to remove a protruding object from the head or give the casualty anything to eat or drink. Further, the first aid provider should be prepared to—

- Clear the airway.
- Control bleeding (external).
- Administer first aid measures for shock.
- Keep the casualty warm.
- Protect the wound.

b. *Unconscious Casualty*. An unconscious casualty does not have control of all of his body's functions and may choke on his tongue, blood, vomitus, or other substances. (Refer to Figure 2-39.)

(1) *Breathing*. The brain requires a constant supply of oxygen. A bluish (or in an individual with dark skin—grayish) color of skin around the lips and nail beds indicates that the casualty is not receiving enough oxygen. Immediate action must be taken to clear the airway, to position the casualty on his side, or to initiate rescue breathing.

(2) *Bleeding*. Bleeding from a head injury usually comes from blood vessels within the scalp. Bleeding can also develop inside the skull or within the brain. In most instances visible bleeding from the head can be controlled by application of the field first aid dressing.

CAUTION

DO NOT attempt to put unnecessary pressure on the wound or attempt to push any brain matter back into the head (skull). **DO NOT** apply a pressure dressing.

c. *Concussion*. If an individual receives a heavy blow to the head or face, he may suffer a brain concussion (an injury to the brain that involves a temporary loss of some or all of the brain's ability to function). For example, the casualty may not breathe properly for a short period of time, or he may become confused and stagger when he attempts to walk. Symptoms of a concussion may only last for a short period of time. However, if a casualty is suspected of having suffered a concussion, he should be transported to an MTF as soon as conditions permit.

d. Convulsions. Convulsions (seizures/involuntary jerking) may occur even after a mild head injury. When a casualty is convulsing, protect him from hurting himself. Take the following measures:

(1) Ease him to the ground if he is standing or sitting.

(2) Support his head and neck.

(3) Maintain his airway.

(4) Protect him from further injury (such as hitting close-by objects).

NOTE

DO NOT forcefully hold the arms and legs if they are jerking because this can lead to broken bones. **DO NOT** force anything between the casualty's teeth—especially if they are tightly clenched because this may obstruct the casualty's airway. Maintain the casualty's airway if necessary.

e. Brain Damage. In *severe* head injuries where brain tissue is protruding, *leave the wound alone*; carefully place a loose moistened dressing (moistened with sterile normal saline if available) and also a first aid dressing over the tissue to protect it from further contamination. ***DO NOT** remove or disturb any foreign matter that may be in the wound.* Position the casualty so that his head is higher than his body. Keep him warm and *seek medical assistance immediately.*

NOTE

If there is an object extending from the wound, **DO NOT** remove the object. Improvise bulky dressings from the cleanest material available and place this material around the protruding object for support, then apply the field dressing.

3-4. Chest Wounds

Blunt trauma, bullet or missile wounds, stab wounds, or falls may cause chest injuries. These injuries can be serious and may cause death quickly if first aid is not administered in a timely manner. A casualty with a chest injury may complain of pain in the chest or shoulder area; he may have difficulty breathing. His chest may not rise normally when he breathes. The injury may cause the casualty to cough up blood and to have a rapid or a weak heartbeat. A casualty with an open chest wound has a punctured chest wall. The sucking sound heard when he breathes is caused by air leaking into his chest cavity. This particular type of wound is dangerous and will collapse the injured lung (Figure 3-1). Breathing becomes difficult for the casualty because the wound is open. The service members life may depend upon how quickly you apply an occlusive dressing over the wound (refer to paragraph 3-5).

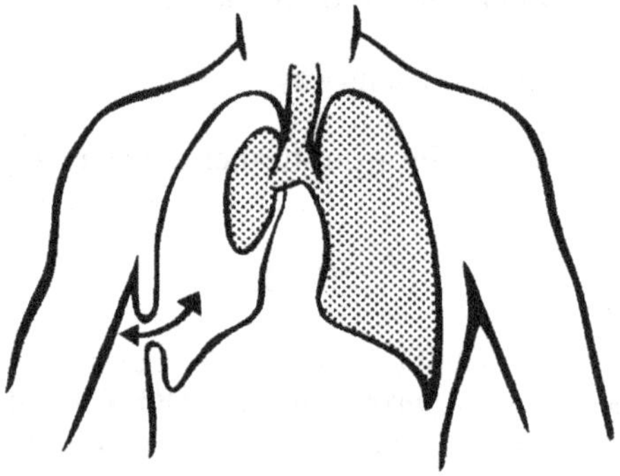

Figure 3-1. Collapsed lung.

3-5. First Aid for Chest Wounds

a. Evaluate the Casualty. Be prepared to perform first aid measures. These measures may include clearing the airway, rescue breathing, treatment for shock, and/or bleeding control.

b. Expose the Wound. If appropriate, cut or remove the casualty's clothing to expose the wound. Remember, **DO NOT** remove clothing that is stuck to the wound because additional injury may result. **DO NOT** attempt to clean the wound.

NOTE

Examine the casualty to see if there is an entry and exit wound. If there are two wounds (entry, exit), perform the same procedure for both wounds. Treat the more serious (heavier bleeding, larger) wound first. It may be necessary to improvise a dressing for the second wound by using strips of cloth, such as a torn T-shirt, or whatever material is available. Also, listen for sucking sounds to determine if the chest wall is punctured.

> **CAUTION**
>
> If there is an object impaled in the wound, **DO NOT** remove it. Apply a dressing around the object and use additional improvised bulky materials/dressings (use the cleanest materials available) to build up the area around the object. Apply a supporting bandage over the bulky materials to hold them in place.

> **CAUTION**
>
> **DO NOT REMOVE** protective clothing in a chemical environment. Apply dressings *over* the protective clothing.

c. *Open the Casualty's Field Dressing Plastic Wrapper.* In cases where there is a sucking chest wound, the plastic wrapper is used with the field dressing to create an occlusive dressing. If a plastic wrapper is not available, or if an additional wound needs to be treated; cellophane, foil, the casualty's poncho, or similar material may be used. The covering should be wide enough to extend 2 inches or more beyond the edges of the wound in all directions.

(1) Tear open one end of the casualty's plastic wrapper covering the field dressing. Be careful not to destroy the wrapper and **DO NOT** touch the inside of the wrapper.

(2) Remove the inner packet (field dressing).

(3) Complete tearing open the empty plastic wrapper using as much of the wrapper as possible to create a flat surface.

d. *Place the Wrapper Over the Wound.* Place the inside surface of the plastic wrapper directly over the wound *when the casualty exhales* and hold it in place (Figure 3-2). The casualty may hold the plastic wrapper in place if he is able.

Figure 3-2. Open chest wound sealed with an occlusive dressing.

e. Apply the Dressing to the Wound.

(1) Use your free hand and shake open the field dressing (Figure 3-3).

Figure 3-3. Shaking open the field dressing.

(2) Place the white side of the dressing on the plastic wrapper covering the wound (Figure 3-4).

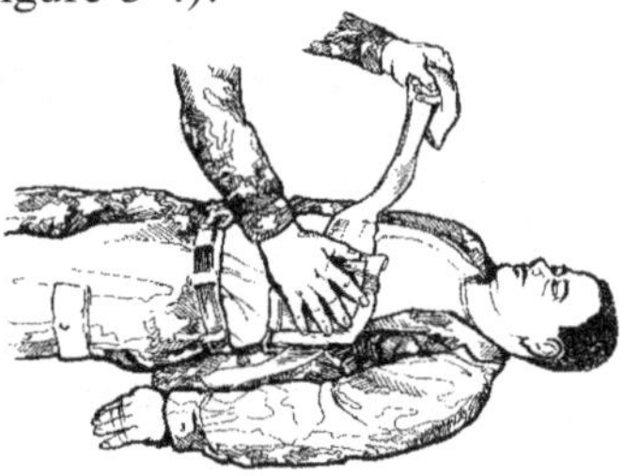

Figure 3-4. Field dressing placed on plastic wrapper.

NOTE

Use the casualty's field dressing, not your own.

(3) Have the casualty breathe normally.

(4) While maintaining pressure on the dressing, grasp one tail of the field dressing with the other hand and wrap it around the casualty's back. If tape is available, tape three sides of the plastic wrapper to the chest wall to provide occlusive type dressing. Leave one side untapped to provide emergency escape for air that may build up in the chest. If tape is not available, secure wrapper on three sides with field dressing leaving the fourth side as a flap.

(5) Wrap the other tail in the opposite direction, bringing both tails over the dressing (Figure 3-5).

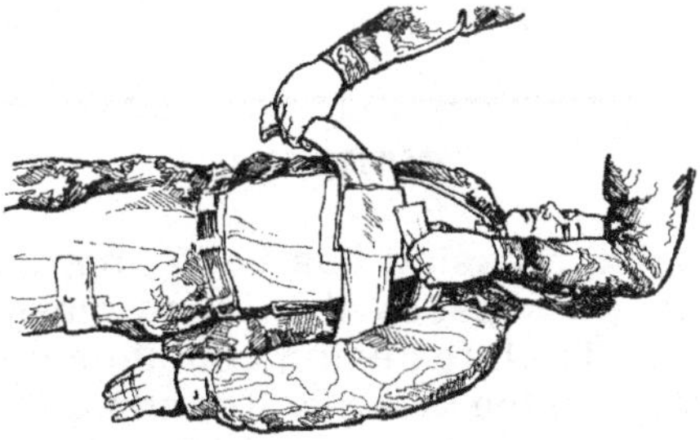

Figure 3-5. Tails of field dressing wrapped around casualty in opposite direction.

(6) Tie the tails into a square knot in the center of the dressing *after* the casualty exhales and *before* he inhales. This will aid in maintaining pressure on the bandage after it has been tied (Figure 3-6). Tie the dressing firmly enough to secure the dressing without interfering with the casualty's breathing.

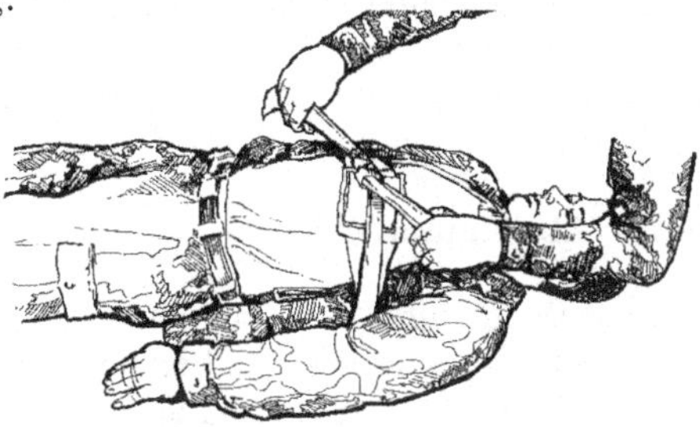

Figure 3-6. Tails of dressing tied into square knot over center of dressing.

NOTE

When practical, apply direct manual pressure over the dressing for 5 to 10 minutes to help control the bleeding.

f. *Position the Casualty*. Position the casualty on his injured side or in a sitting position, whichever makes breathing easier (Figure 3-7).

Figure 3-7. Casualty positioned (lying) on injured side.

g. Seek Medical Assistance. Contact medical personnel.

WARNING

If an occlusive dressing has been improperly placed, air may enter the chest cavity with no means of escape. This causes a life-threatening condition called *tension pneumothorax.* If the casualty's condition (for example, difficulty breathing, shortness of breath, restlessness, or blueness/grayness of the skin) worsens after placing the dressing, quickly lift or remove, and then replace the occlusive dressing.

3-6. Abdominal Wounds

The most serious abdominal wound is one in which an object penetrates the abdominal wall and pierces internal organs or large blood vessels. In these instances, bleeding may be severe and death can occur rapidly.

3-7. First Aid for Abdominal Wounds

a. Evaluate the Casualty. Be prepared to perform basic first aid measures. Always check for both entry and exit wounds. If there are two wounds (entry and exit), treat the wound that appears more serious first (for example, the heavier bleeding, protruding organs, larger wound, and so forth). It may be necessary to improvise dressings for the second wound by using strips of cloth, a T-shirt, or the cleanest material available.

b. Position the Casualty. Place and maintain the casualty on his back with his knees in an upright (flexed) position (Figure 3-8). The knees-up position helps relieve pain, assists in the treatment of shock, prevents further exposure of the bowel (intestines) or abdominal organs, and helps relieve abdominal pressure by allowing the abdominal muscles to relax.

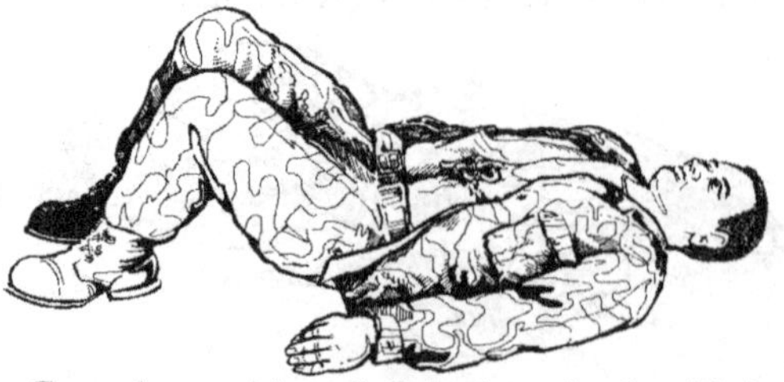

Figure 3-8. Casualty positioned (lying) on back with knees (flexed) up.

c. Expose the Wound.

(1) Remove the casualty's loose clothing to expose the wound. However, **DO NOT** attempt to remove clothing that is stuck to the wound; removing it may cause further injury.

CAUTION

DO NOT REMOVE protective clothing in a chemical environment. Apply dressings *over* the protective clothing.

(2) Gently pick up any organs that may be on the ground. Do this with a clean, dry dressing or with the cleanest available material. Place the organs on top of the casualty's abdomen (Figure 3-9).

Figure 3-9. Protruding organs placed near wound.

NOTE

DO NOT probe, clean, or try to remove any foreign object from the abdomen. **DO NOT** touch with bare hands any exposed organs. **DO NOT** push organs back inside the body.

d. Apply the Field Dressing. Use the casualty's field dressing, not your own. If the field dressing is not large enough to cover the entire wound, the plastic wrapper from the dressing may be used to cover the wound first (placing the field dressing on top). Open the plastic wrapper carefully without touching the inner surface. If necessary, other improvised dressings may be made from clothing, blankets, or the cleanest materials available.

WARNING

If there is an object extending from the wound, DO NOT remove it. Place as much of the wrapper over the wound as possible without dislodging or moving the object. DO NOT place the wrapper over the object.

(1) Grasp the tails in both hands.

(2) Hold the dressing with the white side down directly over the wound. **DO NOT** touch the white (sterile) side of the dressing or allow anything except the wound to come in contact with it.

(3) Pull the dressing open and place it directly over the wound (Figure 3-10). If the casualty is able, he may hold the dressing in place.

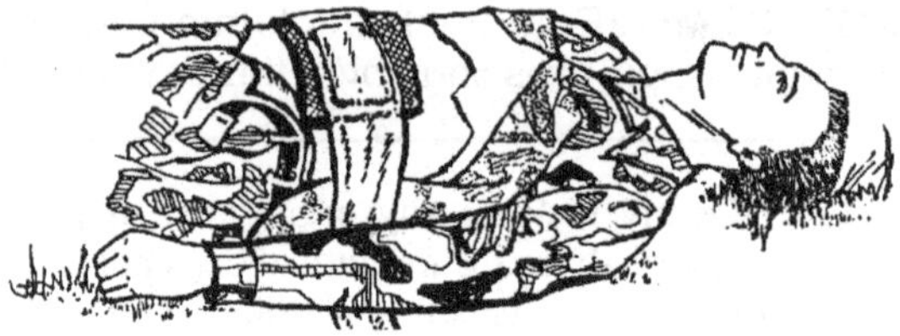

Figure 3-10. Dressing placed directly over the wound.

(4) Hold the dressing in place with one hand and use the other hand to wrap one of the tails around the body.

(5) Wrap the other tail in the opposite direction until the dressing is completely covered. Leave enough of the tail for a knot.

(6) Loosely tie the tails with a square knot at the casualty's side (Figure 3-11).

Figure 3-11. Dressing applied and tails tied with a square knot.

WARNING

When the dressing is applied, DO NOT put pressure on the wound or exposed internal parts, because pressure could cause further injury (vomiting, ruptured intestines, and so forth). Therefore, tie the dressing ties (tails) loosely at casualty's side, not directly over the dressing.

(7) Tie the dressing firmly enough to prevent slipping without applying pressure to the wound site (Figure 3-12).

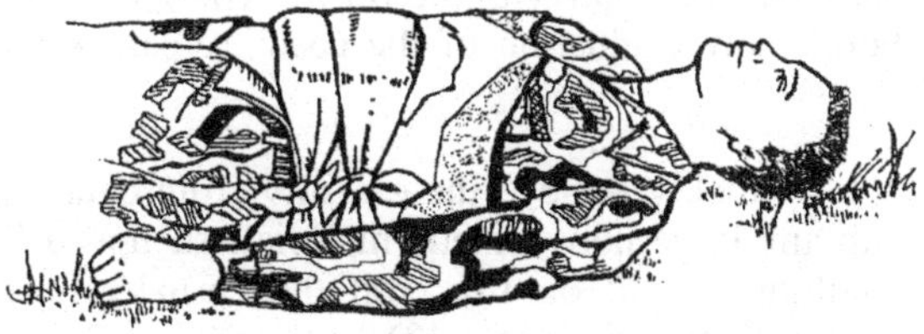

Figure 3-12. Field dressing covered with improvised material and loosely tied.

Field dressings can be covered with improvised reinforcement material (cravats, strips of torn T-shirt, or other cloth) for additional support and protection. Tie improvised bandage on the opposite side of the dressing ties firmly enough to prevent slipping but without applying additional pressure to the wound.

> **CAUTION**
>
> **DO NOT** give casualties with abdominal wounds food or water (moistening the lips is allowed).

e. Seek Medical Assistance. Notify medical personnel.

3-8. Burn Injuries

Burns often cause extreme pain, scarring, or even death. Before administering first aid, you must be able to recognize the type of burn. There are four types of burns:

- Thermal burns caused by fire, hot objects, hot liquids, and gases; or by nuclear blast or fireball.
- Electrical burns caused by electrical wires, current, or lightning.
- Chemical burns caused by contact with wet or dry chemicals or white phosphorus (WP)—from marking rounds and grenades.
- Laser burns (eye [ocular] injury).

3-9. First Aid for Burns

a. Eliminate the Source of the Burn. The source of the burn must be eliminated before any evaluation of the casualty can occur and first aid administered.

(1) Quickly remove the casualty from danger and cover the *thermal burn* with any large nonsynthetic material, such as a field jacket. If the casualty's clothing is still on fire, roll the casualty on the ground to smother (put out) the flames (Figure 3-13).

Figure 3-13. Casualty covered and rolled on ground.

CAUTION

Synthetic materials, such as nylon, may melt and cause further injury.

(2) Remove the *electrical burn* casualty from the electrical source by turning off the electrical current. **DO NOT** attempt to turn off the electricity if the source is not close by. Speed is critical, so **DO NOT** waste unnecessary time. If the electricity cannot be turned off, wrap any *nonconductive* material (*dry* rope, clothing, wood, and so forth) around the casualty's back and shoulders and drag the casualty away from the electrical source (Figure 3-14). **DO NOT** make body-to-body contact with the casualty or touch any wires because you could also become an electrical burn casualty.

Figure 3-14. Casualty removed from electrical source (using nonconductive material).

WARNING

High voltage electrical burns may cause temporary unconsciousness, difficulties in breathing, or difficulties with the heart (heartbeat).

(3) Remove the *chemical* from the *burned casualty*. Remove *liquid* chemicals by flushing with as much water as possible. Remove *dry* chemicals by brushing off loose particles (**DO NOT** use the bare surface of your hand because you could become a chemical burn casualty) and then flush with large amounts of water, if available. If large amounts of water are not available, then **NO** water should be applied because small amounts of water applied to a dry chemical burn may cause a chemical reaction. When WP strikes the skin, smother with a wet cloth or mud. Keep WP covered with a wet material to exclude air; this should help prevent the particles from burning.

(4) Remove the *laser burn* casualty from the source. When removing the casualty from the laser beam source, be careful not to enter the

beam or you may become a casualty. Never look directly at the beam source and if possible, wear appropriate eye protection.

NOTE

After the casualty is removed from the source of the burn, he should be evaluated for conditions requiring basic first aid measures.

b. Expose the Burn. Cut and gently lift away any clothing covering the burned area, without pulling clothing over the burns. Leave in place any clothing that is stuck to the burn. If the casualty's hands or wrists have been burned, remove jewelry if possible without causing further injury (rings, watches, and so forth) and place in his pockets. This prevents the necessity to cut off jewelry since swelling usually occurs as a result of a burn.

CAUTION

DO NOT lift or cut away clothing if in a chemical environment. Apply the dressing directly over the casualty's protective clothing. **DO NOT** attempt to decontaminate skin where blisters have formed.

c. Apply a Field Dressing to the Burn.

(1) Grasp the tails of the casualty's dressing in both hands.

(2) Hold the dressing directly over the wound with the white side down, pull the dressing open, and place it directly over the wound. **DO NOT** touch the white (sterile) side of the dressing or allow anything except the wound to come in contact with it. If the casualty is able, he may hold the dressing in place.

(3) Hold the dressing in place with one hand and use the other hand to wrap one of the tails around the limbs or the body.

(4) Wrap the other tail in the opposite direction until the dressing is completely covered.

(5) Tie the tails into a square knot over the outer edge of the dressing. The dressing should be applied lightly over the burn. Ensure that dressing is applied firmly enough to prevent it from slipping.

NOTE

Use the cleanest improvised dressing material available if a field dressing is not available or if it is not large enough for the entire wound.

d. *Take the Following Precautions*:

- **DO NOT** place the dressing over the face or genital area.

- **DO NOT** break the blisters.

- **DO NOT** apply grease or ointments to the burns.

- For electrical burns, check for both an entry and exit burn from the passage of electricity through the body. Exit burns may appear on any area of the body despite location of entry burn.

- For burns caused by wet or dry chemicals, flush the burns with large amounts of water and cover with a dry dressing.

- For burns caused by WP, flush the area with water, then cover with a wet material, dressing, or mud to exclude the air and keep the WP particles from burning.

- For laser burns, apply a field dressing.

- If the casualty is conscious and not nauseated, give him small amounts of water.

e. *Seek Medical Assistance*. Notify medical personnel.

3-10. Dressings and Bandages

a. *Head Wounds.*

(1) *Position the casualty.*

WARNING

DO NOT move the casualty if you suspect he has sustained a neck, spine, or head injury (which produces any signs or symptoms other than minor bleeding).

- If the casualty has a minor (superficial) scalp wound and is conscious:

  - Have the casualty sit up (unless other injuries prohibit or he is unable to).

  - If the casualty is lying down and is not accumulating fluids or drainage in his throat, elevate his head slightly.

  - If the casualty is bleeding from or into his mouth or throat, turn his head to the side or position him on his side so that the airway will be clear. Avoid putting pressure on the wound and place him on his uninjured side (Figure 3-15).

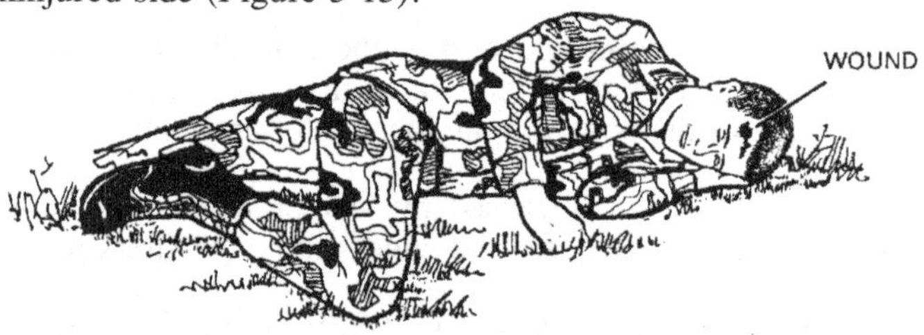

Figure 3-15. Casualty lying on side opposite injury.

- If the casualty is unconscious or has a severe head injury, then suspect and treat him as having a potential neck or spinal injury, *immobilize and **DO NOT** move the casualty.*

NOTE

If the casualty is choking or vomiting or is bleeding from or into his mouth (thus compromising his airway), position him on his uninjured side to allow for drainage and to help keep his airway clear.

WARNING

If it is necessary to turn a casualty with a suspected neck/spine injury; roll the casualty gently onto his side, keeping the head, neck, and body aligned while providing support for the head and neck. DO NOT roll the casualty by yourself but seek assistance. *Move him only if absolutely necessary*, otherwise keep the casualty immobilized to prevent further damage to the neck/spine.

(2) *Expose the wound.* Remove the casualty's helmet (if necessary). In a nuclear, biological, and chemical (NBC) environment, the first aid provider must leave the casualty as much protection (such as protective mask, mission-oriented protective posture [MOPP] overgarments) as possible. What items of protective equipment can be removed is dependent upon the casualty's injuries (where on the body and what type), the MOPP level, integrity of protective equipment (such as tears in the garment or mask seal), availability of chemical protective shelters, and the tactical situation.

WARNING

DO NOT attempt to clean the wound or remove a protruding object.

NOTE

Always use the casualty's field dressing, not your own.

(3) *Apply a dressing to a wound of the forehead or back of head.* To apply a dressing to a wound of the forehead or back of the head—

(*a*) Remove the dressing from the wrapper.

(*b*) Grasp the tails of the dressing in both hands.

(*c*) Hold the dressing (white side down) directly over the wound. **DO NOT** touch the white (sterile) side of the dressing or allow anything except the wound to come in contact with it.

(*d*) Place it directly over the wound.

(*e*) Hold it in place with one hand. If the casualty is able, he may assist.

(*f*) Wrap the first tail horizontally around the head; ensure the tail covers the dressing (Figure 3-16).

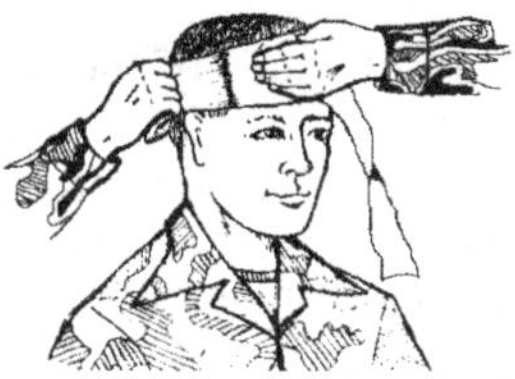

Figure 3-16. First tail of dressing wrapped horizontally around head.

(*g*) Hold the first tail in place and wrap the second tail in the opposite direction, covering the dressing (Figure 3-17).

Figure 3-17. Second tail wrapped in opposite direction.

(*h*) Tie a square knot and secure the tails at the side of the head, making sure they **DO NOT** cover the eyes or ears (Figure 3-18).

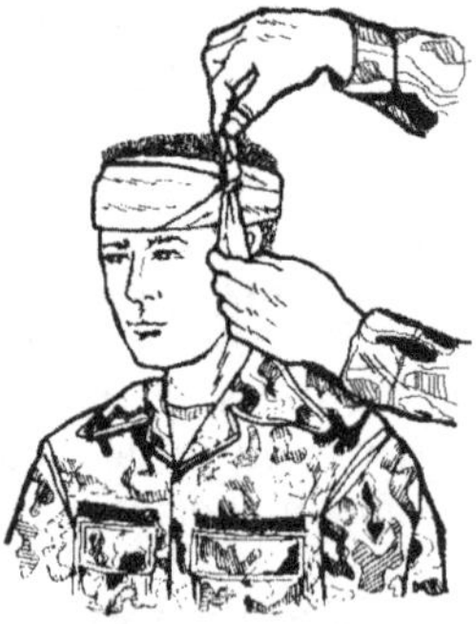

Figure 3-18. Tails tied in square knot at side of head.

(4) *Apply a dressing to a wound on top of the head.* To apply a dressing to a wound on top of the head—

(*a*) Remove the dressing from the wrapper.

(*b*) Grasp the tails of the dressing in both hands.

(*c*) Hold it (white side down) directly over the wound. **DO NOT** touch the white (sterile) side of the dressing or allow anything except the wound to come in contact with it.

(*d*) Place it over the wound (Figure 3-19).

Figure 3-19. Dressing placed over wound.

(*e*) Hold it in place with one hand. If the casualty is able, he may assist.

(*f*) Wrap one tail down under the chin (Figure 3-20), up in front of the ear, over the dressing, and in front of the other ear.

Figure 3-20. One tail of dressing wrapped under chin.

> **WARNING**
>
> **Ensure the tails remain wide and close to the front of the chin to avoid choking the casualty.**

(*g*) Wrap the remaining tail under the chin in the opposite direction and up the side of the face to meet the first tail (Figure 3-21).

Figure 3-21. Remaining tail wrapped under chin in opposite direction.

(*h*) Cross the tails (Figure 3-22), bringing one around the forehead (above the eyebrows) and the other around the back of the head (at the base of the skull) to a point just above and in front of the opposite ear, and tie them using a square knot (Figure 3-23).

Figure 3-22. Tails of dressing crossed with one around forehead.

Figure 3-23. Tails tied in square knot (in front of and above ear).

(5) *Apply a triangular bandage to the head.* To apply a triangular bandage to the head—

(*a*) Turn the base (longest side) of the bandage up and center its base on the center of the forehead, letting the point (apex) fall on the back of the neck (Figure 3-24A).

(*b*) Take the ends behind the head and cross the ends over the apex.

(*c*) Take them over the forehead and tie them (Figure 3-24B).

(*d*) Tuck the apex behind the crossed part of the bandage or secure it with a safety pin, if available (Figure 3-24C).

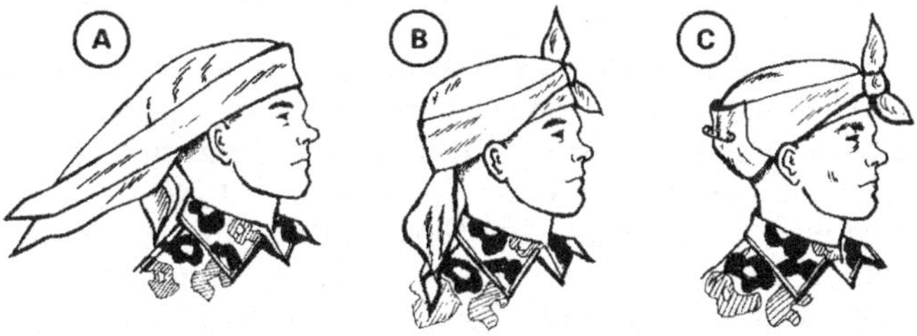

Figure 3-24. Triangular bandage applied to head (Illustrated A—C)

(6) *Apply a cravat bandage to the head.* To apply a cravat bandage to the head—

(*a*) Place the middle of the bandage over the dressing (Figure 3-25A).

(*b*) Cross the two ends of the bandage in opposite directions completely around the head (Figure 3-25B).

(*c*) Tie the ends over the dressing (Figure 3-25C).

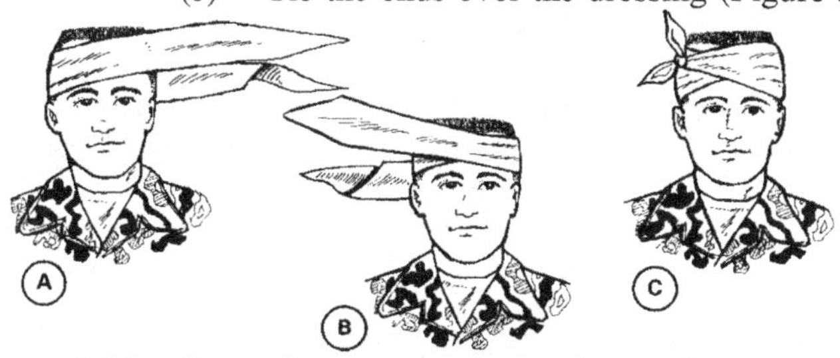

Figure 3-25. Cravat bandage applied to head (Illustrated A—C).

b. Eye Injuries. The eye is a vital sensory organ, and blindness is a severe physical handicap. Timely first aid of the eye may relieve pain and may also help to prevent shock, permanent eye injury, and possible loss of vision. Because the eye is very sensitive, any injury can be easily aggravated if it is improperly handled. Injuries of the eye may be quite severe. Cuts of the eyelids can appear to be very serious, but if the eyeball is not involved, a person's vision usually will not be damaged. However, lacerations (cuts) of the eyeball can cause permanent damage or loss of sight.

(1) *Lacerated/torn eyelids*. Lacerated eyelids may bleed heavily, but bleeding usually stops quickly. Cover the injured eye with a sterile dressing. **DO NOT** put pressure on the wound because you may injure the eyeball. Handle torn eyelids very carefully to prevent further injury. Place any detached pieces of the eyelid on a clean bandage or dressing and immediately send them with the casualty to the medical facility.

(2) *Lacerated eyeball (injury to the globe)*. Lacerations or cuts to the eyeball may cause serious and permanent eye damage. Cover the injury with a loose sterile dressing. **DO NOT** put pressure on the eyeball because additional damage may occur. An important point to remember is that when one eyeball is injured, you should immobilize both eyes. This is done by applying a bandage to both eyes. Because the eyes move together, covering both will lessen the chances of further damage to the injured eye. (However, in hazardous surroundings, leave uninjured eye uncovered to enable casualty to see.)

CAUTION

DO NOT apply pressure when there is a possible laceration of the eyeball. The eyeball contains fluid. Pressure applied over the eye will force the fluid out, resulting in permanent injury. **APPLY PROTECTIVE DRESSING WITHOUT ADDED PRESSURE**.

(3) *Extruded eyeballs*. Service members may encounter casualties with severe eye injuries that include an extruded eyeball (eyeball out-of-socket). In such instances you should gently cover the extruded eye with a loose moistened dressing and also cover the unaffected eye. **DO NOT** bind or exert pressure on the injured eye while applying the dressing. Keep the casualty quiet, place him on his back, treat for shock, and evacuate him immediately.

(4) *Burns of the eyes*. Chemical burns, thermal (heat) burns, and light burns can affect the eyes.

(*a*) *Chemical burns*. Injuries from chemical burns require immediate first aid. Mainly acids or alkalies cause chemical burns. The first aid measures consist of flushing the eyes immediately with large amounts of water for at least 5 to 20 minutes, or as long as necessary to flush out the chemical and, once flushed, bandaging the eyes. If the burn is an acid burn, you should flush the eye for at least 5 to 10 minutes. If the burn is an alkali burn, you should flush the eye for at least 20 minutes. After the eye has been flushed evacuate the casualty immediately.

(*b*) *Thermal burns*. When an individual suffers burns of the face from a fire, the eyes will close quickly due to extreme heat. This reaction is a natural reflex to protect the eyeballs; however, the eyelids remain exposed and are frequently burned. If a casualty receives burns of the eyelids or face—

- **DO NOT** apply a dressing.
- **DO NOT** touch.
- **SEEK** medical assistance immediately.

(*c*) *Light burns*. Exposure to intense light can burn an individual. Infrared rays, eclipse light (if the casualty has looked directly at the sun), or laser burns cause injuries of the exposed eyeball. Ultraviolet rays from arc welding can cause a superficial burn to the surface of the eye. These injuries are generally not painful but may cause permanent damage to the eyes. Immediate first aid is usually not required. Loosely bandaging the eyes may make the casualty more comfortable and protect his eyes from further injury caused by exposure to other bright lights or sunlight.

CAUTION

With impaled objects or significant sized foreign bodies, both eyes are usually bandaged to help secure the foreign body in the injured eye. In a battlefield environment, leave the uninjured eye uncovered so that the casualty can see.

c. *Side-of-Head or Cheek Wound*. Facial injuries to the side of the head or the cheek may bleed profusely (Figure 3-26). Prompt action is necessary to ensure that the airway remains open and also to control the bleeding. It may be necessary to apply a dressing. To apply a dressing—

(1) Remove the dressing from its wrapper.

(2) Grasp the tails in both hands.

(3) Hold the dressing directly over the wound with the white side down and place it directly on the wound (Figure 3-27A). **DO NOT** touch the white (sterile) side of the dressing or allow anything except the wound to come in contact with it.

(4) Hold the dressing in place with one hand (the casualty may assist if able). Wrap the top tail over the top of the head and bring it down in front of the ear (on the side opposite the wound), under the chin (Figure 3-27B) and up over the dressing to a point just above the ear (on the wound side).

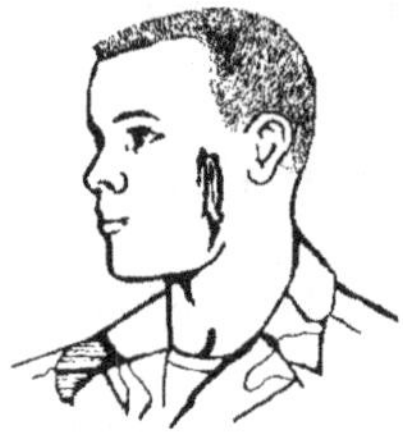

Figure 3-26. Side of head or cheek wound.

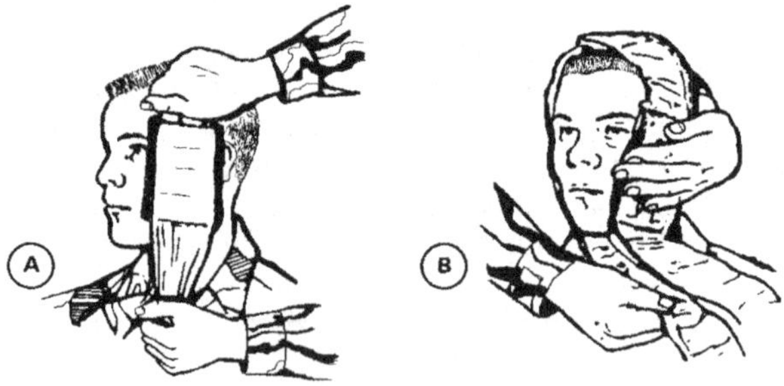

Figure 3-27. Dressing placed directly on wound. Top tail wrapped over top of head, down in front of ear, and under chin (Illustrated A—B).

NOTE

When possible, avoid covering the casualty's ear with the dressing, as this will decrease his ability to hear.

(5) Bring the second tail under the chin, up in front of the ear (on the side opposite the wound), and over the head to meet the other tail (on the wounded side) (Figure 3-28).

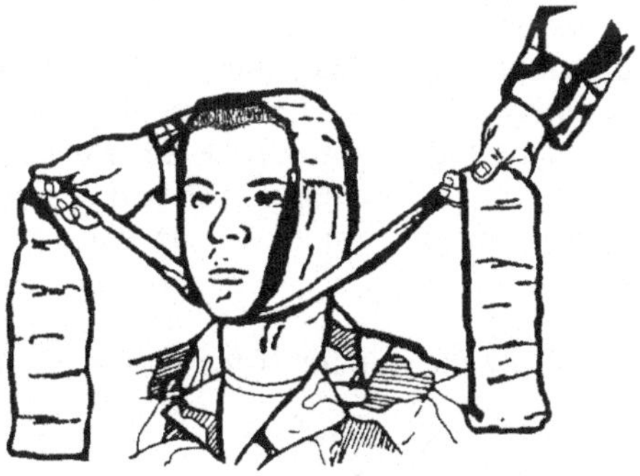

Figure 3-28. Bringing second tail under the chin.

(6) Cross the two tails (on the wound side) (Figure 3-29) and bring one end across the forehead (above the eyebrows) to a point just in front of the opposite ear (on the uninjured side).

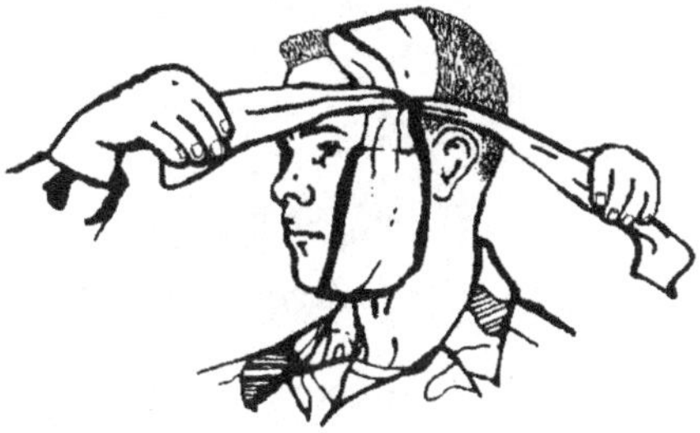

Figure 3-29. Crossing the tails on the side of the wound

(7) Wrap the other tail around the back of the head (at the base of the skull), and tie the two ends just in front of the ear on the uninjured side with a square knot (Figure 3-30).

Figure 3-30. Tying the tails of the dressing in a square knot.

d. Ear Injuries. Lacerated (cut) or avulsed (torn) ear tissue may not, in itself, be a serious injury. Bleeding, or the drainage of fluids from the ear canal, however, may be a sign of a head injury, such as a skull fracture. **DO NOT** attempt to stop the flow from the inner ear canal nor put anything into the ear canal to block it. Instead, you should cover the ear lightly with a dressing. For minor cuts or wounds to the external ear, apply a cravat bandage as follows:

(1) Place the middle of the bandage over the ear (Figure 3-31A).

(2) Cross the ends, wrap them in opposite directions around the head, and tie them (Figures 3-31B and 3-31C).

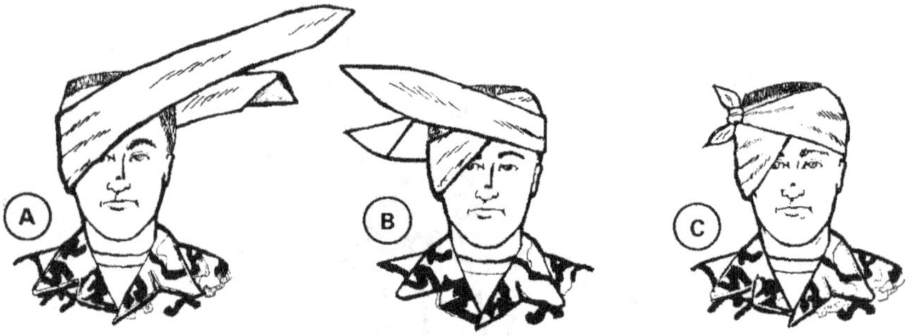

Figure 3-31. Applying cravat bandage to ear (Illustrated A—C).

(3) If possible, place some dressing material between the back of the ear and the side of the head to avoid crushing the ear against the head with the bandage.

e. Nose Injuries. Nose injuries generally produce bleeding. The bleeding may be controlled by placing an ice pack (if available) over the nose, or pinching the nostrils together. The bleeding may also be controlled by placing torn gauze (rolled) between the upper teeth and the lip.

CAUTION

DO NOT attempt to remove objects inhaled into the nose. An untrained person who removes such an object could worsen the casualty's condition and cause permanent injury.

f. Jaw Injuries. Before applying a bandage to a casualty's jaw, remove all loose or free-floating foreign material from the casualty's mouth.

If the casualty is unconscious, check for obstructions in the airway and remove if possible. If there is profuse bleeding in the oral cavity, the cavity may require loose packing with soft bandaging material (for example: Kerlix™ gauze) prior to applying a bandage. Care should be taken to avoid occluding the airway. When applying the bandage, allow the jaw enough freedom to permit passage of air and drainage from the mouth.

(1) *Apply bandages attached to field first aid dressing to the jaw.* After dressing the wound, apply the bandages using the same technique illustrated in Figure 3-32A—C.

NOTE

The dressing and bandaging procedure outlined for the jaw serves a twofold purpose. In addition to stopping the bleeding and protecting the wound, it also immobilizes a fractured jaw.

(2) *Apply a cravat bandage to the jaw.*

(*a*) Place the bandage under the chin and pull its ends upward. Adjust the bandage to make one end longer than the other (Figure 3-32A).

(*b*) Take the longer end over the top of the head to meet the short end at the temple and cross the ends over (Figure 3-32B).

(*c*) Take the ends in opposite directions to the other side of the head and tie them over the part of the bandage that was applied first (Figure 3 32C).

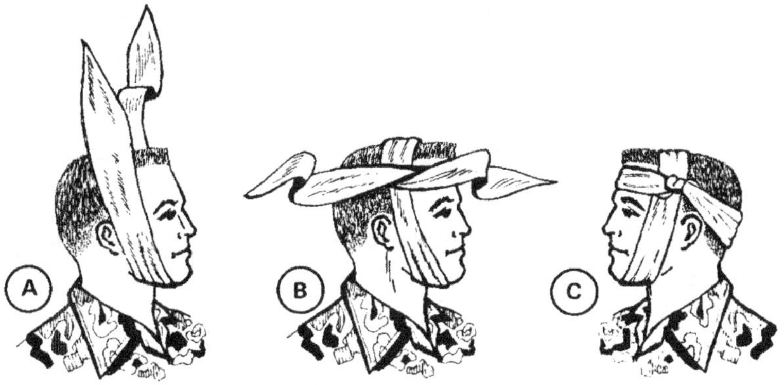

Figure 3-32. Applying a cravat bandage to jaw (Illustrated A—C).

NOTE

The cravat bandage technique is used to immobilize a fractured jaw or to maintain a sterile dressing that does not have tail bandages attached.

3-11. Shoulder Bandage

a. To apply bandages attached to the field first aid dressing—

(1) Take one bandage across the chest and the other across the back and under the arm opposite the injured shoulder.

(2) Tie the ends with a square knot (Figure 3-33).

Figure 3-33. Shoulder bandage.

b. To apply a cravat bandage to the shoulder or armpit—

(1) Make an extended cravat bandage by using two triangular bandages (Figure 3-34A); place the end of the first triangular bandage along the base of the second one (Figure 3-34B).

(2) Fold the two bandages into a single extended bandage (Figure 3-34C).

(3) Fold the extended bandage into a single cravat bandage (Figure 3-34D). After folding, secure the thicker part (overlap) with two or more safety pins (Figure 3-34E).

(4) Place the middle of the cravat bandage under the armpit so that the front end is longer than the back end and safety pins are on the outside (Figure 3-34F).

(5) Cross the ends on top of the shoulder (Figure 3-34G).

(6) Take one of the bandage ends across the back and under the arm on the opposite side and the other end across the chest. Tie the ends (Figure 3-34H).

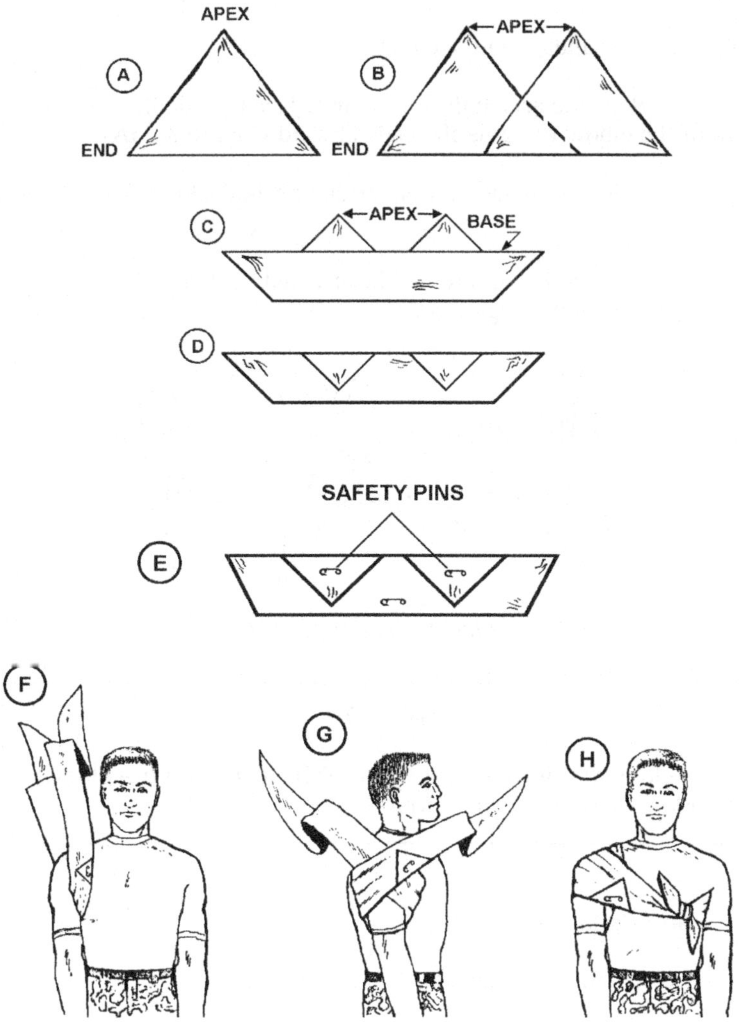

Figure 3-34. Extended cravat bandage applied to shoulder or armpit (Illustrated A—H).

Be sure to place sufficient wadding in the armpit. **DO NOT** tie the cravat bandage too tightly. Avoid compressing the major blood vessels in the armpit.

3-12. Elbow Bandage

To apply a cravat bandage to the elbow—

a. Bend the arm at the elbow and place the middle of the cravat at the point of the elbow bringing the ends upward (Figure 3-35A).

b. Bring the ends across, extending both downward (Figure 3-35B).

c. Take both ends around the arm and tie them with a square knot at the front of the elbow (Figure 3-35C).

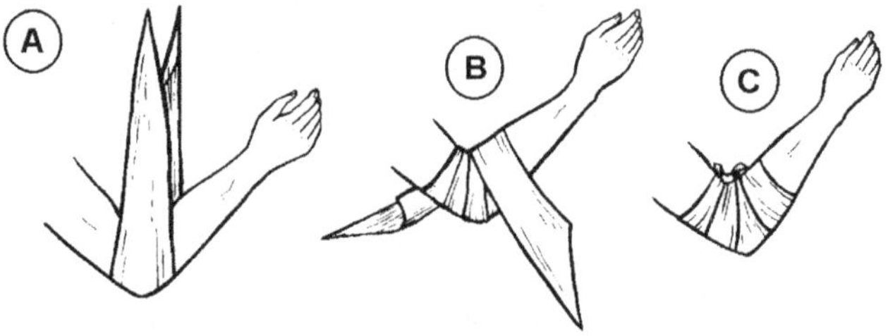

Figure 3-35. Elbow bandage (Illustrated A—C).

> **CAUTION**
>
> If an elbow fracture is suspected, **DO NOT** bend the elbow; bandage it in the position found.

3-13. Hand Bandage

a. To apply a triangular bandage to the hand—

(1) Place the hand in the middle of the triangular bandage with the wrist at the base of the bandage (Figure 3-36A). Ensure that the fingers are separated with absorbent material to prevent chafing and irritation of the skin.

(2) Place the apex over the fingers and tuck any excess material into the pleats on each side of the hand (Figure 3-36B).

(3) Cross the ends on top of the hand, take them around the wrist, and tie them (Figures 3-36C—E) with a square knot.

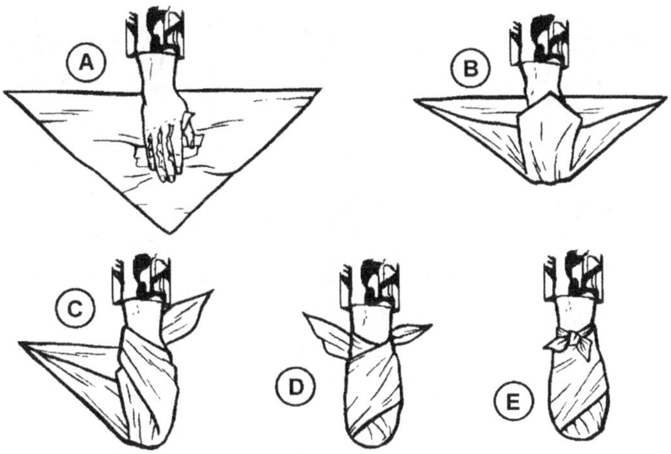

Figure 3-36. Triangular bandage applied to hand (Illustrated A—E).

b. To apply a cravat bandage to the palm of the hand—

(1) Lay the middle of the cravat over the palm of the hand with the ends hanging down on each side (Figure 3-37A).

(2) Take the end of the cravat at the little finger across the back of the hand, extending it upward over the base of the thumb; then bring it downward across the palm (Figure 3-37B).

(3) Take the thumb end across the back of the hand, over the palm, and through the hollow between the thumb and palm (Figure 3-37C).

(4) Take the ends to the back of the hand and cross them; then bring them up over the wrist and cross them again (Figure 3-37D).

(5) Bring both ends down and tie them with a square knot on top of the wrist (Figure 3-37E—F).

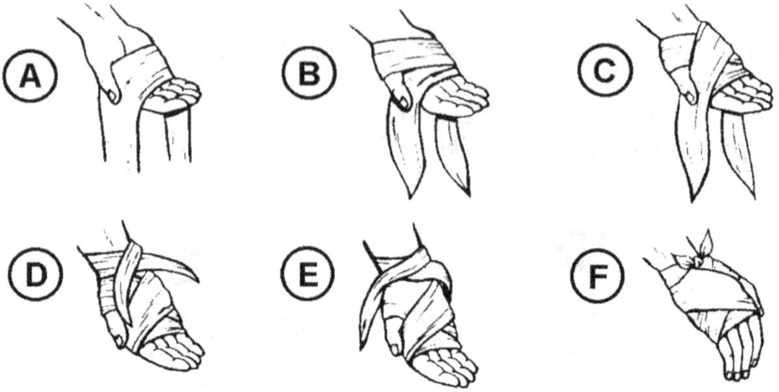

Figure 3-37. Cravat bandage applied to palm of hand (Illustrated A—F).

3-14. Leg (Upper and Lower) Bandage

To apply a cravat bandage to the leg—

a. Place the center of the cravat over the dressing (Figure 3-38A).

b. Take one end around and up the leg in a spiral motion and the other end around and down the leg in a spiral motion, overlapping part of each preceding turn (Figure 3-38B).

c. Bring both ends together and tie them (Figure 3-38C) with a square knot.

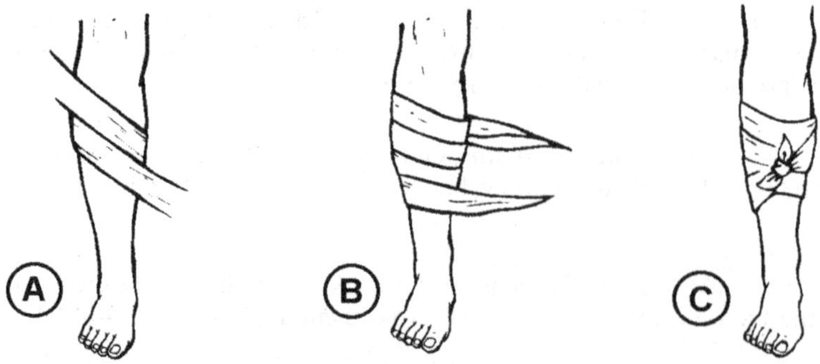

Figure 3-38. Cravat bandage applied to leg (Illustrated A—C).

3-15. Knee Bandage

To apply a cravat bandage to the knee as illustrated in Figure 3-39, use the same technique applied in bandaging the elbow.

> **CAUTION**
>
> If a fracture of the kneecap is suspected, **DO NOT** bend the knee; bandage it in the position found.

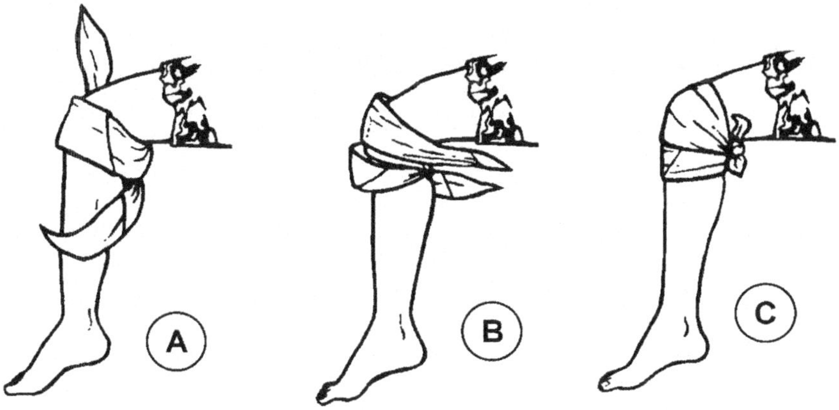

Figure 3-39. Cravat bandage applied to knee (Illustrated A—C).

3-16. Foot Bandage

To apply a triangular bandage to the foot—

a. Place the foot in the middle of the triangular bandage with the heel well forward of the base (Figure 3-40A). Ensure that the toes are separated by absorbent material to prevent chafing and irritation of the skin.

b. Place the apex over the top of the foot and tuck any excess material into the pleats on each side of the foot (Figure 3-40B).

c. Cross the ends on top of the foot, take them around the ankle, and tie them at the front of the ankle (Figure 3-40C—E).

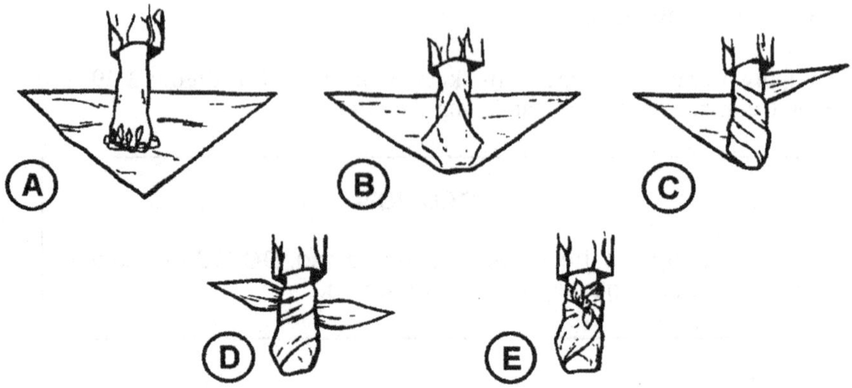

Figure 3-40. Triangular bandage applied to foot (Illustrated A—E).

CHAPTER 4

FIRST AID FOR FRACTURES

4-1. General

A fracture is any break in the continuity of a bone. Fractures can cause total disability or in some cases death by severing vital organs and/or arteries. On the other hand, they can most often be treated so there is a complete recovery. The potential for recovery depends greatly upon the first aid the individual receives before he is moved. First aid includes immobilizing the fractured part in addition to applying lifesaving measures when necessary. The basic splinting principle is to immobilize the joints above and below the fracture.

4-2. Kinds of Fractures

Figure 4-1 depicts types of fractures.

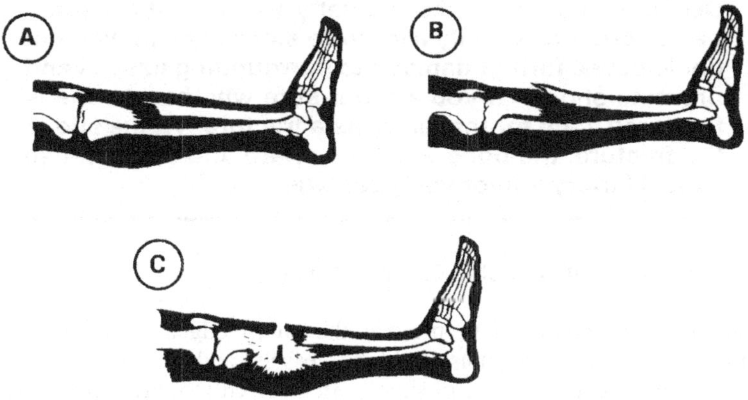

Figure 4-1. Types of fractures (Illustrated A—C).

a. Closed Fracture (Figure 4-1A). A closed fracture is a broken bone that does not break the overlying skin. The tissue beneath the skin may be damaged. A *dislocation* is when a joint, such as a knee, ankle, or shoulder, is not in the proper position. A *sprain* is when the connecting tissues of the joints have been torn. Dislocations and sprains (swelling, possible deformity, and discoloration) should be treated as closed fractures.

b. Open Fracture (Figure 4-1B and 4-1C). An open fracture is a broken bone that breaks (pierces) the overlying skin. The broken bone may come through the skin or a missile such as a bullet or shell fragment may go through the flesh and break the bone.

NOTE

An open fracture is contaminated and subject to infection.

4-3. Signs and Symptoms of Fractures

Indications of a fracture are deformity, tenderness, swelling, pain, inability to move the injured part, protruding bone, bleeding, or discolored skin at the injury site. A sharp pain when the service member attempts to move the part is also a sign of a fracture.

WARNING

DO NOT encourage the casualty to move the injured part in order to identify a fracture since such movement could cause further damage to surrounding tissues and promote shock. If you are not sure whether a bone is fractured, care for the injury as a fracture. At the site of the fracture, the bone ends are sharp and could cause vessel (artery and/or vein) damage.

4-4. Purposes of Immobilizing Fractures

A fracture is immobilized to prevent the sharp edges of the bone from moving and cutting tissue, muscle, blood vessels, and nerves. This reduces pain and helps prevent or control shock. In a closed fracture, immobilization keeps bone fragments from causing an open wound, which can become contaminated and subject to infection.

4-5. Splints, Padding, Bandages, Slings, and Swathes

a. *Splints*. Splints may be improvised from such items as boards, poles, sticks, tree limbs, or cardboard. If nothing is available for a splint, the chest wall can be used to immobilize a fractured arm and the uninjured leg can be used to immobilize (to some extent) the fractured leg.

b. *Padding*. Padding may be improvised from such items as a jacket, blanket, poncho, shelter half, or leafy vegetation.

c. Bandages. Bandages may be improvised from belts, rifle slings, kerchiefs, or strips torn from clothing or blankets. Narrow materials such as wire or cord should not be used to secure a splint in place. The application of wire and/or narrow material to an extremity could cause tissue damage and a tourniquet effect.

d. Slings. A sling is a bandage suspended from the neck to support an upper extremity. If a bandage is not available, a sling can be improvised by using the tail of a coat or shirt or pieces of cloth torn from such items as clothing and blankets. The triangular bandage is ideal for this purpose. Remember that the casualty's hand should be higher than his elbow, and the fingers should be showing at all times. The sling should be applied so that the supporting pressure is on the uninjured side.

e. Swathes. Swathes are any bands (pieces of cloth or load bearing equipment [LBE]) that are used to further immobilize a splinted fracture. Triangular and cravat bandages are often used and are called *swathe bandages*. The purpose of the swathe is to immobilize; therefore, the swathe bandage is placed above and/or below the fracture—not over it.

4-6. Procedures for Splinting Suspected Fractures

Before beginning first aid procedures for a fracture, gather whatever splinting materials are available. Ensure that splints are long enough to immobilize the joint above and below the suspected fracture. If possible, use at least four ties (two above and two below the fracture) to secure the splints. The ties should be square knots and should be tied away from the body on the splint. Distal pulses of the affected extremity should be checked before and after the application of the splint.

a. Evaluate the Casualty. Be prepared to perform any necessary lifesaving measures. Monitor the casualty for development of conditions that may require you to perform necessary lifesaving measures.

WARNING

Unless there is immediate life-threatening danger, such as a fire or an explosion, DO NOT move the casualty with a suspected back or neck injury. Improper movement may cause permanent paralysis or death.

WARNING

In a chemical environment, DO NOT remove any protective clothing. Apply the dressings and splints over the garments.

b. *Locate the Site of the Suspected Fracture.*

(1) Ask the casualty for the location of the injury.

- Does he have any pain?
- Where is it tender?
- Can he move the extremity?

NOTE

With the presence of an obvious deformity, do not make the casualty move extremity.

(2) Look for an unnatural position of the extremity.

(3) Look for a bone sticking out (protruding).

c. *Prepare the Casualty for Splinting the Suspected Fracture.*

(1) Reassure the casualty. Tell him that you will be providing first aid for him and that medical help is on the way.

(2) Loosen any tight or binding clothing.

(3) Remove all jewelry from the injured part and place it in the casualty's pocket. Tell the casualty you are doing this because if the jewelry is not removed and swelling occurs later, he may not be able to get it off and further bodily injury could result.

(4) Boots should not be removed from the casualty unless they are needed to stabilize a neck injury or there is actual bleeding from the foot.

d. *Gather Splinting Materials.* If standard splinting materials (splints, padding, and cravats) are not available, gather improvised materials. If splinting material is not available and the suspected fracture **CANNOT** be splinted, then swathes, or a combination of swathes and slings can be used to immobilize the extremity.

e. Pad the Splints. Pad the splints where they touch any bony part of the body, such as the elbow, wrist, knee, ankle, crotch, or armpit areas. Padding prevents excessive pressure on the area, which could lead to circulation problems.

f. Check the Circulation Below the Site of the Injury.

(1) Note any pale, white, or bluish-gray color of the skin, which may indicate impaired circulation. Circulation can also be checked by depressing the toe or fingernail beds and observing how quickly the color returns. A slower return of color to the injured side when compared with the uninjured side indicates a problem with circulation. The fingernail bed is the method to use to check the circulation in a dark-skinned casualty.

(2) Check the temperature of the injured extremity. Use your hand to compare the temperature of the injured side with the uninjured side. The body area below the injury may be colder to the touch indicating poor circulation.

(3) Question the casualty about the presence of numbness, tightness, cold, or tingling sensations.

WARNING

Casualties with fractures of the extremities may show impaired circulation, such as numbness, tingling, cold or pale to bluish skin tone. These casualties should be evacuated by medical personnel and treated as soon as possible. Prompt medical treatment may prevent possible loss of the limb.

WARNING

If it is an open fracture and the bone is protruding from the skin, *DO NOT ATTEMPT TO PUSH THE BONE BACK UNDER THE SKIN.* Apply a field dressing over the wound to protect the area.

g. Apply the Splint in Place.

(1) Splint the fracture in the position found. **DO NOT** attempt to reposition or straighten the injury. If it is an open fracture, stop the bleeding and protect the wound. Cover all wounds with field dressings before applying a splint. Remember to use the casualty's field dressing, not your own.

(2) Place one splint on each side of the fracture. Make sure that the splints reach, if possible, beyond the joints above and below the fracture.

(3) Tie the splints. Secure each splint in place above and below the fracture site with improvised (or actual) cravats. Improvised cravats, such as strips of cloth, belts, or whatever else you have, may be used. With minimal motion to the injured areas, place and tie the splints with the bandages. Push cravats through and under the natural body curvatures, and then gently position improvised cravats and tie in place. Use square knots. Tie all knots on the splint away from the casualty (Figure 4-2). **DO NOT** tie cravats directly over the suspected fracture site.

Figure 4-2. Square knots tied away from casualty.

h. Check the Splint for Tightness.

(1) **CHECK** to be sure that bandages are tight enough to securely hold splinting materials in place, but not so tight that circulation is impaired.

(2) **RECHECK** the circulation after application of the splint. Check the skin color and temperature. This is to ensure that the bandages holding the splint in place have not been tied too tightly. A fingertip check can be made by inserting the tip of the finger between the bandaged knot and the skin.

(3) **MAKE** any necessary adjustment without allowing the splint to become ineffective.

i. Apply a Sling. An improvised sling may be made from any available nonstretching piece of cloth, such as a battle dress uniform (BDU) shirt or trousers, poncho, or shelter half. Slings may also be improvised using the tail of a coat, belt, or a piece of cloth. Figure 4-3 depicts a shirttail used for support. A trousers belt or LBE may also be used for support (Figure 4-4). A sling should place the supporting pressure on the casualty's uninjured side. The supported arm should have the hand positioned slightly higher than the elbow showing the fingers.

Figure 4-3. Shirttail used for support.

Figure 4-4. Belt used for support.

(1) Insert the splinted arm in the center of the sling (Figure 4-5).

Figure 4-5. Arm inserted in center of improvised sling.

(2) Bring the ends of the sling up and tie them at the side (or hollow) of the neck on the uninjured side (Figure 4-6).

Figure 4-6. Ends of improvised sling tied to side of neck.

(3) Twist and tuck the corner of the sling at the elbow (Figure 4-7).

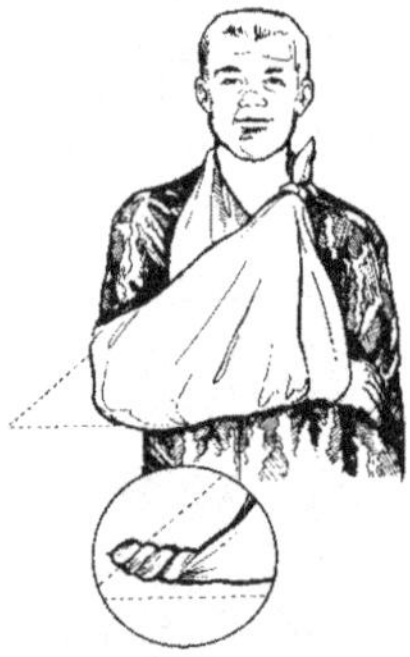

Figure 4-7. Corner of sling twisted and tucked at elbow.

j. Apply a Swathe. You may use any large piece of cloth, service member's belt, or pistol belt, to improvise a swathe.

WARNING

The swathe should not be placed directly on top of the injury, but positioned either above or below the fracture site.

(1) Apply swathes to the injured arm by wrapping the swathe over the injured arm, around the casualty's back, and under the arm on the uninjured side. Tie the ends on the uninjured side (Figure 4-8).

Figure 4-8. Arm immobilized with strip of clothing.

(2) A swathe is applied to an injured leg by wrapping the swathe around both legs and securing it on the uninjured side.

k. Seek Medical Assistance. Notify medical personnel, watch closely for development of life-threatening conditions and/or impaired circulation to the injured extremity. (Refer to Chapter 1 for additional information on life-threatening conditions.)

4-7. Upper Extremity Fractures

Figures 4-9 through 4-17 show how to apply slings, splints, and cravats (swathes) to immobilize and support fractures of the upper extremities. *Although the padding is not visible in some of the illustrations, it is always preferable to apply padding along the injured part for the length of the splint and especially where it touches any bony parts of the body.*

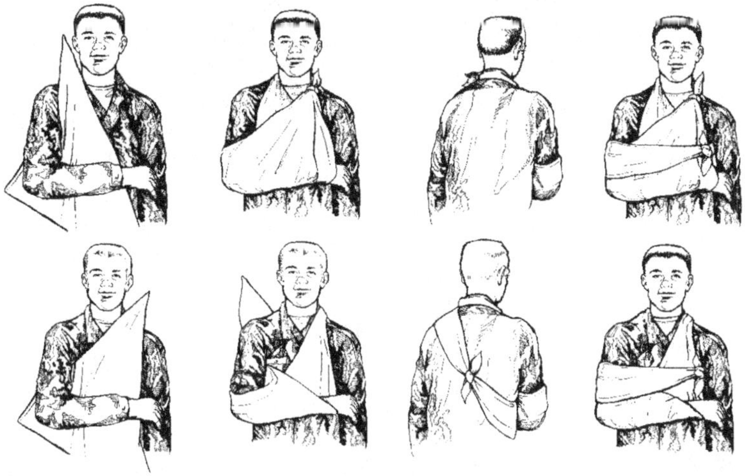

Figure 4-9. Application of triangular bandage to form sling (two methods).

Figure 4-10. Completing sling sequence by twisting and tucking the corner of the sling at the elbow.

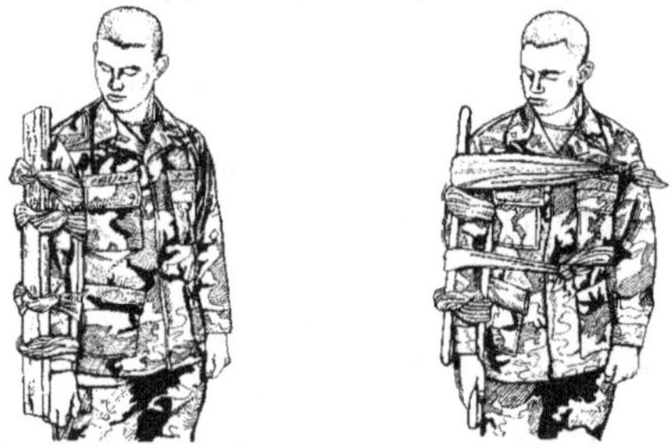

Figure 4-11. Board splints applied to fractured elbow when elbow is not bent (two methods).

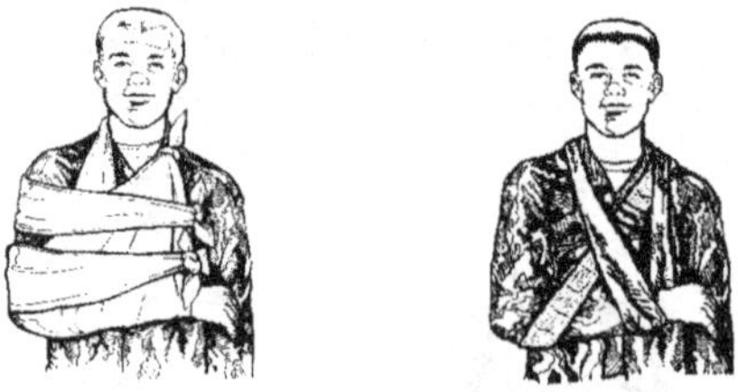

Figure 4-12. Chest wall used as splint for upper arm fracture when no splint is available.

Figure 4-13. Chest wall, sling, and cravat used to immobilize fractured elbow when elbow is bent.

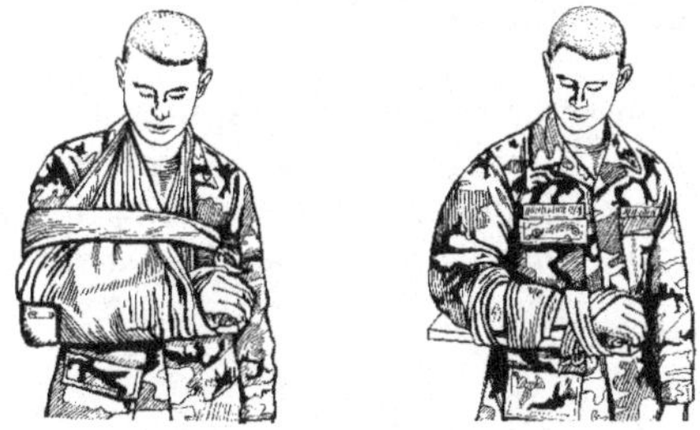

Figure 4-14. Board splint applied to fractured forearm.

Figure 4-15. Fractured forearm or wrist splinted with sticks and supported with tail of shirt and strips of material.

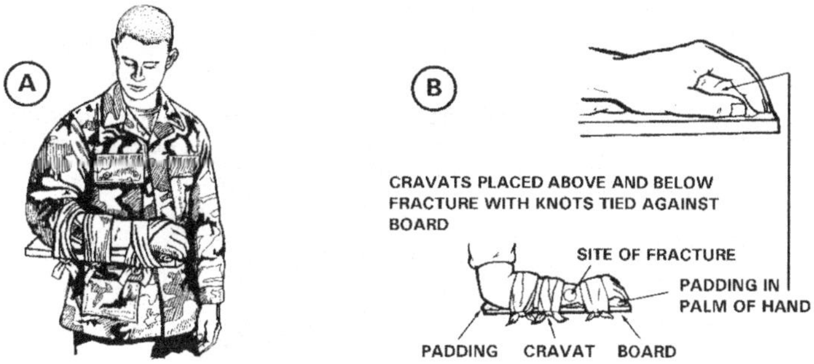

Figure 4-16. Board splint applied to fractured wrist and hand (Illustrated A—B).

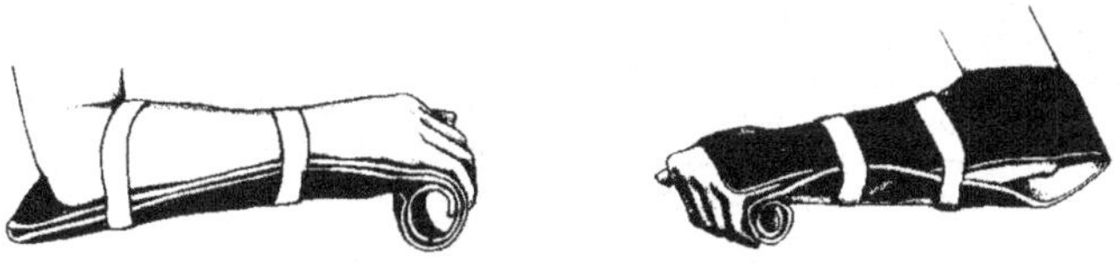

Figure 4-17. SAM® splint applied to fractured wrist or forearm.

4-8. Lower Extremity Fractures

Figures 4-18 through 4-24 show how to apply splints to immobilize fractures of the lower extremities. *Although padding is not visible in some of the illustrations, it is always preferable to apply padding along the injured part for the length of the splint and especially where it touches any bony parts of the body.*

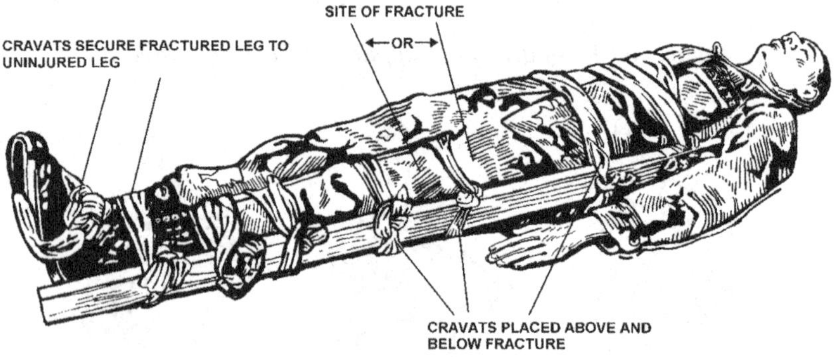

Figure 4-18. Board splints applied to fractured hip or thigh.

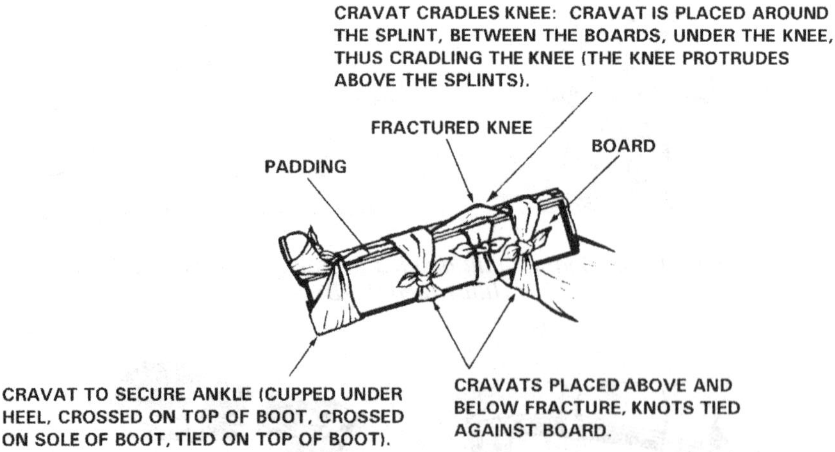

Figure 4-19. Board splint applied to fractured or dislocated knee.

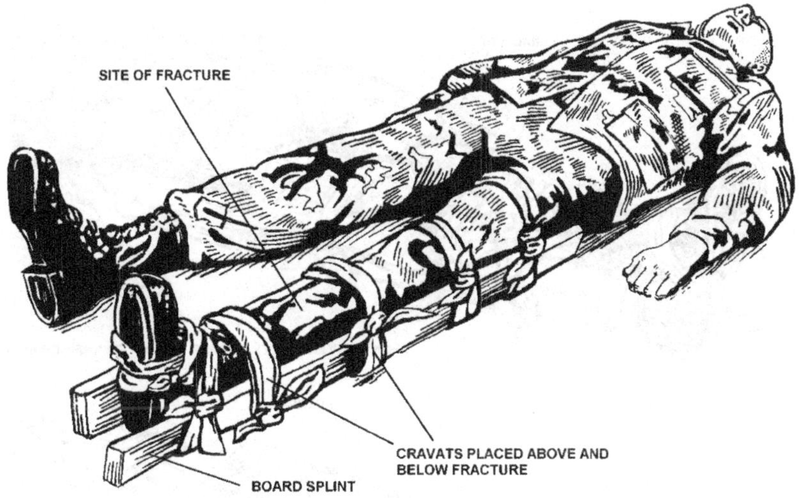

Figure 4-20. Board splints applied to fractured lower leg or ankle.

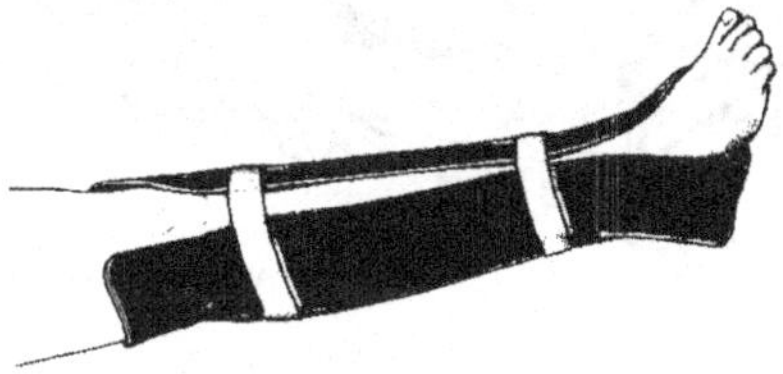

Figure 4-21. SAM® splint applied to fractured lower leg or ankle.

Figure 4-22. Improvised splints applied to fractured lower leg or ankle.

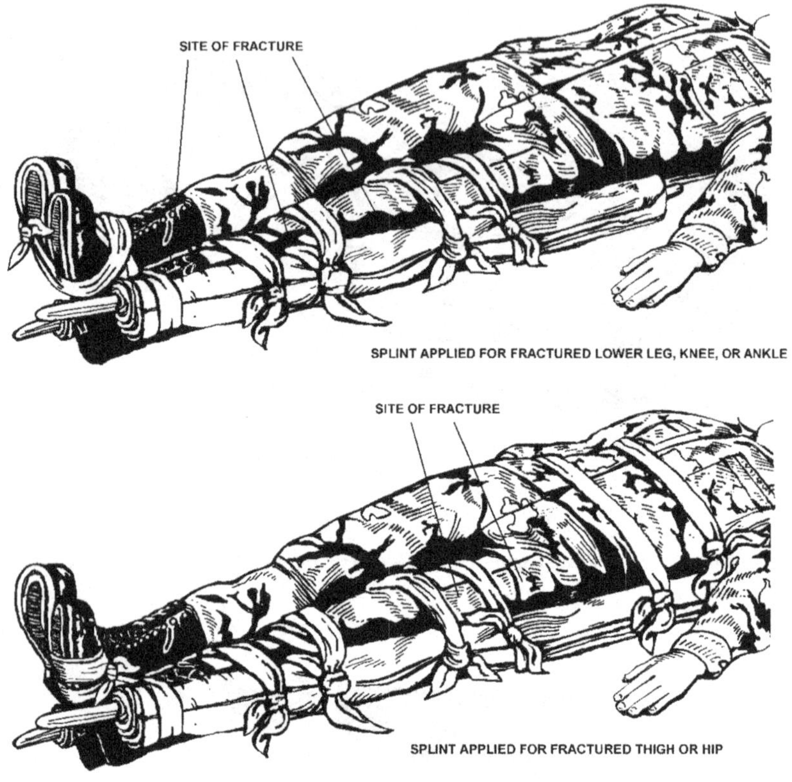

Figure 4-23. Poles rolled in a blanket and used as splints applied to fractured lower extremity.

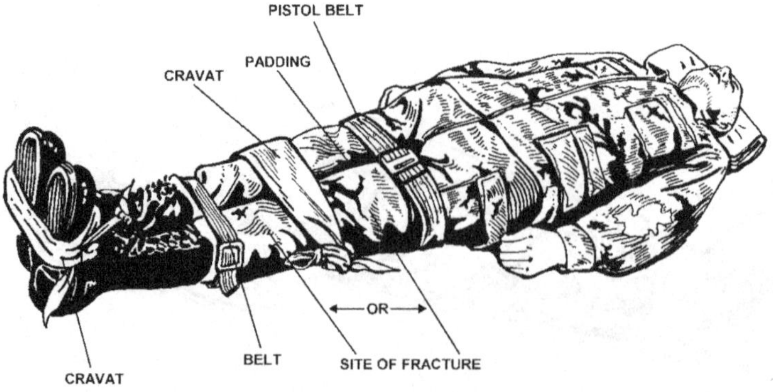

Figure 4-24. Uninjured leg used as splint for fractured leg (anatomical splint).

4-9. Jaw, Collarbone, and Shoulder Fractures

a. Apply a cravat to immobilize a fractured jaw as illustrated in Figure 4-25. Direct all bandaging support to the top of the casualty's head, not to the back of his neck. If incorrectly placed, the bandage will pull the casualty's jaw back and interfere with his breathing.

Figure 4-25. Fractured jaw immobilized.

WARNING

Casualties with lower jaw (mandible) fractures cannot be laid flat on their backs because facial muscles will relax and may cause an airway obstruction.

b. Apply two belts, a sling, and a cravat to immobilize a fractured collarbone, as illustrated in Figure 4-26.

Figure 4-26. Application of belts, sling, and cravat to immobilize a fractured collarbone.

c. Apply a sling and a cravat to immobilize a fractured or dislocated shoulder, using the technique illustrated in Figure 4-27.

Figure 4-27. Application of sling and cravat to immobilize a fractured or dislocated shoulder.

4-10. Spinal Column Fractures

It is often impossible to be sure a casualty has a fractured spinal column. Be suspicious of any back injury, especially if the casualty has fallen or if his back has been sharply struck or bent. If a casualty has received such an injury and does not have feeling in his legs or cannot move them, you can be reasonably sure that he has a severe back injury, which should be managed as a fracture. Remember, that the possibility of a neck fracture or injury to the back should always be suspected, and it is often impossible to be sure if a casualty has a fractured spinal column. If the spine is fractured, bending it can cause the sharp bone fragments to bruise or cut the spinal cord and result in permanent paralysis or death (Figure 4-28A). The spinal column must maintain normal spinal position at the lower back (lumbar region) to help remove pressure from the spinal cord.

a. If the casualty is not to be transported until medical personnel arrive—

- Caution him not to move. Ask him if he is in pain or if he is unable to move any part of his body.

- Leave him in the position in which he is found. **DO NOT** move any part of his body, unless he is in imminent danger.

- If the casualty is lying face up, slip a blanket or other supporting material under the arch of his lower back to help support the spine in a normal position (Figure 4-28B). Take care not to place so much bulky padding as to cause potential damage by causing undo pressure on the spine. If he is lying face down, **DO NOT** put anything under any part of his body.

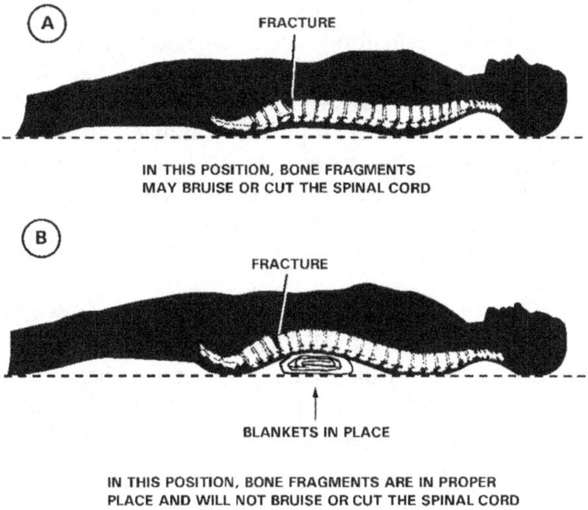

Figure 4-28. Spinal column must maintain a normal spine position.

b. If the casualty must be transported to a safe location before medical personnel arrive and if the casualty is in a—

- Face-up position, transport him by litter or use a firm substitute, such as a wide board or a door longer than his height. Loosely tie the casualty's wrists together over his waistline, using a cravat or a strip of cloth. Tie his feet together to prevent the accidental dropping or shifting of his legs. Lay a folded blanket across the litter where the arch of his back is to be placed. Using a four-man team (Figure 4-29), place the casualty on the litter without bending his spinal column or his neck.

Figure 4-29. Placing face-up casualty with fractured back onto litter.

- The number two man positions himself at the casualty's head. His responsibility is to provide manual in-line (neutral) stabilization of the head and neck. The number three, and four men position themselves on one side of the casualty; all kneel on one knee along the side of the casualty. The number one man positions himself to the opposite side of the casualty (or can be on the same side of number three and four). The numbers two, three, and four men gently place their hands under the casualty. The number one man on the opposite side places his hands under the injured part to assist.

- When all four men are in position to lift, the number two man commands, "**PREPARE TO LIFT**" and then, "**LIFT**." All men, in unison, gently lift the casualty about 8 inches. Once the casualty is lifted, the number one man recovers and slides the litter under the casualty, ensuring that the blanket is in proper position. The number one man then returns to his original lift position (Figure 4-29).

- When the number two man commands, "**LOWER CASUALTY**," all men, in unison, gently lower the casualty onto the litter.

- Facedown position, he must be transported in this same position. The four-man team lifts him onto a regular or improvised litter, keeping the spinal column in a normal spinal position. If a regular litter is used, first place a folded blanket on the litter at the point where the chest will be placed.

4-11. Neck Fractures

A fractured neck is extremely dangerous. Bone fragments may bruise or cut the spinal cord just as they might in a fractured back.

a. If the casualty is not to be transported until medical personnel arrive—

- Caution him not to move. Moving may cause permanent injury or death.

- Leave the casualty in the position in which he is found. If his neck and head (cervical spine) are in an abnormal position, immediately immobilize the neck and head.

- Keep his head still, if the casualty is lying face up, raise his shoulders slightly, and slip a roll of cloth that has the bulk of a bath towel under his neck (Figure 4-31). The roll should be thick enough to arch his neck only slightly, leaving the back of his head on the ground. **DO NOT** bend his neck or head forward. **DO NOT** raise or twist his head. Immobilize the casualty's head (Figure 4-32). Do this by padding heavy objects (such as rocks or his boots filled with dirt, sand, gravel, or rock) and placing them on each side of his head. If it is necessary to use boots, after filling them, tie the top tightly or stuff with pieces of cloth to secure the contents.)

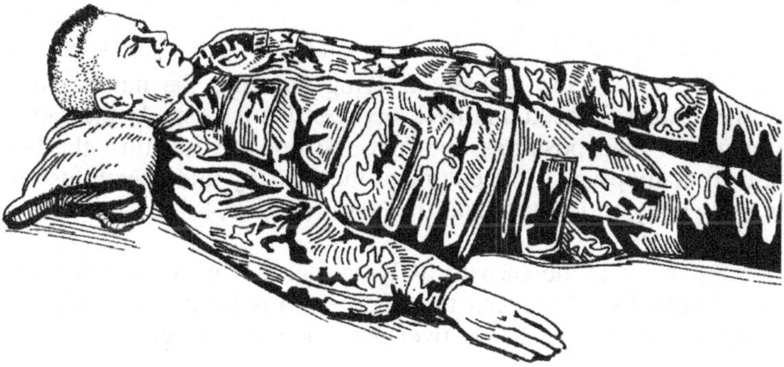

Figure 4-30. Casualty with roll of cloth (bulk) under neck.

Figure 4-31. Immobilization of fractured neck.

- **DO NOT** move him if the casualty is lying face down. Immobilize the head and neck by padding heavy objects and placing them on each side of his head. DO NOT put a roll of cloth under the neck. **DO NOT** bend the neck or head, nor roll the casualty onto his back.

b. If the casualty must be prepared for transportation before medical personnel arrive—

- If the casualty has a fractured neck, at least two persons are needed because the casualty's head and trunk must be moved in unison. The two persons must work in close coordination (Figure 4-32) to avoid bending of the neck.

- A wide board is placed lengthwise beside the casualty. It should extend at least 4 inches beyond the casualty's head and feet (Figure 4-32A).

- If the casualty is lying face up, the number one man steadies the casualty's head and neck between his hands. At the same time, the number two man positions one foot and one knee against the board to prevent it from slipping. He then grasps the casualty underneath his shoulder and hip and gently slides him onto the board (Figure 4-32B).

- If the casualty is lying face down, the number one man steadies the casualty's head and neck between his hands, while the number two man gently rolls the casualty over onto the board (Figure 4-32C).

- The number one man continues to steady the casualty's head and neck. The number two man simultaneously raises the casualty's shoulders slightly, places padding under his neck, and immobilizes the casualty's head (Figures 4-32D—E).

- Any improvised supports are secured in position with a cravat or strip of cloth extended across the casualty's forehead and under the board (Figure 4-32D).

- The board is lifted onto a litter or blanket in order to transport the casualty (Figure 4-32E).

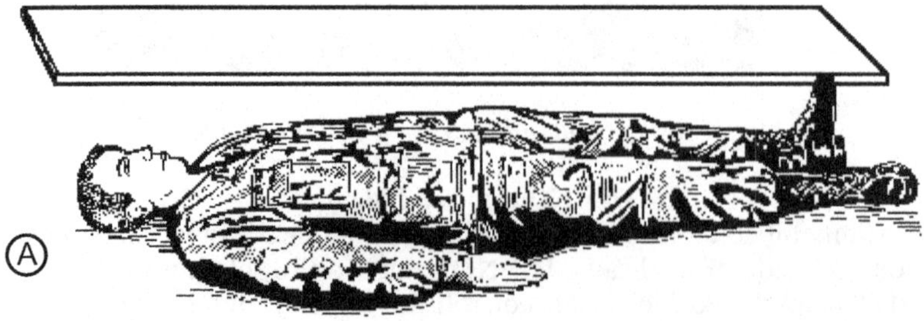

Figure 4-32. Preparing casualty with fractured neck for transportation (Illustrated A—E).

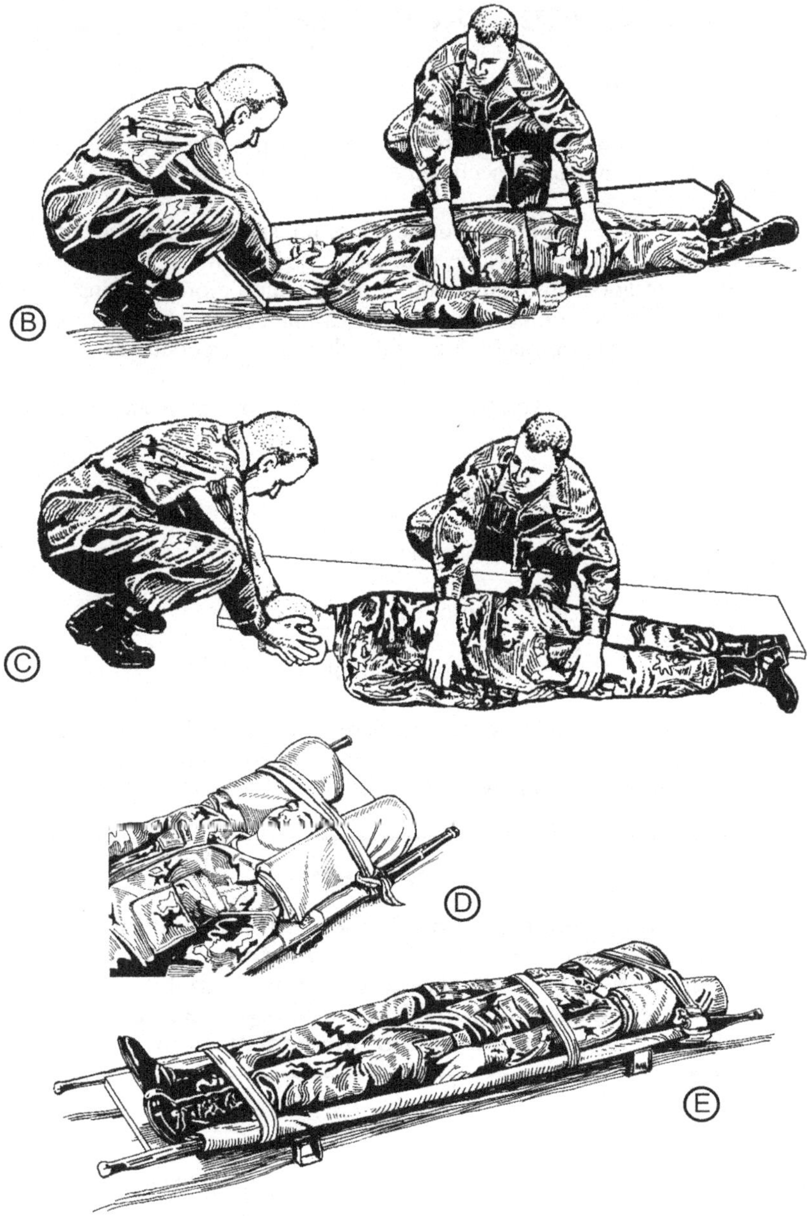

Figure 4-32. Preparing casualty with fractured neck for transportation (Illustrated A—E) (Continued).

CHAPTER 5

FIRST AID FOR CLIMATIC INJURIES

5-1. General

a. It is desirable, but not always possible, for an individual's body to become adjusted (acclimated) to an environment.

(1) The service members physical condition determines the amount of time their bodies need to adjust to the environment. Even those individuals in good physical condition need time before working or training in extremes of hot or cold weather. Climate-related injuries are usually preventable; prevention is both an individual and leadership responsibility.

(2) Several factors contribute to health and well-being in any environment—

- Diet.
- Sleep and rest.
- Exercise.
- Suitable clothing.

(3) Diet should be suited to an individual's needs in a particular climate. A special diet started for any purpose (such as weight reduction) should be done with appropriate medical supervision.

WARNING

Service members should use extreme caution when starting fad diets or taking over-the-counter herbal supplements. Medical records revealed that deaths and severe injuries occurred in individuals using dietary/ herbal supplements without medical monitoring.

NOTE

Weight loss and the use of weight loss supplements should be supervised by a trained health care provider.

(4) Specialized clothing and equipment (such as cold weather gear) for a specific environment should be obtained and used properly.

b. For information on the prevention of heat and cold injuries, refer to FM 21-10/Marine Corps Reference Publication (MCRP) 4-11.1D.

5-2. Heat Injuries

a. Heat injuries are environmental injuries. They may result when a service member—

- Is exposed to extreme heat, such as from the sun or from high temperatures.
- Does not wear proper clothing.
- Is in MOPP gear.
- Is inside closed spaces, such as inside an armored vehicle.
- Wears body armor.

b. Heat injury can be divided into three categories: heat cramps, heat exhaustion, and heatstroke.

c. Each service member must be able to recognize and give first aid for heat injuries.

WARNING

The heat casualty should be continually monitored for development of conditions which may require the performance of necessary basic lifesaving measures.

CAUTION

Do not use salt solutions in first aid procedures for heat injuries.

(1) Check the casualty for signs and symptoms of *cramping*.

- *Signs and symptoms.* Cramping is caused by an imbalance of chemicals (called electrolytes) in the body as a result of excessive sweating. This condition causes the casualty to exhibit:

- Cramping in the extremities (arms and legs).
- Abdominal (stomach) cramps.
- Excessive sweating.

NOTE

Thirst may or may not occur. Cramping can occur without the service member being thirsty.

- *First aid measures.*
  - Move the casualty to a cool, shady area or improvise shade if none is available.
  - Loosen his clothing (if not in a chemical environment).

NOTE

In a chemical environment, transport the heat casualty to a noncontaminated area as soon as the mission permits.

  - Have him slowly drink at least one canteen full of water. (The body absorbs cool water faster than warm or cold water; therefore, cool water is preferred if it is available.)
  - Seek medical assistance should cramps continue.

(2) Check the casualty for signs and symptoms of *heat exhaustion*.

- *Signs and symptoms.* Heat exhaustion is caused by loss of body fluids (dehydration) through sweating without adequate fluid replacement. It can occur in an otherwise fit individual who is involved in physical exertion in any hot environment especially if the service member is not acclimatized to that environment. These signs and symptoms are—
  - Excessive sweating with pale, moist, cool skin.
  - Headache.
  - Weakness.

- Dizziness.
- Loss of appetite.
- Cramping.
- Nausea (with or without vomiting).
- Urge to defecate.
- Chills (gooseflesh).
- Rapid breathing.
- Tingling of hands and/or feet.
- Confusion.

- *First aid measures.*

- Move the casualty to a cool, shady area or improvise shade if none is available.
- Loosen or remove his clothing and boots (unless in a chemical environment); pour water on him and fan him.
- Have him slowly drink at least one canteen of water.
- Elevate his legs.
- If possible, the casualty should not participate in strenuous activity for the remainder of the day.
- Monitor the casualty until the symptoms are gone, or medical assistance arrives.

(3) Check the casualty for signs and symptoms of *heatstroke*.

WARNING

Heatstroke is a medical emergency which may result in death if care is delayed.

- *Signs and symptoms.* A service member suffering from heatstroke has been exposed to high temperatures (such as direct sunlight) or been dressed in protective overgarments, which causes the body temperature to rise. Heatstroke occurs more rapidly in service members who are engaged in work or other physical activity in a high heat environment. Heatstroke is caused by a failure of the body's cooling mechanism which includes a decrease in the body's ability to produce sweat. The casualty's skin is red (flushed), hot, and dry. He may experience weakness, dizziness, confusion, headaches, seizures, nausea, stomach pains or cramps, and his respiration and pulse may be rapid and weak. Unconsciousness and collapse may occur suddenly.

- *First aid measures.* Cool casualty immediately by—

  - Moving him to a cool, shady area or improvising shade if none is available.

  - Loosening or removing his clothing (except in a chemical environment).

  - Spraying or pouring water on him; fanning him to permit the coolant effect of evaporation.

  - Massaging his extremities and skin, which increases the blood flow to those body areas, thus aiding the cooling process.

  - Elevating his legs.

  - Having him slowly drink at least one canteen full of water if he is conscious.

NOTE

Start cooling casualty immediately. Continue cooling while awaiting transportation and during transport to an MTF.

- *Medical assistance.* Seek medical assistance because the casualty should be transported to an MTF as soon as possible. Do not interrupt the cooling process or lifesaving measures to seek help; if someone else is present send them for help. The casualty should be continually monitored for development of conditions that may require the performance of necessary basic lifesaving measures.

d. *Table.* See Table 5-1 for further information.

Table 5-1. Heat Injuries

INJURIES	SIGNS AND SYMPTOMS	FIRST AID[1]
HEAT CRAMPS	THE CASUALTY EXPERIENCES MUSCLE CRAMPS OF THE ARMS, LEGS, AND/OR STOMACH. THE CASUALTY MAY ALSO HAVE EXCESSIVE SWEATING.	1. MOVE THE CASUALTY TO A COOL SHADY AREA OR IMPROVISE SHADE AND LOOSEN CLOTHING.[2] 2. HAVE HIM SLOWLY DRINK AT LEAST ONE CANTEEN FULL OF COOL WATER SLOWLY. 3. MONITOR THE CASUALTY AND GIVE HIM MORE WATER AS TOLERATED.
HEAT EXHAUSTION	THE CASUALTY EXPERIENCES HEAVY SWEATING WITH PALE, MOIST, COOL SKIN; HEADACHE, WEAKNESS, DIZZINESS, AND/OR LOSS OF APPETITE, HEAT CRAMPS, NAUSEA (WITH OR WITHOUT VOMITING), URGE TO DEFECATE, CHILLS (GOOSE-FLESH), RAPID BREATHING, CONFUSION, AND TINGLING OF THE HANDS AND/OR FEET.	1. MOVE THE CASUALTY TO A COOL, SHADY AREA OR IMPROVISE SHADE AND LOOSEN OR REMOVE HIS CLOTHING.[2] 2. POUR WATER ON HIM AND FAN HIM TO PERMIT THE COOLANT EFFECT OF EVAPORATION. 3. HAVE HIM SLOWLY DRINK AT LEAST ONE CANTEEN FULL OF COOL WATER. 4. ELEVATE THE CASUALTY'S LEGS. 5. SEEK MEDICAL ASSISTANCE IF SYMPTOMS CONTINUE; MONITOR UNTIL SYMPTOMS ARE GONE OR MEDICAL ASSISTANCE ARRIVES.
HEATSTROKE[3] (SUNSTROKE)	THE CASUALTY STOPS SWEATING (RED [FLUSHED] HOT, DRY SKIN). HE FIRST MAY EXPERIENCE HEADACHE, DIZZINESS, NAUSEA, FAST PULSE AND RESPIRATION, SEIZURES, AND MENTAL CONFUSION. HE MAY COLLAPSE	1. MOVE THE CASUALTY TO A COOL, SHADY AREA OR IMPROVISE SHADE AND LOOSEN OR REMOVE HIS CLOTHING, REMOVE THE OUTER GARMENTS AND PROTECTIVE

Table 5-1. Heat Injuries (Continued)

INJURIES	SIGNS AND SYMPTOMS	FIRST AID[1]
	AND SUDDENLY BECOME UNCONSCIOUS. **THIS IS A MEDICAL EMERGENCY.**	CLOTHING IF THE SITUATION PERMITS.[2] 2. START COOLING THE CASUALTY IMMEDIATELY. SPRAY OR POUR WATER ON HIM. FAN HIM. MASSAGE HIS EXTREMITIES AND SKIN. 3. ELEVATE HIS LEGS. 4. IF CONSCIOUS, HAVE HIM SLOWLY DRINK AT LEAST ONE CANTEEN FULL OF COOL WATER. 5. SEEK MEDICAL AID. CONTINUE COOLING WHILE AWAITING TRANSPORT AND CONTINUE FIRST AID WHILE EN ROUTE.

LEGEND:

1 THE FIRST AID PROCEDURE FOR HEAT RELATED INJURIES CAUSED BY WEARING INDIVIDUAL PROTECTIVE EQUIPMENT (IPE) IS TO MOVE THE CASUALTY TO A CLEAN AREA AND GIVE HIM WATER TO DRINK.
2 WHEN IN A CHEMICAL ENVIRONMENT, DO NOT LOOSEN OR REMOVE THE CASUALTY'S CLOTHING.
3 CAN BE FATAL IF NOT PROVIDED FIRST AID AND MEDICAL TREATMENT PROMPTLY.

5-3. Cold Injuries

Cold injuries are most likely to occur when conditions are moderately cold, but accompanied by wet or windy conditions. Cold injuries can usually be prevented. Well-disciplined and well-trained service members can be protected even in the most adverse circumstances. They and their leaders must know the hazards of exposure to the cold. They must know the importance of personal hygiene, exercise, care of the feet and hands, and the use of protective clothing.

a. Contributing Factors.

(1) Temperature, humidity, precipitation, and wind greatly increase likelihood of cold injuries, and the service members with wet clothing are at great risk of cold injuries. Riverine operations (river, swamp, and stream crossings) increase likelihood of cold injuries. Low temperatures and low relative humidity (dry cold) promote frostbite. Higher temperatures, together with moisture, promote immersion syndrome. Windchill accelerates the loss of body heat and may aggravate cold injuries.

(2) Relatively stationary activities such as being in an observation post or on guard duty increase the service member's vulnerability to cold injury. Also, a service member is more likely to receive a cold injury if he is—

- In contact with the ground (such as marching, performing guard duty, or engaging in other outside activities).
- Immobile for long periods (such as while riding in an unheated or open vehicle).
- Standing in water, such as in a foxhole.
- Out in the cold for days without being warmed.
- Deprived of an adequate diet and rest.
- Not able to take care of his personal hygiene.

(3) Physical fatigue contributes to apathy, which leads to inactivity, personal neglect, carelessness, and reduced heat production. In turn, these increase the risk of cold injury. Service members with prior cold injuries have a higher-than-normal risk of subsequent cold injury; not necessarily involving the body part previously injured.

(4) Depressed or unresponsive service members are also vulnerable because they are less active. These service members tend to be careless about precautionary measures, especially warming activities, when cold injury is a threat.

(5) Excessive use of alcohol or drugs leading to faulty judgment or unconsciousness in a cold environment increases the risk of becoming a cold injury casualty.

b. *Signs and Symptoms.* Once a service member becomes familiar with the factors that contribute to cold injury, he must learn to recognize cold injury signs and symptoms.

(1) Many service members suffer cold injury without realizing what is happening to them. They may be cold and generally uncomfortable. These service members often do not notice the injured part because it is already numb from the cold.

(2) Superficial cold injury usually can be detected by numbness or tingling sensations. These signs and symptoms often can be relieved simply by loosening boots or other clothing and by exercising to improve circulation. In more advanced cases involving deep cold injury, the service member often is not aware that there is a problem until the affected part feels like a stump or block of wood.

(3) Outward signs of cold injury include discoloration of the skin at the site of injury. In light-skinned persons, the skin first reddens and then becomes pale or waxy white. In dark-skinned persons, grayness in the skin is usually evident. An injured foot or hand feels cold to the touch. Swelling may be an indication of deep injury. Also note that blisters may occur after rewarming the affected parts. Service members should work in pairs (buddy teams) to check each other for signs of discoloration and other symptoms.

c. First Aid Measures. First aid for cold injuries depends on whether they are superficial or deep. Rewarming the affected part using body heat can adequately treat cases of superficial cold injury. (For example, this can be done by covering cheeks with hands, putting fingertips in armpits, or placing the casualty's feet under the clothing of a buddy [next to his belly].) The injured part should **NOT** be massaged, exposed to a fire or stove, rubbed with snow, slapped, chafed, or soaked in cold water. Walking on injured feet should be avoided. Deep cold injury (frostbite) is very serious and requires prompt first aid to avoid or to minimize the loss of parts or all of the fingers, toes, hands, or feet. The sequence for treating cold injuries depends on whether the condition is life-threatening. The first priority in managing cold injuries is to remove the casualty from the cold environment (such as building an improvised shelter). Other injuries the casualty may have are provided first aid simultaneously while waiting for transportation or evacuation. If the casualty is to be transported in a nonmedical vehicle, first aid measures should be continued en route to the MTF.

d. Conditions Caused by Cold. Conditions caused by cold include chilblain, immersion syndrome (immersion foot and trench foot), frostbite, snow blindness, dehydration, and hypothermia.

(1) *Chilblain.*

- *Signs and symptoms.* Chilblain is caused by repeated prolonged exposure of bare skin at temperatures from 60° Fahrenheit (F) to

32°F, or 20°F for acclimated, dry, unwashed skin. The area may be acutely swollen, red, tender, and hot with itchy skin. There may be no loss of skin tissue in untreated cases but continued exposure may lead to infected, ulcerated, or bleeding lesions.

- *First aid measures.* Within minutes, the area usually responds to locally applied body heat. Rewarm the affected part by applying firm steady pressure with your hands, or placing the affected part under your arms or against the stomach of a buddy. **DO NOT** rub or massage affected areas.

NOTE

Medical personnel should evaluate the injury, because signs and symptoms of tissue damage may be slow to appear.

(2) *Immersion syndrome (immersion foot and trench foot).* Immersion foot and trench foot are injuries that result from fairly long exposure of the feet to wet conditions at temperatures from approximately 32°F to 50°F. Inactive feet in damp or wet socks and boots, or tightly laced boots which impair circulation, are even more susceptible to injury. This injury can be very serious; it can lead to loss of toes or parts of the feet. If exposure of the feet has been prolonged and severe, the feet may swell so much that pressure closes the blood vessels and cuts off circulation. Should an immersion injury occur, dry the feet thoroughly and transport the casualty to an MTF by the fastest means possible.

- *Signs and symptoms.* At first, the parts of the affected foot are cold and painless, the pulse is weak, and numbness may be present. Second, the parts may feel hot, and burning and shooting pains may begin. In later stages, the skin is pale with a bluish cast and the pulse decreases. Other signs and symptoms that may follow are blistering, swelling, redness, heat, hemorrhaging (bleeding), and gangrene.

- *First aid measures.* First aid measures are required for all stages of immersion syndrome injury. Rewarm the injured part gradually by exposing it to warm air. Protect it from trauma and secondary infections. Dry, loose clothing or several layers of warm coverings are preferable to extreme heat. Under no circumstances should the injured part be exposed to an open fire. Elevate the injured part to relieve the swelling. Transport the casualty to an MTF as soon as possible. When the part is rewarmed, the casualty often feels a burning sensation and pain. Symptoms may persist for days or weeks even after rewarming.

NOTE

When providing first aid for immersion foot and trench foot— **DO NOT** massage the injured part. **DO NOT** moisten the skin. **DO NOT** apply heat or ice.

(3) *Frostbite*. Frostbite is the injury of tissue caused from exposure to cold, usually below 32°F depending on the windchill factor, duration of exposure, and adequacy of protection. Individuals with a history of cold injury are likely to suffer an additional cold injury. The body parts most easily frostbitten are the cheeks, nose, ears, chin, forehead, wrists, hands, and feet. Frostbite may involve only the skin (superficial), or it may extend to a depth below the skin (deep). Deep frostbite is very serious and requires prompt first aid to avoid or to minimize the loss of parts or all of the fingers, toes, hands, or feet.

- *Signs and symptoms*.
  - Loss of sensation (numb feeling) in any part of the body.
  - Sudden blanching (whitening) of the skin of the affected part, followed by a momentary tingling sensation.
  - Redness of skin in light-skinned service members; grayish coloring in dark-skinned service members.
  - Blisters.
  - Swelling or tender areas.
  - Loss of previous sensation of pain in affected area.
  - Pale, yellowish, waxy-looking skin.
  - Frozen tissue that feels solid (or wooden) to the touch.

CAUTION

Deep frostbite is a very serious injury and requires immediate first aid and subsequent medical treatment to avoid or minimize loss of body parts.

- *First aid measures.*

  - *Face, ears, and nose.* Cover the casualty's affected area with his and/or your bare hands until sensation and color return.

  - *Hands.* Open the casualty's field jacket and shirt. (In a chemical environment, do not loosen or remove the clothing and protective overgarments.) Place the affected hands under the casualty's armpits. Close the field jacket and shirt to prevent additional exposure.

  - *Feet.* Remove the casualty's boots and socks if he does not need to walk any further to receive additional treatment. (Thawing the casualty's feet and forcing him to walk on them will cause additional pain and injury.) Place the affected feet under clothing and against the body of another service member.

WARNING

DO NOT attempt to thaw the casualty's feet or other frozen areas if he will be required to walk or travel to an MTF for additional medical treatment. The possibility of additional injury from walking is less when the feet are frozen than when they are thawed. (However, if possible avoid walking.) Thawing in the field increases the possibilities of infection, gangrene, or other injury.

NOTE

Thawing may occur spontaneously during transportation to the MTF; this cannot be avoided since the body in general must be kept warm.

In all of the above areas, ensure that the casualty is kept warm and that he is covered (to avoid further injury). Seek medical treatment as soon as possible. Reassure the casualty, protect the affected area from further injury by covering it lightly with a blanket or any dry clothing, and seek shelter out of the wind. Remove or loosen constricting clothing (except in a contaminated environment) and increase insulation. Ensure the casualty exercises as much as possible, avoiding trauma to the injured part, and is prepared for pain when thawing occurs. Protect the frostbitten part from additional injury. **DO NOT**—

- Rub the injured part with snow or apply cold water soaks.

- Warm the part by massage or exposure to open fire because the frozen part may be burned due to the lack of feeling.

- Use ointments or other salves.

- Manipulate the part in any way to increase circulation.

- Use alcohol or tobacco because this reduces the body's resistance to cold.

NOTE

Remember, when freezing extends to a depth below the skin, it is a much more serious injury. Extra care is required to reduce or avoid the chances of losing all or part of the toes or feet. This also applies to the fingers and hands.

(4) *Snow blindness*. Snow blindness is the effect that glare from an ice field or snowfield has on the eyes. It is more likely to occur in hazy, cloudy weather than when the sun is shining. Glare from the sun will cause an individual to instinctively protect his eyes. However, in cloudy weather, he may be overconfident and expose his eyes longer than when the threat is more obvious. He may also neglect precautions such as the use of protective eyewear. Waiting until discomfort (pain) is felt before using protective eyewear is dangerous because a deep burn of the eyes may already have occurred.

- *Signs and symptoms*. Symptoms of snow blindness are a sensation of grit in the eyes with pain in and over the eyes, made worse by moving the eyeball. Other signs and symptoms are watering, redness, headache, and increased pain on exposure to light.

- *First aid measures*. First aid measures consist of blindfolding or covering the eyes with a dark cloth which stops painful eye movement. Complete rest is desirable. If further exposure to light is not preventable, the eyes should be protected with dark bandages or the darkest glasses available. Once unprotected exposure to sunlight stops, the condition usually heals in a few days without permanent damage. The casualty should be evacuated to the nearest MTF.

(5) *Dehydration*. Dehydration occurs when the body loses too much fluid. A certain amount of body fluid is lost through normal body processes. A normal daily intake of liquids replaces these losses. When individuals are engaged in any strenuous exercises or activities, fluid is lost

through sweating and this loss creates an imbalance of fluids in the body, and if not matched by rehydration it can contribute to dehydration. The danger of dehydration is as prevalent in cold regions as it is in hot regions. In hot weather, the individual is aware of his body losing fluids through sweat. In cold weather, however, it is extremely difficult to realize that this condition exists since sweating is not as apparent as in a hot environment. The danger of dehydration in cold weather operations is a serious problem. In cold climates, sweat evaporates so rapidly or is absorbed so thoroughly by layers of heavy clothing that it is rarely visible on the skin. Dehydration also occurs during cold weather operations because drinking is inconvenient. Dehydration will weaken or incapacitate a casualty for a few hours, or sometimes, several days. Because rest is an important part of the recovery process, casualties must take care that limited movement during their recuperative period does not enhance the risk of becoming a cold injury casualty.

- *Signs and symptoms.* The symptoms of cold weather dehydration are similar to those encountered in heat exhaustion. The mouth, tongue, and throat become parched and dry, and swallowing becomes difficult. The casualty may have nausea (with or without vomiting) along with extreme dizziness and fainting. The casualty may also feel generally tired and weak and may experience muscle cramps. Focusing the eyes may also become difficult.

- *First aid measures.* The casualty should be kept warm and his clothes should be loosened (if not in a chemical environment) to allow proper circulation. Shelter from wind and cold must be provided. Fluid replacement should begin immediately and the service member transported to an MTF as soon as possible.

(6) *Hypothermia (general cooling).* When exposed to prolonged cold weather a service member may become both mentally and physically numb, thus neglecting essential tasks or requiring more time and effort to achieve them. Under some conditions (particularly cold water immersion), even a service member in excellent physical condition may die in a matter of minutes. The destructive influence of cold on the body is called *hypothermia.* This means bodies lose heat faster than they can produce it. Hypothermia can occur from exposure to temperatures either above or below freezing, especially from immersion in cold water, wet-cold conditions, or from the effect of wind. Physical exhaustion and insufficient food intake may also increase the risk of hypothermia. General cooling of the entire body to a temperature below 95°F is caused by continued exposure to low or rapidly dropping temperatures, cold moisture, snow, or ice. Fatigue, poor physical condition, dehydration, faulty blood circulation, alcohol or other drug use, trauma, and immersion can cause hypothermia. Remember, cold

may affect the body systems slowly and almost without notice. Service members exposed to low temperatures for extended periods may suffer ill effects even if they are well protected by clothing.

- *Signs and symptoms*. As the body cools, there are several stages of progressive discomfort and impairment. A sign that is noticed immediately is shivering. Shivering is an attempt by the body to generate heat. The pulse is faint or very difficult to detect. People with temperatures around 90°F may be drowsy and mentally slow. Their ability to move may be hampered, stiff, and uncoordinated, but they may be able to function minimally. Their speech may be slurred. As the body temperature drops further, shock becomes evident as the person's eyes assume a glassy state, breathing becomes slow and shallow, and the pulse becomes weaker or absent. The person becomes very stiff and uncoordinated. Unconsciousness may follow quickly. As the body temperature drops even lower, the extremities freeze, and a deep (or core) body temperature (below 85°F) increases the risk of irregular heart action. This irregular heart action or heart standstill can result in sudden death.

- *First aid measures*. Except in cases of the most severe hypothermia (marked by coma or unconsciousness and a weak pulse), first aid measures for hypothermia are directed towards protecting the casualty from further loss of body heat. For the casualty who is conscious, first aid measures are directed at rewarming the body evenly and without delay. Provide heat by using a hot water bottle or field expedient or another service member's body heat.

CAUTION

DO NOT expose the casualty to an open fire, as he may become burned.

NOTE

When using a hot water bottle or field expedient (canteen filled with warm water), the bottle or canteen must be wrapped in cloth prior to placing it next to the casualty. This will reduce the chance of burning the casualty's skin.

Always call or send for help as soon as possible and protect the casualty immediately with dry clothing or a sleeping bag. Then, move him to a warm place. Evaluate other injuries and provide first aid as required. First aid measures can be performed while the casualty is waiting transportation or while he is en route. In the case of an accidental breakthrough into ice water, or other hypothermic accident, strip the casualty of wet clothing immediately and bundle him into a sleeping bag. Rescue breathing should be started at once if the casualty's breathing has stopped or is irregular or shallow. Warm liquids (**NOT HOT**) may be given gradually if the casualty is conscious. **DO NOT** force liquids on an unconscious or semiconscious casualty because he may choke. The casualty should be transported on a litter because the exertion of walking may aggravate circulation problems. Medical personnel should immediately treat any hypothermia casualty. Hypothermia is life threatening until normal body temperature has been restored. The first aid measures for a casualty with severe hypothermia are based upon the following principles: attempt to avoid further heat loss, handle the casualty gently, and transport the casualty as soon as possible to the nearest MTF. If at all possible, the casualty should be evacuated by medical personnel.

WARNING

Rewarming a severely hypothermic casualty is extremely dangerous in the field due to the possibility of such complications as rewarming, shock and disturbances in the rhythm of the heartbeat. These conditions require treatment by medical personnel.

NOTE

Resuscitation of casualties with hypothermic complications is difficult if not impossible to do outside of an MTF setting.

CAUTION

The casualty is unable to generate his own body heat. Therefore, merely placing him in a blanket or sleeping bag is not sufficient.

e. *Table.* See Table 5-2 for further information.

Table 5-2. Injuries Caused by Cold and Wet Conditions

INJURIES	SIGNS/SYMPTOMS	FIRST AID
CHILBLAIN	RED SWOLLEN, HOT, TENDER, ITCHING SKIN. CONTINUED EXPOSURE MAY LEAD TO INFECTED (ULCERATED OR BLEEDING) SKIN LESIONS.	1. AREA USUALLY RESPONDS TO LOCALLY APPLIED REWARMING (BODY HEAT). 2. DO NOT RUB OR MASSAGE AREA. 3. SEEK MEDICAL AID.
IMMERSION SYNDROME (IMMERSION FOOT/TRENCH FOOT)	AFFECTED PARTS ARE COLD, NUMB, AND PAINLESS. PARTS MAY THEN BE HOT, WITH BURNING AND SHOOTING PAINS. ADVANCED STAGE: SKIN PALE WITH BLUISH CAST; PULSE DECREASES; BLISTERING, SWELLING, HEAT, HEMORRHAGING, AND GANGRENE MAY FOLLOW.	1. GRADUAL REWARMING BY EXPOSURE TO WARM AIR. 2. DO NOT MASSAGE OR MOISTEN SKIN. 3. PROTECT AFFECTED PARTS FROM TRAUMA. 4. DRY FEET THOROUGHLY, AVOID WALKING. 5. SEEK MEDICAL AID.
FROSTBITE	LOSS OF SENSATION (NUMB FEELING) IN ANY PART OF THE BODY. SUDDEN BLANCHING (WHITENING) OF THE SKIN OF THE AFFECTED PART, FOLLOWED BY A MOMENTARY TINGLING SENSATION. REDNESS OF SKIN IN LIGHT-SKINNED SERVICE MEMBERS; GRAYISH COLORING IN DARK-SKINNED SERVICE MEMBERS. BLISTERS. SWELLING OR TENDER AREAS. LOSS OF PREVIOUS SENSATION OF PAIN IN THE AFFECTED AREA. PALE YELLOWISH, WAXY-LOOKING SKIN. FROZEN TISSUE THAT FEELS SOLID (WOODEN) TO THE TOUCH.	1. WARM THE AREA AT THE FIRST SIGN OF FROSTBITE, USING FIRM, STEADY PRESSURE OF THE HAND, UNDERARM, OR ABDOMEN. 2. FACE, EARS, NOSE: COVER AREA WITH HANDS (CASUALTY'S OWN OR BUDDY'S). 3. HANDS: OPEN FIELD JACKET AND PLACE CASUALTY'S HANDS AGAINST HIS BODY, THEN CLOSE THE JACKET TO PREVENT HEAT LOSS. 4. FEET: REMOVE THE CASUALTY'S BOOTS AND SOCKS AND PLACE HIS FEET AGAINST THE BODY OF ANOTHER SERVICE MEMBER.

Table 5-2. Injuries Caused by Cold and Wet Conditions (Continued)

INJURIES	SIGNS/SYMPTOMS	FIRST AID
		5. WARNING: DO NOT ATTEMPT TO THAW THE CASUALTY'S FEET OR OTHER FROZEN AREAS IF HE WILL BE REQUIRED TO WALK OR TRAVEL TO AN MTF FOR ADDITIONAL TREATMENT. THE POSSIBILITY OF INJURY FROM WALKING IS LESS WHEN THE FEET ARE FROZEN THAN WHEN THEY HAVE BEEN THAWED. (HOWEVER, IF POSSIBLE AVOID WALKING.) THAWING IN THE FIELD INCREASES THE POSSIBILITY OF INFECTION, GANGRENE, OR INJURY. 6. LOOSEN OR REMOVE CONSTRICTING CLOTHING AND REMOVE ANY JEWELRY. 7. INCREASE INSULATION (COVER WITH BLANKET OR OTHER DRY MATERIAL). ENSURE CASUALTY EXERCISES AS MUCH AS POSSIBLE, AVOIDING TRAUMA TO INJURED PART.
SNOW BLINDNESS	EYES MAY FEEL SCRATCHY. WATERING, REDNESS, HEADACHE, AND INCREASED PAIN WITH EXPOSURE TO LIGHT CAN OCCUR.	1. COVER THE EYES WITH A DARK CLOTH. 2. SEEK MEDICAL AID.
DEHYDRATION	SIMILAR TO HEAT EXHAUSTION (REFER TO TABLE 5-1).	1. KEEP WARM. 2. CASUALTY NEEDS FLUID REPLACEMENT, REST, AND PROMPT MEDICAL AID.
HYPOTHERMIA	CASUALTY IS COLD. SHIVERING. CORE TEMPERATURE IS LOW. CONSCIOUSNESS MAY BE	**MILD HYPOTHERMIA** 1. REWARM BODY

Table 5-2. Injuries Caused by Cold and Wet Conditions (Continued)

INJURIES	SIGNS/SYMPTOMS	FIRST AID
	ALTERED. UNCOORDINATED MOVEMENTS MAY OCCUR. SHOCK AND COMA MAY RESULT AS BODY TEMPERATURE DROPS.	EVENLY AND WITHOUT DELAY. (NEED TO PROVIDE HEAT SOURCE; CASUALTY'S BODY UNABLE TO GENERATE HEAT.) 2. KEEP DRY, PROTECT FROM THE ELEMENTS. 3. WARM (NOT HOT) LIQUIDS MAY BE GIVEN GRADUALLY (TO CONSCIOUS CASUALTIES ONLY). 4. BE PREPARED TO START BASIC LIFE SUPPORT MEASURES FOR THE CASUALTY. 5. SEEK MEDICAL TREATMENT IMMEDIATELY. **SEVERE HYPOTHERMIA** 1. STABILIZE THE TEMPERATURE. 2. ATTEMPT TO AVOID FURTHER HEAT LOSS. 3. HANDLE THE CASUALTY GENTLY. 4. EVACUATE TO THE NEAREST MTF AS SOON AS POSSIBLE. **5. WARNING: HYPOTHERMIA IS A MEDICAL EMERGENCY. PROMPT MEDICAL ATTENTION IS NECESSARY.**

CHAPTER 6

FIRST AID FOR BITES AND STINGS

6-1. General

Snakebites, insect bites, or stings can cause intense pain and/or swelling. If not treated promptly and correctly, they can cause serious illness or death. The severity of a snakebite depends upon: whether the snake is poisonous or nonpoisonous, the type of snake, the location of the bite, and the amount of venom injected. Bites from humans and other animals, such as dogs, cats, bats, raccoons, and rats, can cause severe bruises and infection and tears or lacerations of tissue. Awareness of the potential sources of injuries can reduce or prevent them from occurring. Knowledge and prompt application of first-aid measures can lessen the severity of injuries from bites and stings and keep the service member from becoming a serious casualty.

6-2. Types of Snakes

a. Nonpoisonous Snakes. There are approximately 130 different varieties of nonpoisonous snakes in the United States. They have oval-shaped heads and round eyes. Unlike poisonous snakes, discussed below, nonpoisonous snakes do not have fangs with which to inject venom. Figure 6-1 depicts the characteristics of a nonpoisonous snake.

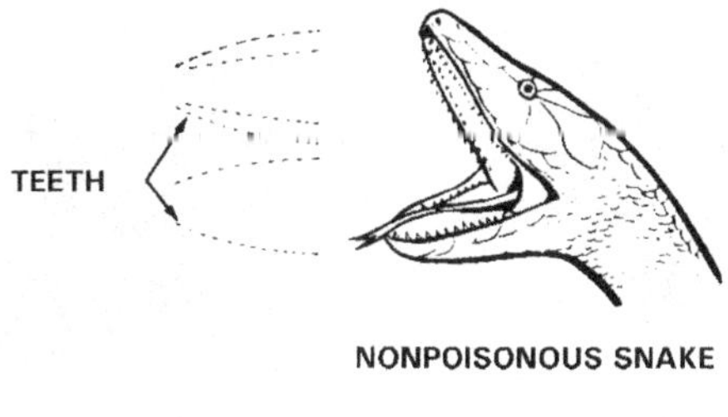

Figure 6-1. Characteristics of nonpoisonous snake.

b. Poisonous Snakes. Poisonous snakes are found throughout the world, primarily in tropical to moderate climates. Within the United States, there are four kinds: rattlesnakes, copperheads, water moccasins (cottonmouth), and coral snakes. Poisonous snakes in other parts of the world include sea snakes, the fer-de-lance, the bushmaster, and the tropical rattlesnake in tropical Central America; the Malayan pit viper in the tropical Far East; the cobra in Africa and Asia; the mamba (or black mamba) in central and southern Africa; and the krait in India and Southeast Asia. Refer to Figure 6-2 for characteristics of a poisonous pit viper.

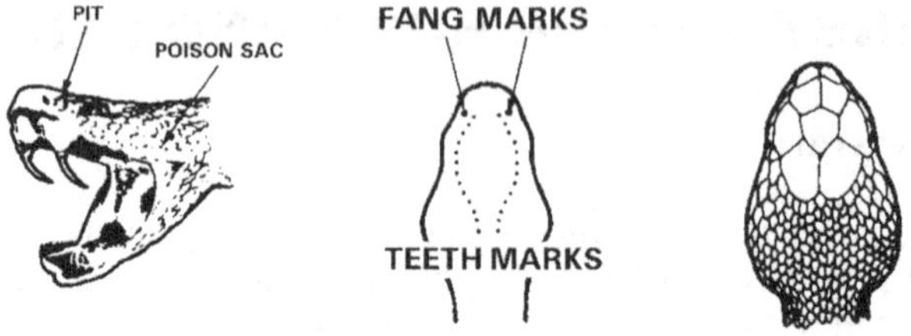

Figure 6-2. Characteristics of poisonous pit viper.

c. Pit Vipers (Poisonous). Figure 6-3 depicts a variety of poisonous snakes.

Figure 6-3. Poisonous snakes.

(1) Rattlesnakes, bushmasters, copperheads, fer-de-lance, Malayan pit vipers, and water moccasins (cottonmouth) are called pit vipers because of the small, deep pits between the nostrils and eyes on each side of the head (Figure 6-2). In addition to their long, hollow fangs, these snakes have other identifying features: thick bodies, slit-like pupils of the eyes, and flat, almost triangular-shaped heads. Color markings and other identifying characteristics, such as rattles or a noticeable white interior of the mouth (cottonmouth), also help distinguish these poisonous snakes. Further

identification is provided by examining the bite pattern of the wound for signs of fang entry. Occasionally there will be only one fang mark, as in the case of a bite on a finger or toe where there is no room for both fangs, or when the snake has broken off a fang.

(2) The casualty's condition provides the best information about the seriousness of the situation, or how much time has passed since the bite occurred. Pit viper bites are characterized by severe burning pain. Discoloration and swelling around the fang marks usually begins within 5 to 10 minutes after the bite. If only minimal swelling occurs within 30 minutes, the bite will almost certainly have been from a nonpoisonous snake or possibly from a poisonous snake which did not inject venom. The venom destroys blood cells, causing a general discoloration of the skin. Blisters and numbness in the affected area follow this reaction. Other signs, which can occur, are weakness, rapid pulse, nausea, shortness of breath, vomiting, and shock.

d. Corals, Cobras, Kraits, and Mambas. Corals (Figure 6-4), cobras (Figure 6-5), kraits, and mambas all belong to the same group even though they are found in different parts of the world. All four inject their venom through short, grooved fangs, leaving a characteristic bite pattern.

Figure 6-4. Coral snake.

(1) The small coral snake, found in the Southeastern US, is brightly colored with bands of red, yellow (or almost white), and black completely encircling the body. Other nonpoisonous snakes have the same coloring, but on the coral snake found in the US, the red ring always touches the yellow ring. To know the difference between a harmless snake and the coral snake found in the United States, remember the following:

"Red on yellow will kill a fellow,
Red on black, venom will lack."

Figure 6-5. Cobra snake.

(2) The venom of corals, cobras, kraits, and mambas produces symptoms different from those of pit vipers. Because there is only minimal pain and swelling, many people believe that the bite is not serious. Delayed reactions in the nervous system normally occur between 1 to 7 hours after the bite. Symptoms include blurred vision, drooping eyelids, slurred speech, drowsiness, and increased salivation and sweating. Nausea, vomiting, shock, respiratory difficulty, paralysis, convulsions, and coma will usually develop if the bite is not treated promptly.

e. Sea Snakes. Sea snakes (Figure 6-6) are found in the warm water areas of the Pacific and Indian oceans, along the coasts, and at the mouths of some larger rivers. Their venom is **VERY** poisonous, but their fangs are only 1/4 inch long. The first aid outlined for land snakes also applies to sea snakes.

Figure 6-6. Sea snake.

6-3. Snakebites

a. Poisonous snakes **DO NOT** always inject venom when they bite or strike a person. However, all snakes may carry tetanus (lockjaw); anyone bitten by a snake, whether poisonous or nonpoisonous, should immediately seek medical attention.

- Poison is injected from the venom sacs through grooved or hollow fangs. Depending on the species, these fangs are either long or short. Pit vipers have long hollow fangs. These fangs are folded against the roof of the mouth and extend when the snake strikes. This allows them to strike quickly and then withdraw. Cobras, coral snakes, kraits, mambas, and sea snakes have short, grooved fangs. These snakes are less effective in their attempts to bite, since they must chew after striking to inject enough venom (poison) to be effective. Figure 6-7 depicts the characteristics of a poisonous snakebite.

- In the event you are bitten, attempt to identify and/or kill the snake. Take it to medical personnel for inspection/identification. This provides valuable information to medical personnel who deal with snakebites. **TREAT ALL SNAKEBITES AS POISONOUS.**

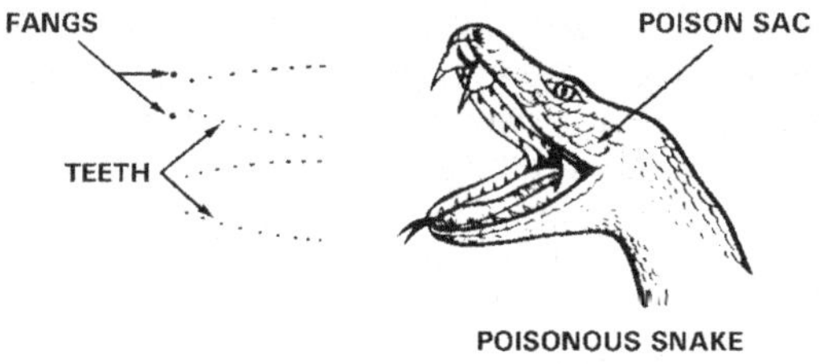

Figure 6-7. Characteristics of poisonous snakebite.

b. The venoms of different snakes cause different effects. Pit viper venom (hemotoxin [blood toxin]) destroys tissue and blood cells. Cobras, adders, and coral snakes inject powerful venom (neurotoxin [nerve toxin]) which affect the central nervous system, causing respiratory paralysis. Water moccasins and sea snakes have venom that is both hemotoxic and neurotoxic.

c. The identification of poisonous snakes is very important since medical treatment will be different for each type of venom. *Unless it can be positively identified, the snake should be killed and saved.* When this is not possible or when doing so is a serious threat to others, identification may sometimes be difficult since many venomous snakes resemble harmless varieties. When dealing with snakebite problems in foreign countries, seek advice, professional or otherwise, which may help identify species in the particular area of operations.

d. Get the casualty to an MTF as soon as possible and with minimum movement. Until evacuation or treatment is possible, have the casualty lie quietly and not move any more than necessary. If the casualty has been bitten on an extremity, **DO NOT** elevate the limb; keep the extremity level with the body. Keep the casualty comfortable and reassure him. If the casualty is alone when bitten, he should go to the medical facility himself rather than wait for someone to find him. Unless the snake has been positively identified, attempt to kill it and send it with the casualty. Be sure that retrieving the snake does not endanger anyone or delay transporting the casualty.

(1) If the bite is on an arm or leg, place a constricting band (narrow cravat [swathe], or narrow gauze bandage) one to two fingerbreadths above and below the bite (Figure 6-8). If the bite is on the hand or foot, place a single band above the wrist or ankle. The band should be tight enough to stop the flow of blood near the skin, but not tight enough to interfere with circulation. In other words, it should not have a tourniquet-like affect. If no swelling is seen, place the bands about 1 inch from either side of the bite. If swelling is present, put the bands on the unswollen part at the edge of the swelling. If the swelling extends beyond the band, move the band to the new edge of the swelling. (If possible, leave the old band on, place a new one at the new edge of the swelling, and then remove and save the old one in case the process has to be repeated.)

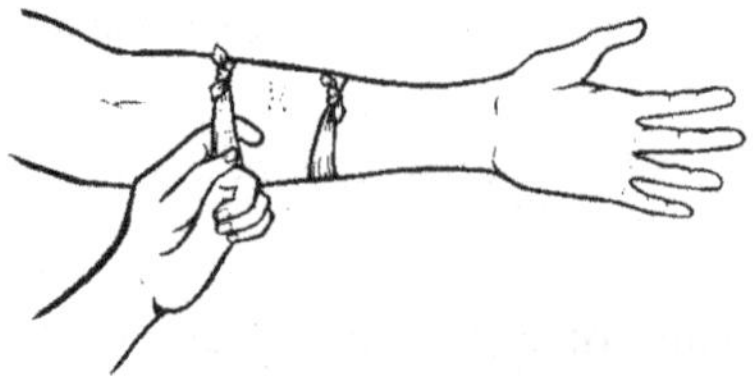

Figure 6-8. Constricting band.

CAUTION

DO NOT attempt to cut open the bite nor suck out the venom. If the venom should seep through any damaged or lacerated tissues in your mouth, you could immediately lose consciousness or even die.

(2) If the bite is located on an arm or leg, immobilize it at a level below the heart. **DO NOT** elevate an arm or leg even with or above the level of the heart.

CAUTION

When a splint is used to immobilize the arm or leg, take **EXTREME** care to ensure the splinting is done properly and does not bind. Watch it closely and adjust it if any changes in swelling occur.

(3) When possible, clean the area of the bite with soap and water. **DO NOT** use ointments of any kind.

(4) **NEVER** give the casualty food, alcohol, stimulants (coffee or tea), drugs, or tobacco.

(5) Remove rings, watches, or other jewelry from the affected limb.

6-4. Human or Animal Bites

Human or other land animal bites may cause lacerations or bruises. In addition to damaging tissue, bites always present the possibility of infection.

a. Human Bites. Human bites that break the skin may become seriously infected since the mouth is heavily contaminated with bacteria. Medical personnel **MUST** treat all human bites.

b. Animal Bites. Land animal bites can result in both infection and disease. Tetanus, rabies, and various types of fevers can follow an untreated animal bite. Because of these possible complications, the animal causing the bite should, if possible, be captured or killed (without damaging its head) so that it can be tested for disease.

c. First Aid.

(1) Cleanse the wound thoroughly with soap.

(2) Flush it well with water.

(3) Cover it with a sterile dressing.

(4) Immobilize the injured arm or leg, if appropriate.

(5) Transport the casualty immediately to an MTF.

NOTE

If unable to capture or kill the animal, provide medical personnel with any information that will help identify it.

6-5. Marine (Sea) Animals

With the exception of sharks and barracuda, most marine animals will not deliberately attack. The most frequent injuries from marine animals are wounds by biting, stinging, or puncturing. Wounds inflicted by marine animals can be very painful, but are rarely fatal.

a. Sharks, Barracuda, and Alligators. Wounds from these marine animals can involve major trauma as a result of bites and lacerations. Bites from large marine animals are potentially the most life threatening of all injuries from marine animals. Major wounds from these animals can be treated by controlling the bleeding, preventing shock, giving basic life support, splinting the injury, and by securing prompt medical aid.

b. Turtles, Moray Eels, and Corals. These animals normally inflict minor wounds. Treat by cleansing the wound(s) thoroughly and by splinting if necessary.

c. Jellyfish, Portuguese Man-of-War, Anemones, and Others. This group of marine animals inflict injury by means of stinging cells in their tentacles. Contact with the tentacles produces burning pain with a rash and small hemorrhages on the skin. Shock, muscular cramping, nausea, vomiting, and respiratory distress may also occur. Gently remove the clinging tentacles with a towel and wash or treat the area. Use diluted ammonia or alcohol, meat tenderizer, and talcum powder. If symptoms become severe or persist, seek medical assistance.

d. Spiny Fish, Urchins, Stingrays, and Cone Shells. These animals inject their venom by puncturing the skin with their spines. General signs and symptoms include swelling, nausea, vomiting, generalized cramps, diarrhea, muscular paralysis, and shock. Deaths are rare. Treatment consists of soaking the wounds in hot water (when available) for 30 to 60 minutes. This inactivates the heat sensitive toxin. In addition, further first aid measures (controlling bleeding, applying a dressing, and so forth) should be carried out as necessary.

CAUTION

Be careful not to scald the casualty with water that is too hot because the pain of the wound will mask the normal reaction to heat.

6-6. Insect (Arthropod) Bites and Stings

An insect bite or sting can cause great pain, allergic reaction, inflammation, and infection. If not treated correctly, some bites/stings may cause serious illness or even death. When an allergic reaction is not involved, first aid is a simple process. In any case, medical personnel should examine the casualty at the earliest possible time. It is important to properly identify the spider, bee, or creature that caused the bite/sting, especially in cases of allergic reaction.

a. Types of Insects. The insects found throughout the world that can produce a bite or sting are too numerous to mention in detail. Commonly encountered stinging or biting insects include brown recluse spiders (Figure 6-9), black widow spiders (Figure 6-10), tarantulas (Figure 6-11), scorpions (Figure 6-12), urticating caterpillars, bees, wasps, centipedes, conenose beetles (kissing bugs), ants, and wheel bugs. Upon being reassigned, especially to overseas areas, take the time to become acquainted with the types of insects to avoid.

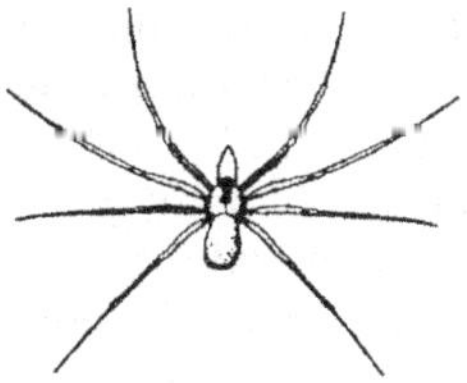

Figure 6-9. Brown recluse spider.

Figure 6-10. Black widow spider.

Figure 6-11. Tarantula.

Figure 6-12. Scorpion.

b. Signs and Symptoms. Discussed in paragraphs (1) and (2) below are the most common effects of insect bites/stings. They can occur alone or in combination with the others.

(1) *Less serious.* Commonly seen signs/symptoms are pain, irritation, swelling, heat, redness, and itching. Hives or wheals (raised areas of the skin that itch) may occur. These are the least severe of the allergic reactions that commonly occur from insect bites/stings. They are usually dangerous only if they affect the air passages (mouth, throat, nose, and so forth), which could interfere with breathing. The bites/stings of bees, wasps, ants, mosquitoes, fleas, and ticks are usually not serious and normally produce mild and localized symptoms. A tarantula's bite is usually no worse than that of a bee sting. Scorpions are rare and their stings (except for a specific species found only in the Southwest desert) are painful but usually not dangerous.

(2) *Serious.* Emergency allergic or hypersensitive reactions sometimes result from the stings of bees, wasps, and ants. Many people are allergic to the venom of these particular insects. Bites or stings from these insects may produce more serious reactions, to include generalized itching and hives, weakness, anxiety, headache, breathing difficulties, nausea, vomiting, and diarrhea. Very serious allergic reactions (called *anaphylactic shock*) can lead to complete collapse, shock, and even death. Spider bites (particularly from the black widow and brown recluse spiders) can also be

serious. Venom from the black widow spider affects the nervous system. This venom can cause muscle cramps, a rigid, nontender abdomen, breathing difficulties, sweating, nausea, and vomiting. The brown recluse spider generally produces local rather than system-wide problems; however, local tissue damage around the bite can be severe and can lead to an ulcer and even gangrene.

c. First Aid. There are certain principles that apply regardless of what caused the bite/sting. Some of these are—

- If there is a stinger present (for example, from a bee), remove the stinger by scraping the skin's surface with a fingernail or knife. **DO NOT** squeeze the sac attached to the stinger because it may inject more venom.

- Wash the area of the bite/sting with soap and water (alcohol or an antiseptic may also be used) to help reduce the chances of an infection and remove traces of venom.

- Remove jewelry from bitten extremities because swelling may occur.

- In most cases of insect bites the reaction will be mild and localized; use ice or cold compresses (if available) on the site of the bite/sting. This will help reduce swelling, ease the pain, and slow the absorption of venom. Meat tenderizer (to neutralize the venom) or calamine lotion (to reduce itching) may be applied locally. If necessary, seek medical assistance.

- In more serious reactions (severe and rapid swelling, allergic symptoms, and so forth) treat the bite/sting like you would treat a snakebite; that is, apply constricting bands above and below the site.

- Be prepared to perform basic life-support measures, such as rescue breathing.

- Reassure the casualty and keep him calm.

- In serious reactions, attempt to capture the insect for positive identification; however, be careful not to become a casualty yourself.

- If the reaction to the bite/sting appears serious, seek medical assistance.

WARNING

Insect bites/stings may cause *anaphylactic shock* (a shock caused by a severe allergic reaction). This is a life-threatening event and a TRUE MEDICAL EMERGENCY. Be prepared to perform the basic life-support measures and to immediately transport the casualty to an MTF.

NOTE

Be aware that some allergic or hypersensitive individuals may carry identification or emergency insect bite treatment kits. If the casualty is having an allergic reaction and has such a kit, administer the medication in the kit according to the instructions which accompany the kit.

d. Supplemental Information. For additional information concerning biting insects, see FM 21-10.

6-7. First Aid for Bites and Stings

See the table below for information on bites and stings.

Table 6-1. First Aid Measures for Bites and Stings

TYPES	FIRST AID MEASURES
SNAKEBITE	1. MOVE CASUALTY AWAY FROM THE SNAKE. 2. REMOVE JEWELRY FROM THE AFFECTED AREA, IF APPLICABLE. 3. REASSURE CASUALTY AND KEEP HIM QUIET. 4. APPLY CONSTRICTING BAND, 1-2 FINGERBREADTHS FROM THE BITE. YOU SHOULD BE ABLE TO INSERT A FINGER BETWEEN THE BAND AND THE SKIN. *a. ARM OR LEG BITE.* PLACE ONE BAND ABOVE AND ONE BAND BELOW THE BITE SITE. *b. HAND OR FOOT BITE.* PLACE ONE BAND ABOVE THE WRIST OR ANKLE.

Table 6-1. First Aid Measures for Bites and Stings

TYPES	FIRST AID MEASURES
	5. IMMOBILIZE THE AFFECTED PART IN A POSITION BELOW THE LEVEL OF THE HEART. 6. KILL THE SNAKE (IF POSSIBLE, WITHOUT DAMAGING ITS HEAD OR ENDANGERING YOURSELF) AND SEND IT WITH THE CASUALTY. 7. SEEK MEDICAL ASSISTANCE IMMEDIATELY.
BROWN RECLUSE SPIDER OR BLACK WIDOW SPIDER BITE	1. KEEP CASUALTY QUIET. 2. REMOVE ALL JEWELRY FROM AFFECTED PART, IF APPLICABLE. 3. WASH THE AREA. 4. APPLY ICE OR FREEZE PACK, IF AVAILABLE. 5. SEEK MEDICAL ASSISTANCE.
TARANTULA BITE OR SCORPION STING OR ANT BITE	1. WASH THE AREA. 2. REMOVE ALL JEWELRY FROM AFFECTED PART, IF APPLICABLE. 3. APPLY ICE OR FREEZE PACK, IF AVAILABLE. 4. APPLY BAKING SODA, CALAMINE LOTION, OR MEAT TENDERIZER (IF AVAILABLE) TO BITE SITE TO RELIEVE PAIN AND ITCHING. 5. IF THE SITE OF THE BITE IS ON THE FACE, NECK (POSSIBLE AIRWAY PROBLEMS), OR GENITAL AREA, OR IF LOCAL REACTION SEEMS SEVERE, OR IF THE STING IS BY THE DANGEROUS TYPE OF SCORPION FOUND IN THE SOUTHWEST UNITED STATES DESERT, KEEP THE CASUALTY AS QUIET AS POSSIBLE. SEEK MEDICAL ASSISTANCE.
BEE STING	1. IF THE STINGER IS PRESENT, REMOVE BY SCRAPING WITH A KNIFE OR FINGERNAIL. DO NOT SQUEEZE VENOM SAC ON STINGER; MORE VENOM MAY BE INJECTED. 2. REMOVE ALL JEWELRY FROM AFFECTED PART, IF APPLICABLE.

Table 6-1. First Aid Measures for Bites and Stings

TYPES	FIRST AID MEASURES
	3. WASH THE AREA. 4. APPLY ICE OR FREEZE PACK, IF AVAILABLE. 5. IF ALLERGIC SIGNS OR SYMPTOMS APPEAR, BE PREPARED TO PERFORM BASIC LIFE SUPPORT MEASURES. SEEK IMMEDIATE MEDICAL ASSISTANCE.

CHAPTER 7

FIRST AID IN A NUCLEAR, BIOLOGICAL, AND CHEMICAL ENVIRONMENT

7-1. General

American forces have not been exposed to NBC weapons/agents on the battlefield since World War I. In future conflicts and wars we can expect the use of such agents. Nuclear, biological, and chemical weapons will rapidly degrade unit effectiveness by forcing troops to wear protective clothing and by creating confusion and fear. Through training in protective procedures and first aid, units can maintain their effectiveness on the integrated battlefield.

7-2. First Aid Materials

You may be issued the following materials to protect, decontaminate, and use as first aid for NBC exposure. You must know how to use the items; some items are described in *a* through *d* below. It is equally important that you know when to use them.

a. Nerve Agent Pyridostigmine Pretreatment (*NAPP*). You may be issued a blister pack of pretreatment tablets when your commander directs. The NAPP is a pretreatment; it is not an antidote. It improves the effectiveness of the nerve agent antidote. When ordered to take the pretreatment you must take one tablet every 8 hours, mission permitting. This must be taken prior to exposure to nerve agents, since it may take several hours to develop adequate blood levels.

NOTE

Commanders must follow investigational new drug protocols for use of the NAPP.

b. M291 Skin Decontaminating Kit. The M291 Skin Decontaminating Kit (Figure 7-1) contains six packets of XE-555 decontaminant resin.

WARNING

For external use only. May be slightly irritating to the eyes. Keep decontaminating powder out of eyes. Use water to wash toxic agent out of eyes.

c. *Nerve Agent Antidote Kit, MARK I.* Each service member is issued three MARK Is for use in first aid for nerve agent poisoning (Figure 7-2 and paragraph 7-6).

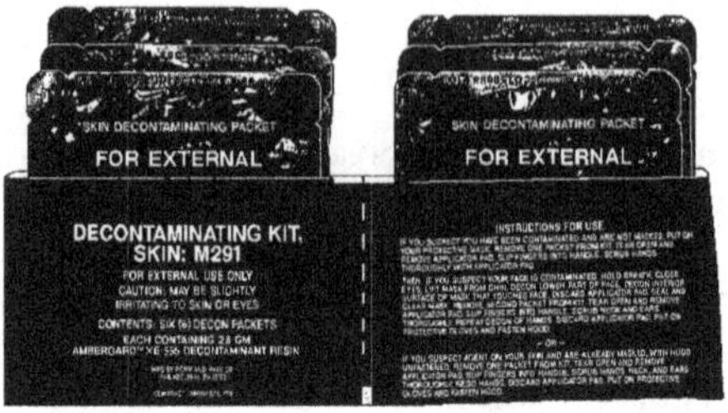

Figure 7-1. M291 Skin Decontamination Kit.

d. *Antidote Treatment, Nerve Agent, Autoinjector.* A new nerve agent antidote injection device, Antidote Treatment, Nerve Agent, Autoinjector (ATNAA) is currently under development that will replace the MARK I. The ATNAA is a multichambered device with the atropine and pralidoxime chloride in separate chambers. Both antidotes will be administered through a single needle.

7-3. Classification of Chemical and Biological Agents

a. Chemical agents are classified according to the primary physiological effects they produce, such as blistering, choking, vomiting, and incapacitating agents.

b. Biological warfare agents are classified according to the effect they have on man. The effects include their ability to incapacitate and cause death. Most biological warfare agents are delivered as aerosols that effect the respiratory tract; some can be delivered by releasing infected insects, by contaminating food and water, and by injection (injecting material in individuals by terrorist, not mass exposure). These agents are found in living organisms such as fungi, bacteria, and viruses.

WARNING

Swallowing water or food contaminated with nerve, blister, and other chemical agents and with some biological agents can be fatal. NEVER consume water or food that is suspected of being contaminated until it has been tested and found safe for consumption by medical personnel.

7-4. Conditions for Masking Without Order or Alarm

a. Once an attack with a chemical or biological agent is detected or suspected, or information is available that such an agent is about to be used, you must **STOP BREATHING** and mask immediately. **DO NOT WAIT** to receive an order or alarm under the following circumstances:

- Your position is hit by artillery missiles, rockets that produce vapors, smoke, and mists, and aerial sprays.
- Smoke or vapor cloud from an unknown source is present or approaching.
- A suspicious odor, liquid, or solid is present.
- A chemical or biological warfare agent attack is occurring.
- You are entering an area known or suspected of being contaminated.
- When casualties are being received from an area where chemical or biological agents have reportedly been used.
- You have one or more of the following symptoms:
  - An unexplained runny nose.
  - A sudden unexplained headache.
  - A feeling of choking or tightness in the chest or throat.
  - Dimness of vision.
  - Irritation of the eyes.
  - Difficulty in or increased rate of breathing without obvious reasons.
  - Sudden feeling of depression.
  - Dread, anxiety, or restlessness.
  - Dizziness or light-headedness.

- Slurred speech.
- Unexplained laughter or unusual behavior is noted in others.
- Numerous unexplained ill personnel.
- Service members suddenly collapsing without evident cause.
- Animals or birds exhibiting unusual behavior or suddenly dying.

b. For further information on protection and masking procedures, refer to FM 3-4, FM 4-02.7, FM 8-284, and FM 8-285.

7-5. First Aid for a Chemical Attack

Your field protective mask gives protection against biological and chemical warfare agents as well as radiological fallout. With practice you can mask in 9 seconds or less, or put on your mask with hood within 15 seconds.

a. Stop breathing. Don your mask, seal it properly, and clear and check it; then resume breathing. Give the alarm, and continue the mission. Keep your mask on until the "all clear" signal has been given.

NOTE

Keep your mask on until the area is no longer hazardous and you are told to unmask.

b. If symptoms of nerve agent poisoning (paragraph 7-7) appear, immediately give yourself one MARK I or ATNAA.

CAUTION

Do not inject a nerve agent antidote until you are sure you need it.

c. If your eyes and face become contaminated, you must immediately try to get under cover. You need shelter to prevent further contamination while performing decontamination procedures on your face. If no overhead cover is available, put your poncho over your head before beginning the decontamination process. Then you put on the remaining protective clothing. If vomiting occurs, the mask should be lifted momentarily and drained—with your eyes closed and while holding your breath—then replaced, cleared, and sealed.

d. If nerve agents are used, mission permitting, watch for persons needing nerve agent antidotes and immediately follow procedures outlined in paragraph 7-8*b* or *c*.

e. Decontaminate your skin immediately and clothing and equipment as soon as the mission permits.

7-6. Background Information on Nerve Agents

a. Nerve agents are among the deadliest of chemical agents. Nerve agents enter the body by inhalation, by ingestion, and through the skin. Depending on the route of entry and the amount, nerve agents can produce injury or death within minutes. Nerve agents can achieve their effects with small amounts. Nerve agents are absorbed rapidly, and the effects are felt immediately upon entry into the body. You will be issued three MARK Is or three ATNAAs and one Convulsant Antidote for Nerve Agent (CANA). Each MARK I consists of one atropine autoinjector and one pralidoxime chloride (2 PAM Cl) autoinjector (Figure 7-2A). Each ATNAA consist of a multichambered autoinjector with the atropine and pralidoxime chloride in separate chambers (Figure 7-2C). The CANA is a single autoinjector with flanges (Figure 7-2B). Procedures for use of both the MARK I and ATNAA are described below. You will use either the MARK I or the ATNAA in self-aid and buddy aid as issued.

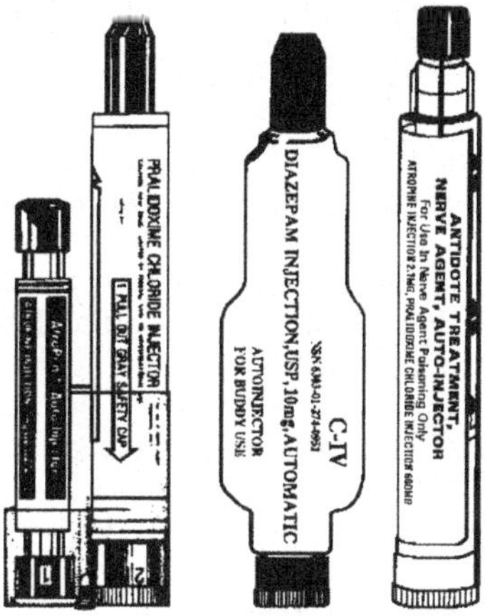

A B C

Figure 7-2. Nerve Agent Antidote Kit, MARK I, CANA, and ATNAA.

b. When you have the signs and symptoms of nerve agent poisoning, you should immediately put on the protective mask and then inject yourself with one set of the MARK I or ATNAA. Do not administer the CANA. You should inject yourself in the outer (lateral) thigh muscle (Figure 7-3) or if you are thin, in the upper outer (lateral) part of the buttocks (Figure 7-4).

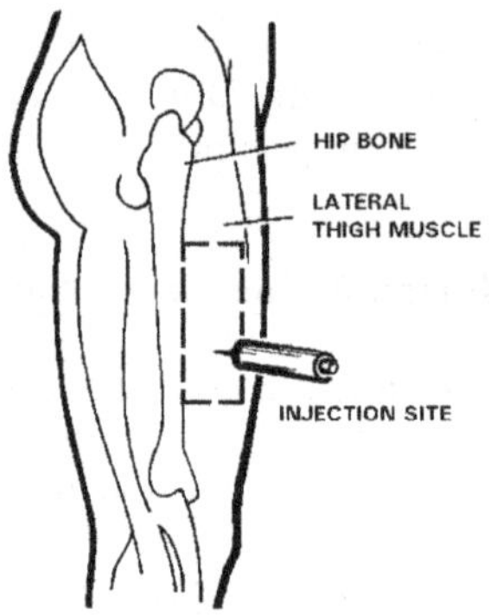

Figure 7-3. Thigh injection site.

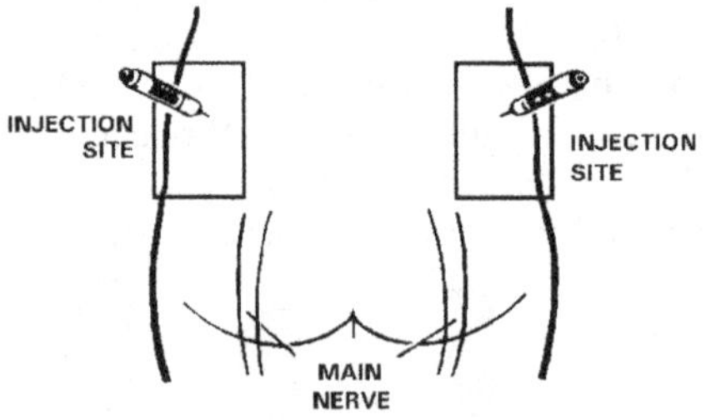

Figure 7-4. Buttocks injection site.

c. Also, you may come upon an unconscious chemical agent casualty who will be unable to care for himself and who will require first aid. You should be able to successfully—

(1) Mask him if he is unmasked.

(2) Inject him, if necessary, with all of **HIS** autoinjectors.

(3) Decontaminate his skin.

(4) Seek medical assistance.

7-7. Signs and Symptoms of Nerve Agent Poisoning

The symptoms of nerve agent poisoning are grouped as **MILD**—those that you recognize and for which you can perform self-aid, and **SEVERE**—those which require buddy aid.

a. MILD Signs and Symptoms.

- Unexplained runny nose.
- Unexplained sudden headache.
- Sudden drooling.
- Difficulty seeing (dimness of vision and miosis).
- Tightness in the chest or difficulty in breathing.
- Localized sweating and muscular twitching in the area of contaminated skin.
- Stomach cramps.
- Nausea.
- Tachycardia followed by bradycardia. (*Tachycardia* is an abnormally rapid heartbeat with a heart rate of over 100 beats per minute. *Bradycardia* is a slow heart rate of less than 60 beats per minute.)

b. SEVERE Signs and Symptoms.

- Strange or confused behavior.
- Wheezing, dyspnea (difficulty in breathing), and coughing.
- Severely pinpointed pupils.
- Red eyes with tearing.
- Vomiting.
- Severe muscular twitching and general weakness.
- Involuntary urination and defecation.

- Convulsions.
- Unconsciousness.
- Respiratory failure.
- Bradycardia.

7-8. First Aid for Nerve Agent Poisoning

First aid for nerve agent poisoning consists of administering the MARK I or ATNAA and CANA.

a. Injection Site. The injection site for administering the antidotes is normally in the outer thigh muscle. The thigh injection site is the area about a hand's width above the knee to a hand's width below the hip joint (Figure 7-3). It is important that the injection be given into a large muscle area. If the individual is thinly built, then the injections should be administered into the upper outer quarter (quadrant) of the buttock (Figure 7-4). Injecting in the buttocks of a thinly built individual avoids injury to the thighbone.

b. Self-Administer MARK I. If you experience any or all of the nerve agent **MILD** symptoms (paragraph 7-7*a*), you must **IMMEDIATELY** put on your protective mask and self-administer one MARK I (Figure 7-2A). Follow the procedure given in Table 7-1. The MARK I is carried in your protective mask carrier, pocket of the MOPP overgarment, or other location as specified in your unit tactical standing operating procedure (TSOP). (In cold weather, the MARK I should be stored in an inside pocket of your clothing to protect the antidote from freezing. A frozen MARK I cannot be immediately used to provide you with antidote, when needed. (However, the MARK I can still be used after complete thawing.)

Table 7-1. Self Aid for Nerve Agent Poisoning

MARK I*	ATNAA*
STEP 1. OBTAIN ONE MARK I.**	STEP 1. OBTAIN ONE ATNAA.**
STEP 2. CHECK INJECTION SITE.	STEP 2. CHECK INJECTION SITE.
STEP 3. HOLD MARK I AT EYE LEVEL WITH NONDOMINANT HAND WITH THE LARGE INJECTOR ON TOP (FIGURE 7-5A).	STEP 3. HOLD ATNAA WITH DOMINANT HAND (FIGURE 7-12A).

Table 7-1. Self Aid for Nerve Agent Poisoning (Continued)

MARK I*	ATNAA*
STEP 4. GRASP SMALL INJECTOR (ATROPINE) (FIGURE 7-5B) AND REMOVE FROM CLIP (FIGURE 7-5C).	STEP 4. GRASP SAFETY CAP WITH NONDOMINANT HAND AND REMOVE FROM INJECTOR (FIGURE 7-12B).
STEP 5. CLEAR HARD OBJECTS FROM INJECTION SITE.	STEP 5. CLEAR HARD OBJECTS FROM INJECTION SITE.
STEP 6. INJECT ATROPINE AT INJECTION SITE APPLYING EVEN PRESSURE TO THE INJECTOR (FIGURE 7-6 OR 7-7). HOLD IN PLACE FOR 10 SECONDS.	STEP 6. INJECT ATNAA AT INJECTION SITE APPLYING EVEN PRESSURE TO THE INJECTOR (FIGURE 7-14 OR 7-15). HOLD IN PLACE FOR 10 SECONDS.
STEP 7. HOLD USED INJECTOR WITH NONDOMINANT HAND.	STEP 7. BEND NEEDLE OF USED INJECTOR BY PRESSING ON A HARD SURFACE TO FORM A HOOK.
STEP 8. GRASP THE LARGE (2 PAM Cl) INJECTOR (FIGURE 7-8B) AND PULL IT FROM CLIP (FIGURE 7-8C). DROP CLIP TO GROUND.	STEP 8. ATTACH USED INJECTOR TO BLOUSE POCKET FLAP OF BDO/JSLIST (FIGURE 7-16).
STEP 9. INJECT 2 PAM Cl AT INJECTION SITE APPLYING EVEN PRESSURE TO THE INJECTOR (FIGURE 7-6 OR 7-7). HOLD IN PLACE FOR 10 SECONDS.	STEP 9. MASSAGE INJECTION SITE, MISSION PERMITTING.
STEP 10. BEND THE NEEDLES OF ALL USED INJECTORS BY PRESSING ON A HARD SURFACE TO FORM A HOOK.	
STEP 11. ATTACH ALL USED INJECTORS TO BLOUSE POCKET FLAP OF BDO/JSLIST (FIGURE 7-9).	
STEP 12. MASSAGE INJECTION SITE, MISSION PERMITTING.	

* USE STEPS LISTED FOR TYPE OF ANTIDOTE DEVICE ISSUED.

** ONLY ADMINISTER ONE MARK I OR ATNAA AS SELF-AID. DO NOT SELF-ADMINISTER CANA.

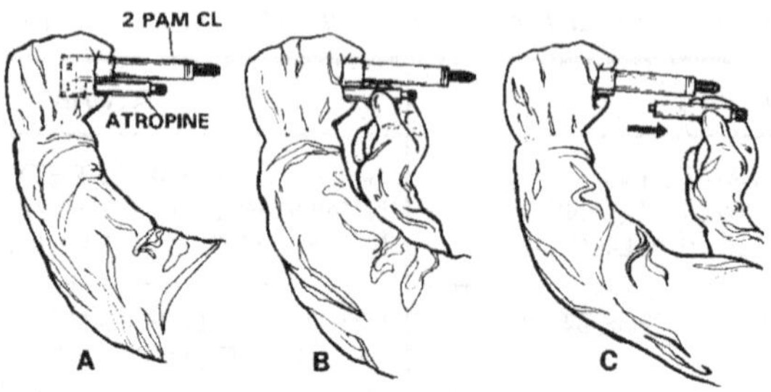

Figure 7-5. Removing the atropine autoinjector from the MARK I clip.

CAUTION

DO NOT cover or hold the needle end with your hand, thumb, or fingers—you might accidentally inject yourself. An accidental injection into the hand **WILL NOT** deliver an effective dose of the antidote, especially if the needle goes through the hand.

Figure 7-6. Thigh injection site for self-aid.

NOTE

If you are thinly built, inject yourself into the upper outer quadrant of the buttock (Figure 7-7). There is a nerve that crosses the buttocks; hitting this nerve can cause paralysis. Therefore, you must only inject into the *upper outer quadrant* of the buttock.

Figure 7-7. Buttocks injection site for self-aid.

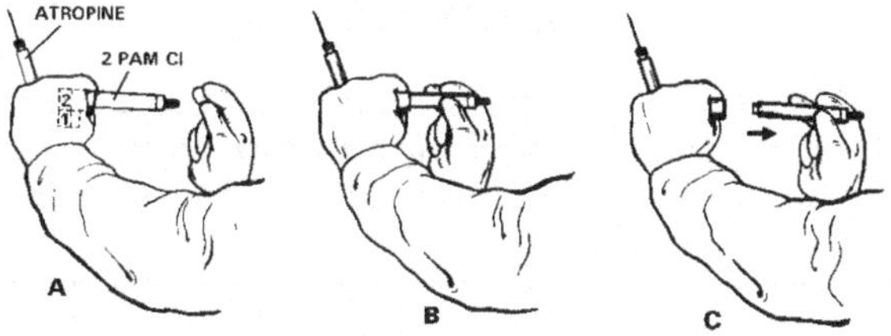

Figure 7-8. Removing the 2 PAM Cl autoinjector from the MARK I clip.

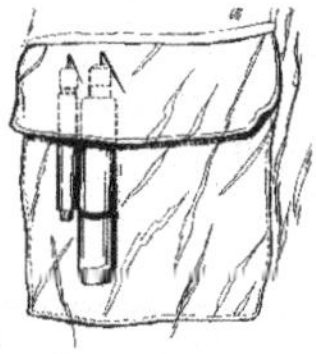

Figure 7-9. One set of used MARK I autoinjectors attached to pocket flap.

NOTES

1. **DO NOT** give yourself another set of injections. If you are able to walk without assistance, know who you are, and where you are, you **WILL NOT** need the second set of injections. (If not needed, giving yourself a second set of MARK I injections or ATNAA may create a nerve agent antidote overdose, which could cause incapacitation [inability to perform mission or defend yourself].)

2. If you continue to have symptoms of nerve agent poisoning, seek someone else (a buddy) to check your symptoms and administer the additional sets of injections, if required.

c. Buddy Evaluation and Buddy Aid. Service members may seek assistance after self-aid (self-administering one MARK I or ATNAA) or may become incapacitated after self-aid. A buddy must evaluate the individual to determine if additional antidotes are required to counter the effects of the nerve agent. Also, service members may experience **SEVERE** symptoms of nerve agent poisoning (paragraph 7-7*b*); they will not be able to treat themselves. In either case, other service members must perform buddy aid as quickly as possible. Before initiating buddy aid, determine if one set of MARK I autoinjectors has already been used so that no more than three sets of the antidote are administered. Buddy aid also includes administering the CANA with the third MARK I or ATNAA to prevent convulsions. Follow the procedures indicated in Table 7-2.

WARNING

Squat, DO NOT kneel, when masking the casualty or administering the nerve agent antidote to the casualty. Kneeling may force the chemical agent into or through your protective clothing.

CAUTION

DO NOT use your own MARK I, ATNAA, or CANA on a casualty. If you use your own, you may not have any antidote if needed for self-aid.

WARNING

DO NOT inject into areas close to the hip, knee, or thigh-bone.

Table 7-2. Buddy Aid/Combat Lifesaver Aid for Nerve Agent Casualty.

MARK I*	ATNAA*	CANA**
STEP 1. MASK THE CASUALTY AND POSITION HIM ON HIS SIDE (SWIMMER'S POSITION).	STEP 1. MASK THE CASUALTY AND POSITION HIM ON HIS SIDE (SWIMMER'S POSITION).	STEP 1. OBTAIN BUDDY'S CANA.
STEP 2. POSITION YOURSELF NEAR THE CASUALTY'S THIGH.	STEP 2. POSITION YOURSELF NEAR THE CASUALTY'S THIGH.	STEP 2. CHECK INJECTION SITE.

Table 7-2. Buddy Aid/Combat Lifesaver Aid for Nerve Agent Casualty (Continued).

MARK I*	ATNAA*	CANA**
STEP 3. OBTAIN BUDDY'S THREE OR REMAINING MARK Is.	STEP 3. OBTAIN BUDDY'S THREE OR REMAINING ATNAAs.	STEP 3. HOLD CANA IN A CLOSED FIST WITH DOMINANT HAND (FIGURE 7-12A).
STEP 4. CHECK INJECTION SITE.	STEP 4. CHECK INJECTION SITE.	STEP 4. GRASP SAFETY CAP WITH NONDOMINANT HAND AND REMOVE FROM INJECTOR (FIGURE 7-12B).
STEP 5. HOLD MARK I WITH NONDOMINANT HAND (FIGURE 7-5A).	STEP 5. HOLD ATNAA IN A CLOSED FIST WITH DOMINANT HAND (FIGURE 7-12A).	STEP 5. CLEAR HARD OBJECTS FROM INJECTION SITE.
STEP 6. GRASP SMALL INJECTOR (ATROPINE) AND REMOVE FROM CLIP (FIGURE 7-5B).	STEP 6. GRASP SAFETY CAP WITH NONDOMINANT HAND AND REMOVE FROM INJECTOR (FIGURE 7-12B).	STEP 6. INJECT CANA AT INJECTION SITE BY APPLYING EVEN PRESSURE TO THE INJECTOR, NOT A JABBING MOTION (FIGURE 7-14 OR 7-15). HOLD IN PLACE FOR 10 SECONDS.
STEP 7. CLEAR HARD OBJECTS FROM INJECTION SITE.	STEP 7. CLEAR HARD OBJECTS FROM INJECTION SITE.	STEP 7. BEND NEEDLE OF INJECTOR BY PRESSING ON A HARD SURFACE TO FORM A HOOK.
STEP 8. INJECT ATROPINE AT INJECTION SITE BY APPLYING EVEN PRESSURE TO THE INJECTOR, NOT A JABBING MOTION (FIGURE 7-10 OR 7-11). HOLD IN PLACE FOR 10 SECONDS.	STEP 8. INJECT ATNAA AT INJECTION SITE BY APPLYING EVEN PRESSURE TO THE INJECTOR, NOT A JABBING MOTION (FIGURE 7-14 OR 7-15). HOLD IN PLACE FOR 10 SECONDS.	STEP 8. ATTACH USED INJECTOR TO BLOUSE POCKET FLAP OF BDO/JSLIST (FIGURE 7-16).
STEP 9. HOLD USED INJECTOR BETWEEN LITTLE FINGER AND RING FINGER OF NONDOMINANT HAND (FIGURE 7-5A).	STEP 9. BEND NEEDLE OF INJECTOR BY PRESSING ON A HARD SURFACE TO FORM A HOOK.	STEP 9. MASSAGE INJECTION SITE, MISSION PERMITTING.
STEP 10. PULL LARGE INJECTOR (2 PAM Cl) FROM CLIP (FIGURE 7-5C). DROP CLIP TO GROUND.	STEP 10. ATTACH ALL USED INJECTORS TO BLOUSE POCKET FLAP OF BDO/JSLIST (FIGURE 7-16).	

Table 7-2. Buddy Aid/Combat Lifesaver Aid for Nerve Agent Casualty (Continued).

MARK I*	ATNAA*	CANA**
STEP 11. INJECT 2 PAM CI AT INJECTION SITE BY APPLYING EVEN PRESSURE TO THE INJECTOR, NOT A JABBING MOTION (FIGURE 7-10 OR 7-11). HOLD IN PLACE FOR 10 SECONDS.	STEP 11. MASSAGE INJECTION SITE, MISSION PERMITTING.	
STEP 12. REPEAT STEPS ABOVE FOR REMAINING MARK Is.		
STEP 13. BEND THE NEEDLES OF ALL USED INJECTORS BY PRESSING ON A HARD SURFACE TO FORM A HOOK.		
STEP 14. ATTACH ALL USED INJECTORS TO BLOUSE POCKET FLAP OF BDO/JSLIST (FIGURE 7-13).		
STEP 15. MASSAGE INJECTION SITE, MISSION PERMITTING.		

* USE STEPS LISTED FOR TYPE OF ANTIDOTE DEVICE ISSUED.
** CANA IS USED IN BUDDY AID/CLS AID ONLY. DO NOT USE IN SELF-AID.

NOTE

If the casualty is thinly built, inject the antidote into the buttock. Only inject the antidote into the upper outer portion of the casualty's buttock (Figure 7-11). This avoids hitting the nerve that crosses the buttocks (Figure 7-4). Hitting this nerve can cause paralysis.

Figure 7-10. Injecting the casualty's thigh (Mark I or CANA).

Figure 7-11. Injecting the casualty's buttocks (Mark I or CANA).

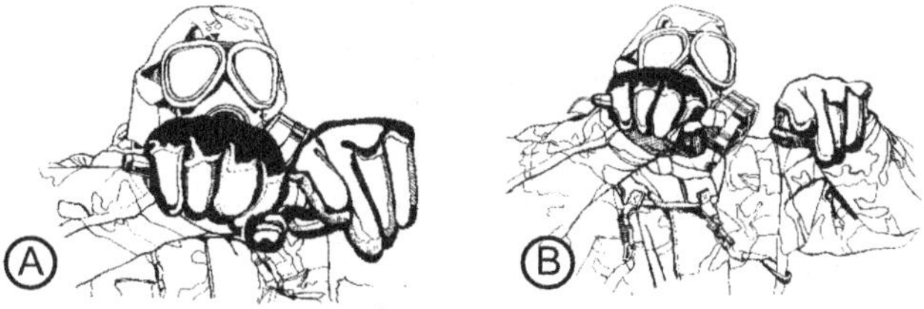

Figure 7-12. Preparing CANA or ATNAA for injection.

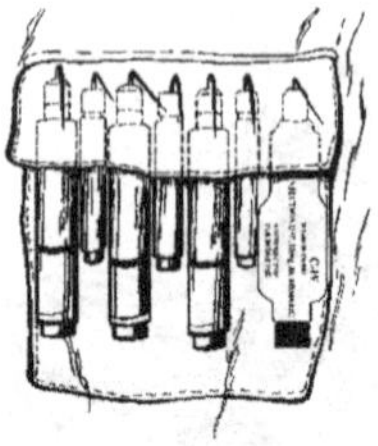

Figure 7-13. Three sets of used MARK I autoinjectors and one CANA autoinjector attached to pocket flap.

d. Self-Administer Antidote Treatment Nerve Agent Autoinjector. If you experience any or all of the nerve agent **MILD** symptoms (paragraph 7-7*b*), you must **IMMEDIATELY** self-administer one ATNAA following the procedure given Table 7-1.

NOTE

If you are thinly-built, inject yourself into the upper outer quarter (quadrant) of the buttock (Figure 7-15). There is a nerve that crosses the buttocks; hitting this nerve can cause paralysis. Therefore, you must only inject into the upper outer quarter (quadrant) of the buttocks.

Figure 7-14. Self-administration of ATNAA (thigh).

Figure 7-15. Self-administration of ATNAA (buttock).

NOTE

If you continue to have symptoms of nerve agent poisoning, seek someone else (a buddy) to check your symptoms and administer your remaining sets of injections, if required.

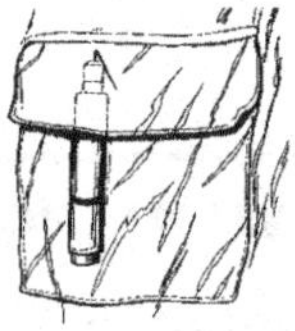

Figure 7-16. Used ATNAA attached to clothing.

e. Buddy Assistance. Service members may seek assistance after self-aid (self-administering one ATNAA) or may become incapacitated after self-aid. A buddy must evaluate the individual to determine if additional antidotes are required to counter the effects of the nerve agent. Also, service members may experience **SEVERE** symptoms of nerve agent poisoning (paragraph 7-7*b*); they will not be able to treat themselves. In either case, other service members must perform buddy aid as quickly as possible. Before initiating buddy aid, determine if one ATNAA has already been used so that no more than three ATNAA are administered. Buddy aid also includes administering the CANA with the third ATNAA to prevent convulsions. Follow the procedures indicated in Table 7-2.

WARNING

Squat, DO NOT kneel, when masking the casualty or administering the nerve agent antidotes to the casualty. Kneeling may force any chemical agent on your overgarment into or through your protective clothing.

Figure 7-17. Buddy injecting casualty's outer thigh (ATNAA or CANA).

NOTE

If the casualty is thinly built, inject the antidote into the buttocks (Figure 7-18). Only inject the antidote into the upper outer portion of the casualty's buttocks. This avoids hitting the nerve that crosses the buttocks (Figure 7-4). Hitting this nerve can cause paralysis.

WARNING

DO NOT inject into areas close to the hip, knee, or thighbone.

Figure 7-18. Buddy injecting casualty's buttocks (ATNAA or CANA).

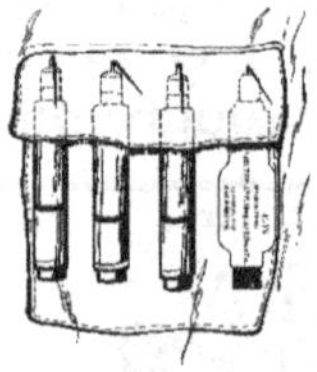

Figure 7-19. Three used ATNAAs and one CANA autoinjector attached to clothing.

f. Combat Lifesaver.

(1) The combat lifesaver must check to verify if the individual has received three sets of MARK I or ATNAAs. If not, the combat lifesaver performs first aid as described for buddy aid above. If the

individual has received the initial three sets of MARK I, then the combat lifesaver may administer additional atropine injections at approximately 15 minute intervals until atropinization is achieved (that is a heart rate above 90 beats per minute, reduced bronchial secretions, and reduced salivations). Administer additional atropine at intervals of 30 minutes to 4 hours to maintain atropinization or until the casualty is placed under the care of medical personnel. Check the heart rate by lifting the casualty's mask hood and feeling for a pulse at the carotid artery. Request medical assistance as soon as the tactical situation permits.

(2) The CLS should administer additional CANA to casualties suffering convulsions. Administer a second, and if needed, a third CANA at 5 to 10 minute intervals for a maximum of three injections (30 milligrams diazepam). Follow the steps and procedures described in buddy aid for administering the CANA. **DO NOT** give more than two additional injections for a total of three (one self-aid plus two by the CLS).

7-9. Blister Agents

Blister agents (vesicants) include mustard (H and HD), nitrogen mustards (HN), lewisite (L), and other arsenicals, mixtures of mustards and arsenicals, and phosgene oxime (CX). Blister agents may act on the eyes, mucous membranes, lungs, and skin. They burn and blister the skin or any other body parts they contact. Even relatively low doses may cause serious injury. Blister agents damage the respiratory tract (nose, sinuses, and windpipe) when inhaled and cause vomiting and diarrhea when absorbed. Lewisite and CX cause immediate pain on contact. However, mustard agents are deceptive as there is little or no pain at the time of exposure. Thus, in some cases, signs of injury may not appear for several hours after exposure.

a. Protective Measures. Your protective mask with hood and protective overgarment provide protection against blister agents. If it is known or suspected that blister agents are being used, **STOP BREATHING**, put on your mask and your protective overgarment.

CAUTION

Large drops of liquid vesicants on the protective overgarment ensemble may penetrate it if allowed to stand for an extended period. Remove large drops as soon as possible.

b. Signs and Symptoms of Blister Agent Poisoning.

(1) Immediate and intense pain upon contact with L, LH (lewisite and mustard) mixture, and CX. No initial pain upon contact with mustard.

(2) Inflammation and blisters (burns) resulting in tissue destruction. The severity of a chemical burn is directly related to the concentration of the agent and the duration of contact with the skin. The longer the agent is in contact with the tissue, the more serious the injury will be.

(3) Vomiting and diarrhea. Exposure to high concentrations of vesicants may cause vomiting or diarrhea.

(4) Death. The blister agent vapors absorbed during ordinary field exposure will probably not cause enough internal body (systemic) damage to result in death. However, death may occur from prolonged exposure to high concentrations of vapor or from extensive liquid contamination over wide areas of the skin, particularly *when decontamination is neglected or delayed.*

c. First Aid Measures.

(1) Use your M291 Skin Decontaminating Kit to decontaminate your skin and use water to flush contaminated eyes. Decontamination of vesicants must be done immediately (within 1 minute is best).

(2) If blisters form, cover them loosely with a field dressing and secure the dressing.

CAUTION

Blisters are actually burns. **DO NOT** attempt to decontaminate the skin where blisters have formed, as the agent has already been absorbed.

(3) If you receive blisters over a wide area of the body, you are considered seriously burned. Seek medical assistance immediately.

(4) If vomiting occurs, the mask should be lifted momentarily and drained—while the eyes are closed and the breath is held—and replaced, cleared, and sealed.

(5) Remember, if vomiting or diarrhea occurs after having been exposed to blister agents, seek medical assistance immediately.

7-10. Choking Agents (Lung-Damaging Agents)

Chemical agents that attack lung tissue, primarily causing fluid buildup (pulmonary edema), are classified as choking agents (lung-damaging agents). This group includes phosgene (CG), diphosgene (DP), chlorine (Cl), and chloropicrin (PS). Of these four agents, CG is the most dangerous and is more likely to be employed by the enemy in future conflict.

a. Protective Measures. Your protective mask gives adequate protection against choking agents.

b. Signs and Symptoms. During and immediately after exposure to choking agents (depending on agent concentration and length of exposure), you may experience some or all of the following signs and symptoms:

- Tears (lacrimation).
- Coughing.
- Choking.
- Tightness of chest.
- Nausea and vomiting.
- Headaches.

c. Self-Aid.

(1) The protective mask should be put on immediately when any of the conditions described in *b* above exist. Another indication of a CG attack is an odor like newly mown hay; however, **DO NOT** rely upon odor as indication of a chemical attack.

(2) If some CG is inhaled, normal combat duties should be continued unless there is difficulty in breathing, nausea, vomiting, or more than the usual shortness of breath during exertion. If any of the above symptoms occur and the mission permits, remain at quiet rest until medical evacuation is accomplished.

d. Death. With ordinary field exposure to choking agents, death will probably not occur. *However, prolonged exposure to high concentrations of the vapor and neglect or delay in masking can be fatal.*

7-11. Cyanogen (Blood) Agents

Cyanogen agents interfere with proper oxygen utilization in the body. Hydrogen cyanide (AC) and cyanogen chloride (CK) are the primary agents in this group.

a. Protective Measures. Your protective mask with a fresh filter gives adequate protection against field concentrations of cyanogen agent vapor. The protective overgarments, as well as the mask, are needed when exposed to liquid AC.

b. Signs and Symptoms. During and immediately after exposure to cyanogen agents (depending on agent concentration and length of exposure), you may experience some or all of the following signs and symptoms:

- Tearing (lacrimation).
- Eye, nose, and throat irritation.
- Sudden stimulation of breathing (unable to hold breath).
- Nausea.
- Coughing.
- Tightness of chest.
- Headache.
- Light-headedness (dizziness).
- Unconsciousness.

c. First Aid.

(1) *Hydrogen cyanide.* During any chemical attack, if you get a sudden stimulation of breath or detect an odor like bitter almonds, **PUT ON YOUR MASK IMMEDIATELY**. Speed is absolutely essential since this agent acts so rapidly that within a few seconds its effects will make it impossible for service members to put on their mask by themselves. Stop breathing until the mask is on, if at all possible. This may be very difficult since the agent strongly stimulates respiration.

(2) *Cyanogen chloride.* Put your mask on immediately if you experience any irritation of the eyes, nose, or throat. Service members

who are unable to mask should be masked by the nearest service member (buddy).

d. *Medical Assistance.* If you suspect that you have been exposed to blood agents, seek medical assistance immediately.

7-12. Incapacitating Agents

An incapacitating agent is a chemical agent which produces temporary, disabling conditions which persist for hours to days after exposure. Unlike riot control agents, which usually are momentary or fleeting in action, incapacitating agents have a persistent effect. It is likely that smoke-producing munitions or aerosols will disseminate such agents, thus making breathing their means of entry into the body. The protective mask is, therefore, essential.

a. There are no specific first aid measures to relieve the symptoms of incapacitating agents. Supportive first aid and physical restraint may be indicated. If the casualty is stuporous or comatose, be sure that respiration is unobstructed; then turn him on his side in case vomiting should occur. Complete cleansing of the skin with soap and water should be done as soon as possible; or, the M291 Skin Decontaminating Kit can be used if washing is impossible. Remove weapons and other potentially harmful items from service members who are suspected of having these symptoms. Harmful items include cigarettes, matches, medications, and small items that might be swallowed accidentally. Delirious (confused) persons have been known to attempt to eat items bearing only a superficial resemblance to food.

b. Incapacitating agents (anticholinergic drugs BZ type) may produce alarming dryness and coating of the lips and tongue; however, there is usually no danger of immediate dehydration. Fluids should be given sparingly, if at all, because of the danger of vomiting and because of the likelihood of temporary urinary retention due to paralysis of bladder muscles.

c. If the body temperature is elevated and mucous membranes are dry, immediate and vigorous cooling (as for heatstroke) is indicated. Methods that can be used to cool the skin are spraying with cool water and air circulation (fanning); applying alcohol soaked cloths and air circulation; and providing maximum exposure to air in a shaded area, along with maximum air circulation. Such cases are usually a result of anticholinergic poisoning. Rapid evacuation should be accomplished since medical treatment with the appropriate medication may be lifesaving.

CAUTION

DO NOT use **ice** for cooling the skin.

d. Reassurance and a firm, but friendly, attitude by individuals providing first aid will be beneficial if the casualty appears to comprehend what is being said. Conversation is a waste of time if the service member is incoherent or cannot understand what is being said. In such cases, the less said, the better it is—these casualties will benefit more from prompt and vigorous restraint and evacuation to an MTF.

7-13. Incendiaries

Incendiaries can be grouped as WP, thickened gasoline, metal, and oil and metal. You must learn to protect yourself against these incendiaries.

a. White phosphorus is used primarily as a smoke producer but can be used for its incendiary effect to ignite field expedients and combustible materials. The burns from WP are usually multiple, deep, and variable in size. When particles of WP get on the skin or clothing, they continue to burn until deprived of air. They also have a tendency to stick to a surface and must be brushed off or picked out.

(1) If burning particles of WP strike and stick to your clothing, quickly take off the contaminated clothing before the WP burns through to the skin.

(2) If burning WP strikes your skin, smother the flame with water, a wet cloth, or mud.

NOTE

Since WP is soluble in oil, **DO NOT** use grease, oily ointments, or eye ointments to smother the flame.

(3) Keep the WP particles covered with a wet material to exclude air until you can remove them or have them removed from your skin.

(4) Remove the WP particles from the skin by brushing them with a wet cloth and by picking them out with a knife, bayonet, stick, or other available object.

(5) Seek medical assistance when the mission permits.

b. Thickened fuel mixtures (napalm) have a tendency to cling to clothing and body surfaces, thereby producing prolonged exposure and severe burns. The first aid for these burns is the same as for other heat burns. The heat and irritating gases given off by these combustible mixtures may cause lung damage, which must be treated by medical personnel.

c. Metal incendiaries pose special problems. Thermite particles on the skin should be immediately cooled with water and then removed. The first aid for these burns is the same as for other heat burns. Particles of magnesium on the skin burn quickly and deeply. Like other metal incendiaries, they must be removed. Ordinarily, medical personnel should do the complete removal of these particles as soon as possible. Immediate medical treatment is required.

d. Oil and metal incendiaries have much the same effect on contact with the skin and clothing as those discussed (*b* and *c* above). First aid measures for burns are discussed in Chapter 3.

7-14. Biological Agents and First Aid

a. Biological attacks can result in combat ineffectiveness by introducing disease-causing organisms into a troop population.

b. Once a disease is identified, first aid or medical treatment is initiated, depending on the seriousness of the disease. First aid measures are concerned with observable symptoms of the disease such as diarrhea or vomiting.

7-15. Toxins

Toxins are alleged to have been used in past conflicts. Witnesses and victims have described the agent as toxic rain (or yellow rain) because it was reported to have been released from aircraft as a yellow powder or liquid that covered ground, structures, vegetation, and people.

a. *Signs and Symptoms.* The occurrence of the symptoms from toxins may appear in a period of a few minutes to several hours depending on the particular toxin, the service member's susceptibility, and the amount of toxin inhaled, ingested, or deposited on the skin. Symptoms from toxins usually involve the central nervous system but are often preceded by less prominent symptoms, such as nausea, vomiting, diarrhea, cramps, or stomach irritation and burning sensation. Typical neurological symptoms often develop rapidly in severe cases; for example, visual disturbances, inability to swallow, speech difficulty, lack of muscle coordination, and sensory abnormalities (numbness of mouth, throat, or extremities). Yellow rain (mycotoxins) also may have hemorrhagic symptoms, which could include any or all of the following:

- Dizziness.
- Severe itching or tingling of the skin.
- Formation of multiple, small, hard blisters.
- Coughing up blood.
- Shock (which could result in death).

b. *Self-Aid.* Upon recognition of an attack employing toxins, you must immediately take the following actions:

(1) Stop breathing, put on your protective mask with hood, and then resume breathing. Next, put on your protective clothing.

(2) Should severe itching of the face become unbearable, quickly—

- Loosen the cap on your canteen.
- Take and hold a deep breath and lift your mask.
- While holding your breath, close your eyes and flush your face with generous amounts of water.

CAUTION

DO NOT rub or scratch your eyes. Try not to let the water run onto your clothing or protective overgarment.

- Put your protective mask back on, seat it properly, clear it, and check it for a seal; then resume breathing.
- Decontaminate your skin by bathing with soap and water as soon as the mission permits.

- Change clothing and decontaminate your protective mask using soap and water. Replace the filters if directed.

(3) If vomiting occurs, the mask should be lifted momentarily and drained—while the eyes are closed and the breath is held—and replaced, cleared, and sealed.

c. *Medical Assistance.* If you suspect that you have been exposed to toxins, you should seek medical assistance immediately.

7-16. Nuclear Detonation

a. Three types of injuries may result from a nuclear detonation. These are thermal, blast, and radiation injuries. Many times the casualty will have a combination of these types of injuries. First aid for thermal and blast injuries is provided based on observable injuries, such as burns, hemorrhage, or fractures.

b. The signs and symptoms of radiation illness in the initial phase include the rapid onset of nausea, vomiting, and malaise (tiredness). The only first aid procedure for radiological casualties is decontamination.

CHAPTER 8

FIRST AID FOR PSYCHOLOGICAL REACTIONS

8-1. General

Psychological first aid is as natural and reasonable as physical first aid and is just as familiar. When you were hurt as a child, the understanding attitude of your parents did as much as the psychological effect of a bandage. Later, your disappointment or grief was eased by supportive words from a friend. Certainly, taking a walk and talking things out with a friend are familiar ways of dealing with an emotional crisis. The same natural feelings that make us want to help a person who is injured make us want to give a helping hand to a buddy who is upset. *Psychological first aid* really means nothing more complicated than assisting people with emotional distress whether it results from physical injury, disease, or excessive stress. Emotional distress is not always as visible as a wound or a broken bone. However, overexcitement, severe fear, excessive worry, deep depression, misdirected irritability, and anger are signs that stress has reached the point of interfering with effective coping. The more noticeable the symptoms become, the more urgent the need for you to be of help and the more important it is for you to know *how* to help.

8-2. Importance of Psychological First Aid

You must know how to give psychological first aid to be able to help yourself, your buddies, and your unit in order to keep performing the mission. Psychological first aid measures are simple and easy to understand. Your decision of what to do depends upon your ability to observe the service member and understand his needs. Making the best use of resources requires ingenuity on your part. A stress reaction resulting in poor judgment can cause injury or even death to yourself or others on the battlefield. It can be even more dangerous if other persons are affected by the judgment of an emotionally upset service member. If it is detected early enough, the affected service member stands a good chance of remaining in his unit as an effective member. If it is not detected early and if the service member becomes more emotionally upset, he may become a threat to himself and to others.

8-3. Situations Requiring Psychological First Aid

- Psychological first aid (buddy aid) is most needed at the first sign that a service member cannot perform the mission because of emotional distress. Stress is inevitable in combat, in hostage and terrorist situations, and in civilian disasters such as floods, hurricanes, or industrial accidents. Most emotional reactions to such situations are temporary, and the service member can still carry on with encouragement. Painful or disruptive symptoms may last for minutes, hours, or days. However, if the stress symptoms are seriously disabling, they may be psychologically contagious and endanger not only the emotionally upset service member but also the entire unit.

- Sometimes people continue to function well during a disastrous event, but suffer from emotional scars which impair their job performance or quality of life at a later time. Painful memories and dreams may recur for months and years and still be considered a normal reaction. However, if the memories are so painful that the person must avoid all situations which arouse them, becomes socially withdrawn, or shows symptoms of anxiety, depression, or substance abuse, he needs treatment. Experience with police, firemen, emergency medical technicians, and others who deal with disasters has proved that the routine application of psychological first aid to all the participants, including those who have functioned well, greatly reduces the likelihood of future serious post-traumatic stress disorders (PTSDs).

8-4. Interrelationship of Psychological and Physical First Aid

Psychological first aid should go hand in hand with physical first aid. The discovery of a physical injury or cause for an inability to function does not rule out the possibility of a psychological injury (or vice versa). The person suffering from pain, shock, fear of serious injury, or fear of death does not respond well to joking, indifference, or tearful-tearful attention. Fear and anxiety may take as high a toll of the service member's strength as does the loss of blood.

8-5. Goals of Psychological First Aid

The goals of psychological first aid are to—

- Be supportive; assist the service member in dealing with his stress reaction.

- Prevent, and if necessary control, behavior harmful to himself and to others.

- Return the service member to duty as soon as possible after dealing with the stress reaction.

8-6. Respect for Others' Feelings

a. Accept the service member you are trying to help without censorship or ridicule. Respect his right to his own feelings. Even though your feelings, beliefs, and behavior are different, DO NOT blame or make light of him for the way he feels or acts. Your purpose is to help him in this tough situation, not to be his critic. A person DOES NOT WANT to be upset and worried. When he seeks help, he needs and expects consideration of his fears, not abrupt dismissal or ridicule.

b. Realize that people are the products of a wide variety of factors. All people DO NOT react the same way to the same situations. Each individual has complex needs and motivations, both conscious and unconscious, that are uniquely his own. Often the one thing that finally causes the person to become overloaded by a stressful situation is not the stressor itself, but some other problem.

8-7. Emotional and Physical Disability

a. Accept emotional disability as being just as real as physical disability. If a service member's ankle is seriously sprained in a fall, no one expects him to run right away. A service member's emotions may be temporarily strained by the overwhelming stress of battle or other traumatic incident. DO NOT demand that he pull himself together immediately and carry on without a break. Some individuals can pull themselves together immediately, but others cannot. The service member whose emotional stability has been disrupted has a disability just as real as the service member who has sprained his ankle. There is an unfortunate tendency in many people to regard as real only what they can see, such as a wound or bleeding. Some people tend to assume that damage involving a person's mind and emotions is just imagined, that he is not really sick or injured, and that he could overcome his trouble by using his will power.

b. The terms *it's all in your head, snap out of it,* and *get control of yourself* are often used by people who believe they are being helpful. Actually, these terms are expressions of hostility because they show lack of understanding. They only emphasize weakness and inadequacy. Such terms are of no use in psychological first aid.

c. Every physically injured person has some emotional reaction to the fact that he is injured.

(1) It is normal for an injured person to feel upset. The more severe the injury, the more insecure and fearful he becomes, especially

if the injury is to a body part which is highly valued. For example, an injury to the eyes or the genitals, even though relatively minor, is likely to be extremely upsetting. An injury to some other part of the body may be especially disturbing to an individual for his own particular reason. For example, an injury of the hand may be a terrifying blow to a surgeon or an injury to the eye of a pilot.

(2) An injured service member always feels less secure, more anxious, and more afraid not only because of what has happened to him but because of what he imagines may happen as a result of his injury. This fear and insecurity may cause him to be irritable, uncooperative, or unreasonable. As you help him, always keep in mind that such behavior has little or nothing to do with you personally. He needs your patience, reassurance, encouragement, and support.

8-8. Combat and Other Operational Stress Reactions

Stress reaction is a temporary emotional disorder or inability to function, experienced by a previously normal service member as a reaction to the overwhelming or cumulative stress of combat. Stress reaction gets better with reassurance, rest, physical replenishment, and activities that restore confidence. All service members are likely to feel stress reaction under conditions of intense and/or prolonged stress. They may even become stress reaction casualties, unable to perform their mission for hours or days. Other combat and operational stress reactions (COSRs) may result in negative behavior, but are not termed *stress reaction,* as they need more intensive treatment. These negative COSRs may result in misconduct stress behaviors such as drug and alcohol abuse, criminal acts, looting, desertion, and self-inflicted wounds. These harmful COSRs can often be prevented by good psychological first aid. Service members who commit misconduct stress behaviors may require disciplinary action rather than medical treatment.

8-9. Reactions to Stress

Most service members react to stressful incidents after the situation has passed. All service members feel some fear. This fear may be greater than they have experienced at any other time, or they may be more aware of their fear. In such a situation, they should not be surprised if they feel shaky or become sweaty, nauseated, or confused. These reactions are normal and are not a cause for concern. However, some reactions, either short- or long-term, will cause problems if left unchecked. See paragraph 8-13 for more information.

a. Emotional Reactions.

(1) The most obvious combat stress reaction (CSR) is inefficient performance. This can be demonstrated by—

- Slow thinking (or reaction time).
- Difficulty recognizing priorities and seeing what needs to be done.
- Difficulty getting started.
- Indecisiveness and having trouble focusing attention.
- Tendency to do familiar tasks and be preoccupied with familiar details. (This can reach the point where the person is very passive, such as just sitting or wandering about not knowing what to do.)

(2) A less common reaction may be uncontrolled emotional outbursts; this can be demonstrated by crying, screaming, or laughing. Some service members will react in the opposite way. They will be very withdrawn and silent and try to isolate themselves from everyone. These service members should be encouraged to remain with their assigned unit. Uncontrolled reactions may appear by themselves or in any combination (the person may be crying uncontrollably one minute and then laughing the next). In this state, the person is restless and cannot keep still. He may run about, apparently without purpose. Inside, he feels a great rage or fear and his physical acts may show this. In his anger he may indiscriminately strike out at others.

b. Loss of Adaptability.

(1) In a desperate attempt to get away from the danger, which has overwhelmed him, a service member may panic and become confused. His mental ability may be so impaired he cannot think clearly or even follow simple commands. His judgment may be faulty and he may not be aware of his actions, such as standing up in his fighting position during an attack.

(2) In other cases, overwhelming stress may produce symptoms that are often associated with head injuries. For example, the service member may appear dazed or be found wandering around aimlessly. He may appear confused and disoriented and may seem to have a complete or partial loss of memory. In such cases, especially when no eyewitnesses can provide evidence that the service member has NOT suffered a head injury, it is necessary for him to be rapidly medically evacuated. **DO NOT** allow the

service member to expose himself to further personal danger until the cause of the problem has been determined.

c. Sleep Disturbance and Repetition of Dreams. A person who has been overwhelmed by stress often has difficulty sleeping. The service member may experience nightmares related to the stressors. Remember that nightmares, in themselves, are not considered abnormal when they occur soon after a period of intensive stress. As time passes, the nightmares usually become less frequent and less intense. In extreme cases, a service member, even when awake, may think repeatedly of the incident, feel as though it is happening again, and act out parts of his stress over and over again. For some persons, this repetitious reexperiencing of the stressful event may be necessary for eventual recovery; therefore, it should not be discouraged or viewed as abnormal. For the person reexperiencing the event, such reaction may be disruptive. The service member needs to be encouraged to *ventilate* about the incident. Ventilation is a technique where the service member is given the opportunity to talk extensively, often repetitiously about the experience.

8-10. Severe Stress or Stress Reaction

You do not need specialized training to recognize severe stress or stress reaction that will cause problems for the service member, the unit, or the mission. Reactions that are less severe, however, are more difficult to detect. To determine whether a person needs help, you must observe him to see whether he is doing something meaningful, performing his duties, taking care of himself, behaving in an unusual fashion, or acting out of character.

8-11. Application of Psychological First Aid

The emotionally disturbed service member has built a barrier against fear. He does this for his own protection, although he is probably not aware that he is doing it. If he finds that he does not have to be afraid and that there are normal, understandable things about him, he will feel safer in dropping this barrier. Persistent efforts to make him realize that you want to understand him will be reassuring, especially if you remain calm. Nothing can cause an emotionally disturbed person to become even more fearful than feeling that others are afraid of him. Try to remain calm. Familiar things, such as a cup of coffee, the use of his name, attention to a minor wound, being given a simple job to do, or the sight of familiar people and activities, will add to his ability to overcome his fear. He may not respond well if you get excited, angry, or abrupt.

a. *Ventilation.* After the service member becomes calmer, he is likely to have dreams about the stressful event. He also may think about it when he is awake or even repeat his personal reaction to the event. One benefit of this natural pattern is that it helps him master the stress by going over it just as one masters the initial fear of parachuting from an aircraft by doing it over and over again. Eventually, it is difficult to remember how frightening the event was initially. In giving first aid to the emotionally disturbed service member, you should let him follow this natural pattern. Encourage him to talk. Be a good listener. Let him tell, in his own words, what actually happened. If home front problems or worries have contributed to the stress, it will help him to talk about them. Your patient listening will prove to him that you are interested in him, and by describing his personal problem, he can work at mastering his fear. If he becomes overwhelmed in the telling, suggest a cup of coffee or a break. Whatever you do, assure him that you will listen again as soon as he is ready. Do try to help put the service member's perception of what happened back into realistic perspective; but DO NOT argue about it.

b. *Activity.*

(1) A person who is emotionally disturbed as the result of a combat action is a casualty of anxiety and fear. He is disabled because he has become temporarily overwhelmed by his anxiety. A good way to control fear is through activity. Almost all service members, for example, experience a considerable sense of anxiety and fear while they are poised, awaiting the opening of a big offensive; but this is normally relieved, and they actually feel better once they begin to move into action. They take pride in effective performance and pleasure in knowing that they are good service members, perhaps being completely unaware that overcoming their initial fear was their first major accomplishment.

(2) Useful activity is very beneficial to the emotionally disturbed service member who is not physically incapacitated. After you help a service member get over his initial fear, help him to regain some self-confidence. Make him realize his job is continuing by finding him something useful to do. Encourage him to be active. Get him to help load trucks, clean up debris, or dig fighting positions. If possible, get him back to his usual duty. Seek out his strong points and help him apply them. Avoid having him just sit around. You may have to provide direction by telling him what to do and where to do it. The instructions should be clear and simple and should be repeated. A person who has panicked is likely to argue. Respect his feelings, but point out more immediate, obtainable, and demanding needs. Channel his excessive energy and, above all, DO NOT argue. If you cannot get him interested in doing more profitable work, it may be necessary to enlist aid in controlling his overactivity before it spreads to the group and

results in more panic. Prevent the spread of such infectious feelings by restraining and segregating if necessary.

(3) Involvement in activity helps a service member in three ways; he—

- Forgets himself.
- Has an outlet for his excessive tensions.
- Proves to himself he can do something useful.

c. *Rest.* There are times, particularly in combat, when physical exhaustion is a principal cause for emotional reactions. A unit sleep plan should be established and implemented. When possible, service members should be given a safe and relatively comfortable area in which to sleep. Examples would be an area away from heavy traffic, noise, and congestion or a place that is clean and dry and protected from environmental conditions. The more uninterrupted sleep a service member gets the better he will be able to function in the tactical environment.

d. *Hygiene.* Field hygiene is an important ingredient in a service member's morale. A service member who is dirty and unkempt will not function as well as a service member who has had the opportunity to bathe and put on clean, dry clothing. During combat, unit leaders should stress the importance of personal hygiene. Good personal hygiene not only improves morale, it also is a preventive measure against disease and nonbattle injury (DNBI).

e. *Group Activity.* You have probably already noticed that a person works, faces danger, and handles serious problems better if he is a member of a closely-knit group. Each service member in the team supports the other team members. Esprit de corps is built because the service members have the same interests, goals, and mission, and as a result they are more productive; furthermore, they are less worried because everyone is involved. It is this spirit that takes a strategic hill in battle. It is so powerful that it is one of the most effective tools you have in your *psychological first aid bag*. Getting the service member back into the team or squad activities will reestablish his sense of belonging and security and will go far toward making him a useful member of the unit.

8-12. Reactions and Limitations

Up to this point the discussion has been primarily about the feelings of the emotionally distressed service member. What about your feelings toward him? Whatever the situation, you will have emotional reactions (conscious or unconscious) toward this service member. Your reactions can either help or hinder your ability to help him. When you are tired or worried, you may very easily become impatient with him if he is unusually slow or exaggerates. You may even feel resentful toward him. At times when many physically wounded lie about you, it will be especially natural for you to resent disabilities that you cannot see. Physical wounds can be seen and easily accepted. Emotional reactions are more difficult to accept as injuries. On the other hand, will you tend to be overly sympathetic? Excessive sympathy for an incapacitated person can be as harmful as negative feelings in your relationship with him. He needs strong help, but not your sorrow. To overwhelm him with pity will make him feel even more inadequate. You must expect your buddy to recover, to be able to return to duty, and to become a useful service member again. This expectation should be displayed in your behavior and attitude as well as in what you say. If he can see your calmness, confidence, and competence, he will be reassured and will feel a sense of greater security.

8-13. Stress Reactions

See Tables 8-1, 8-2, and 8-3 for more information.

Table 8-1. Mild Stress Reaction

PHYSICAL SIGNS*	EMOTIONAL SIGNS*
1. TREMBLING, TEARFUL 2. JUMPINESS, NERVOUSNESS 3. COLD SWEAT, DRY MOUTH 4. POUNDING HEART, DIZZINESS 5. INSOMNIA, NIGHTMARES 6. NAUSEA, VOMITING, DIARRHEA 7. FATIGUE 8. THOUSAND-YARD STARE 9. DIFFICULTY THINKING, SPEAKING, AND COMMUNICATING	1. ANXIETY, INDECISIVENESS 2. IRRITABLE, COMPLAINING 3. FORGETFUL, UNABLE TO CONCENTRATE 4. EASILY STARTLED BY NOISE, MOVEMENT 5. GRIEF, TEARFUL 6. ANGER, BEGINNING TO LOSE CONFIDENCE IN SELF AND UNIT

SELF- AND BUDDY AID
1. CONTINUE MISSION PERFORMANCE, FOCUS ON IMMEDIATE MISSION. 2. EXPECT SERVICE MEMBER TO PERFORM ASSIGNED DUTIES. 3. REMAIN CALM AT ALL TIMES; BE DIRECTIVE AND IN CONTROL. 4. LET SERVICE MEMBER KNOW HIS REACTION IS NORMAL, AND THAT THERE IS NOTHING SERIOUSLY WRONG WITH HIM. 5. KEEP SERVICE MEMBER INFORMED OF THE SITUATION, OBJECTIVES, EXPECTATIONS, AND SUPPORT. CONTROL RUMORS. 6. BUILD SERVICE MEMBER'S CONFIDENCE, TALK ABOUT SUCCEEDING. 7. KEEP SERVICE MEMBER PRODUCTIVE (WHEN NOT RESTING) THROUGH RECREATIONAL ACTIVITIES, EQUIPMENT MAINTENANCE.

8. ENSURE SERVICE MEMBER MAINTAINS GOOD PERSONAL HYGIENE.
9. ENSURE SERVICE MEMBER EATS, DRINKS, AND SLEEPS AS SOON AS POSSIBLE.
10. LET SERVICE MEMBER TALK ABOUT HIS FEELINGS. DO NOT "PUT DOWN" HIS FEELINGS OF GRIEF OR WORRY. GIVE PRACTICAL ADVICE AND PUT EMOTIONS INTO PERSPECTIVE.

* MOST OR ALL OF THESE SIGNS ARE PRESENT IN MILD STRESS REACTION. THEY CAN BE PRESENT IN ANY NORMAL SERVICE MEMBER IN COMBAT YET HE CAN STILL DO HIS JOB.

Table 8-2. More Serious Stress Reaction

PHYSICAL SIGNS*	EMOTIONAL SIGNS*
1. CONSTANTLY MOVES AROUND 2. FLINCHING OR DUCKING AT SUDDEN SOUNDS 3. SHAKING, TREMBLING (WHOLE BODY OR ARMS) 4. CANNOT USE PART OF BODY, NO PHYSICAL REASON (HAND, ARM, LEGS) 5. CANNOT SEE, HEAR, OR FEEL (PARTIAL OR COMPLETE LOSS) 6. PHYSICAL EXHAUSTION, CRYING 7. FREEZING UNDER FIRE, OR TOTAL IMMOBILITY 8. VACANT STARES, STAGGERS, SWAYS WHEN STANDS 9. PANIC RUNNING UNDER FIRE	1. RAPID AND/OR INAPPROPRIATE TALKING 2. ARGUMENTATIVE, RECKLESS MOVEMENTS/ACTIONS 3. INATTENTIVE TO PERSONAL HYGIENE 4. INDIFFERENT TO DANGER 5. MEMORY LOSS 6. SEVERE STUTTERING, MUMBLING, OR CANNOT SPEAK AT ALL 7. INSOMNIA, NIGHTMARES 8. SEEING OR HEARING THINGS THAT DO NOT EXIST 9. RAPID EMOTIONAL SHIFTS 10. SOCIAL WITHDRAWAL 11. APATHETIC 12. HYSTERICAL OUTBURSTS 13. FRANTIC OR STRANGE BEHAVIOR

TREATMENT PROCEDURES**

1. IF A SERVICE MEMBER'S BEHAVIOR ENDANGERS THE MISSION, SELF, OR OTHERS, DO WHATEVER IS NECESSARY TO CONTROL HIM.
2. IF THE SERVICE MEMBER IS UPSET, CALMLY TALK HIM INTO COOPERATING.
3. IF CONCERNED ABOUT THE SERVICE MEMBER'S RELIABILITY:
    - UNLOAD HIS WEAPON.
    - TAKE WEAPON IF SERIOUSLY CONCERNED.
    - PHYSICALLY RESTRAIN HIM ONLY WHEN NECESSARY FOR SAFETY OR TRANSPORTATION.
4. REASSURE EVERYONE THAT THE SIGNS ARE PROBABLY JUST STRESS REACTION AND WILL QUICKLY IMPROVE.
5. IF STRESS REACTION SIGNS CONTINUE:
    - GET THE SERVICE MEMBER TO A SAFER PLACE.
    - DO NOT LEAVE THE SERVICE MEMBER ALONE, KEEP SOMEONE HE KNOWS WITH HIM.
    - NOTIFY SENIOR NONCOMMISSIONED OFFICER (NCO) OR OFFICER.
    - HAVE THE SERVICE MEMBER EXAMINED BY MEDICAL PERSONNEL.

Table 8-2. More Serious Stress Reaction (Continued)

TREATMENT PROCEDURES**

6. GIVE THE SERVICE MEMBER EASY TASKS TO DO WHEN NOT SLEEPING, EATING, OR RESTING.
7. ASSURE THE SERVICE MEMBER HE WILL RETURN TO FULL DUTY IN 24 HOURS; AND, RETURN HIM TO NORMAL DUTIES AS SOON AS HE IS READY.

* THESE SIGNS ARE PRESENT IN ADDITION TO THE SIGNS OF MILD STRESS REACTION.

** DO THESE PROCEDURES IN ADDITION TO THE SELF- AND BUDDY AID CARE.

Table 8-3. Preventive Measures to Combat Stress Reaction

1. WELCOME NEW MEMBERS INTO YOUR TEAM, GET TO KNOW THEM QUICKLY. IF YOU ARE NEW, BE ACTIVE IN MAKING FRIENDS.
2. BE PHYSICALLY FIT (STRENGTH, ENDURANCE, AND AGILITY).
3. KNOW AND PRACTICE LIFESAVING SELF- AND BUDDY AID.
4. PRACTICE RAPID RELAXATION TECHNIQUES (FM 22-51).
5. HELP EACH OTHER OUT WHEN THINGS ARE TOUGH AT HOME OR IN THE UNIT.
6. KEEP INFORMED; ASK YOUR LEADER QUESTIONS, IGNORE RUMORS.
7. WORK TOGETHER TO GIVE EVERYONE FOOD, WATER, SHELTER, HYGIENE, AND SANITATION.
8. SLEEP WHEN MISSION AND SAFETY PERMIT; LET EVERYONE GET TIME TO SLEEP.
   - SLEEP ONLY IN SAFE PLACES AND BY STANDING OPERATING PROCEDURE (SOP).
   - IF POSSIBLE, SLEEP 6 TO 9 HOURS PER DAY.
   - TRY TO GET AT LEAST 4 HOURS SLEEP PER DAY.
   - GET GOOD SLEEP BEFORE GOING ON SUSTAINED OPERATIONS.
   - CATNAP WHEN YOU CAN, BUT ALLOW TIME TO WAKE UP FULLY.
   - CATCH UP ON SLEEP AFTER GOING WITHOUT.

APPENDIX A

FIRST AID CASE AND KITS, DRESSINGS, AND BANDAGES

A-1. First Aid Case with Field Dressings and Bandages

Every service member is issued a first aid case (Figure A-1A) with a field first aid dressing encased in a plastic wrapper (Figure A-1B). He carries it at all times for his use. The field first aid dressing is a standard sterile (germ-free) compress or pad with bandages attached (Figure A-1C). This dressing is used to cover the wound, to protect against further contamination, and to stop bleeding (pressure dressing). When a service member administers first aid to another person, he must remember to use the wounded person's dressing; he may need his own later. The service member must check his first aid case regularly and replace any used or missing dressing. The field first aid dressing may normally be obtained from his unit supply.

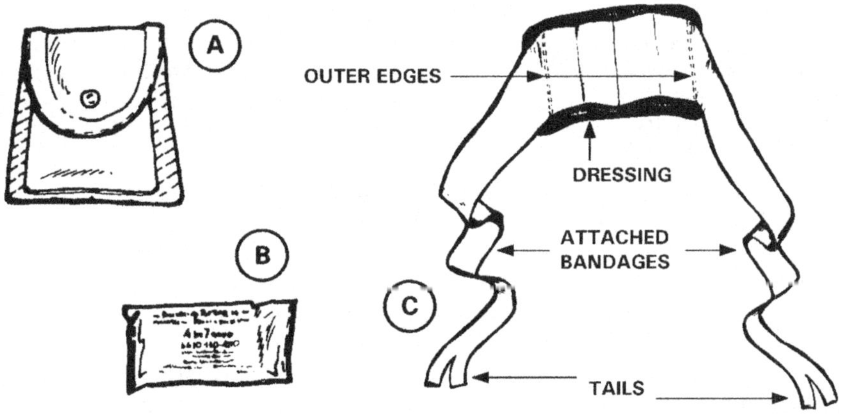

Figure A-1. Field first aid case and dressing (Illustrated A—C).

A-2. General Purpose First Aid Kits

General-purpose first aid kits are listed in the common table of allowances (CTA) 8-100. The operators, crew, and passengers carry these kits on Army vehicles, aircraft, and boats for use. Individuals designated by unit TSOP to be responsible for these kits are required to check them regularly and replace all items used. The general-purpose kit and its contents can be obtained through the unit supply system.

NOTE

Periodically check the dressings (for holes or tears in the packaging) and the medicines (for expiration date) that are in the first aid kits. If necessary, replace defective or outdated items.

A-3. Dressings

Dressings are sterile pads or compresses used to cover wounds. They usually are made of gauze or cotton wrapped in gauze (Figure A-1C). In addition to the standard field first aid dressing, other dressings such as sterile gauze compresses and small sterile compresses on adhesive strips may be available under CTA 8-100.

A-4. Standard Bandages

a. Standard bandages are made of gauze or muslin and are used over a sterile dressing to secure the dressing in place, to close off its edge from dirt and germs, and to create pressure on the wound and control bleeding. A bandage can also support an injured part or secure a splint.

b. Tailed bandages may be attached to the dressing as indicated on the field first aid dressing (Figure A-1C).

A-5. Triangular and Cravat (Swathe) Bandages

a. Triangular and cravat (or swathe) bandages (Figure A-2) are fashioned from a triangular piece of muslin (37 by 37 by 52 inches) provided in the general-purpose first aid kit. If it is folded into a strip, it is called a cravat. Two safety pins are packaged with each bandage. These bandages are valuable in an emergency since they are easily applied.

b. To improvise a triangular bandage, cut a square of available material, slightly larger than 3 feet by 3 feet, and *fold it diagonally*. If two bandages are needed, cut the material along the diagonal fold.

c. A cravat can be improvised from such common items as T-shirts, other shirts, bed linens, trouser legs, scarfs, or any other item made of pliable and durable material that can be folded, torn, or cut to the desired size.

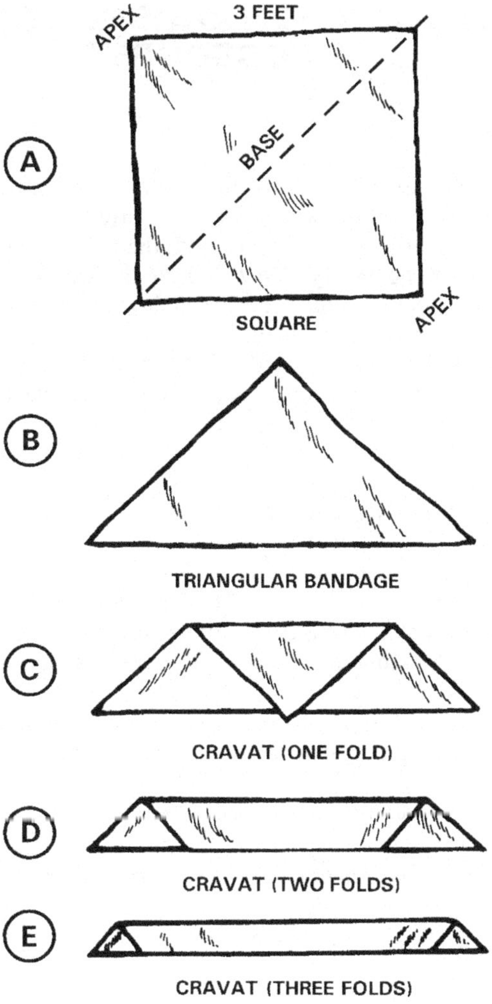

Figure A-2. Triangular and cravat bandages (Illustrated A—E).

APPENDIX B

RESCUE AND TRANSPORTATION PROCEDURES

B-1. General

A basic principle of first aid is to evaluate the casualty's injuries and administer first aid before moving him. However, adverse situations or conditions may jeopardize the lives of both the rescuer and the casualty if this is done. It may be necessary *first to rescue* the casualty before first aid can be effectively or safely given. The life and/or the well-being of the casualty will depend as much upon the manner in which he is *rescued and transported,* as it will upon the first aid and medical treatment he receives. Rescue actions must be done quickly and safely. Careless or rough handling of the casualty during rescue operations can aggravate his injuries.

B-2. Principles of Rescue Operations

a. When faced with the necessity of rescuing a casualty who is threatened by hostile action, fire, water, or any other immediate hazard, DO NOT take action without first determining the extent of the hazard and your ability to handle the situation. DO NOT become a casualty.

b. The rescuer must evaluate the situation and analyze the factors involved. This evaluation involves three major steps:

- Identify the task.
- Evaluate circumstances of the rescue.
- Plan the action.

B-3. Considerations

a. First determine if a rescue attempt is actually needed. It is a waste of time, equipment, and personnel to rescue someone not in need of rescuing. It is also a waste to look for someone who is not lost or needlessly risk the lives of the rescuer(s). In planning a rescue, attempt to obtain the following information:

- Who, what, where, when, why, and how the situation happened?

- How many casualties are involved and the nature of their injuries?

- What is the tactical situation?

- What are the terrain features and the location of the casualties?

- Will there be adequate assistance available to aid in the rescue/evacuation?

- Can first aid and/or medical treatment be provided at the scene; will the casualties require movement to a safer location?

- What specialized equipment will be required for the rescue operation?

- Is the rescue area contaminated? Will decontamination equipment and materiel be required for casualties, rescue personnel, and rescue equipment?

- How much time is available?

b. The time element can play a significant role in how the rescue is attempted. If the casualties are in imminent danger of losing their lives (such as near a burning vehicle or in a burning building) the time available will be relatively short and will sometimes cause a rescuer to compromise planning stages and/or the first aid which can be given. However, if the casualty is in a relatively secure area and his physical condition is strong, more deliberate planning can take place. A realistic estimate of time available must be made as quickly as possible to determine action time remaining. The key elements are the casualty's physical and mental condition, the tactical situation, and the environment.

B-4. Plan of Action

a. The casualty's ability to endure is of primary importance in estimating the time available. Age, physical condition, and extent of wounds and/or injuries will differ from casualty to casualty. Therefore, to determine the time available, you will have to consider—

- Endurance time of the casualty.

- Extent of injuries.

- Type of situation.
- Personnel and/or equipment availability.
- Weather.
- Terrain (natural and man-made).
- Environment (contaminated or uncontaminated).

b. In respect to terrain, you must consider altitude and visibility. In some cases, the casualty may be of assistance because he knows more about the particular terrain or situation than you do. Maximum use of secure/reliable trails or roads is essential.

c. When taking weather into account, ensure that blankets and/or rain gear are available. Even a mild rain can complicate a normally simple rescue. In high altitudes and/or extreme cold and gusting winds, the time available is critically shortened. Be prepared to provide shelter and warmth for the casualty as well as the rescuers.

B-5. Proper Handling of Casualties

a. You may have saved the casualty's life through the application of appropriate first aid measures. However, his life can be lost through rough handling or careless transportation procedures. Before you attempt to move the casualty—

- Evaluate the type and extent of his injuries.
- Ensure that dressings over wounds are adequately reinforced.
- Ensure that fractured bones are properly immobilized and supported to prevent them from cutting through muscle, blood vessels, and skin.

b. Based upon your evaluation of the type and extent of the casualty's injury and your knowledge of the various manual carries, you must select the best possible method of manual transportation. If the casualty is conscious, tell him how he is to be transported. This will help allay his fear of movement and gain his cooperation and confidence.

c. Buddy aid for chemical agent casualties includes those actions required to prevent an incapacitated casualty from receiving additional injury from the effects of chemical hazards. If a casualty is physically unable to decontaminate himself or administer the proper chemical agent antidote, the casualty's buddy assists him and assumes responsibility for his care. Buddy-aid includes—

- Administering the proper chemical agent antidote.
- Decontaminating the incapacitated casualty's exposed skin.
- Ensuring that his protective ensemble remains correctly emplaced.
- Maintaining respiration.
- Controlling bleeding.
- Providing other standard first aid measures
- Transporting the casualty out of the contaminated area.

B-6. Positioning the Casualty

The first step in any manual carry is to position the casualty to be lifted. If he is conscious, he should be told how he is to be positioned and transported. This helps lessen his fear of movement and to gain his cooperation. It may be necessary to roll the casualty onto his abdomen, or his back, depending upon the position in which he is lying and the particular carry to be used.

a. To roll a casualty onto his abdomen, kneel at the casualty's uninjured side.

(1) Place his arms above his head; cross his ankle which is farther from you over the one that is closer to you.

(2) Place your hands on the shoulder which is farther from you; place your other hand in the area of his hip or thigh (Figure B-1).

(3) Roll him gently toward you onto his abdomen (Figure B-2).

b. To roll a casualty onto his back, follow the same procedure described in *a* above, except gently roll the casualty onto his back, rather than onto his abdomen.

Figure B-1. Positioning the casualty.

Figure B-2. Rolling casualty onto his abdomen.

B-7. Medical Evacuation and Transportation of Casualties

a. Medical evacuation of the sick and wounded (with en route medical care) is the responsibility of medical personnel who have been provided special training and equipment. Therefore, unless a good reason for you to transport a casualty arises, wait for some means of medical evacuation to be provided. When the situation is urgent and you are unable to obtain medical assistance or know that no medical evacuation assets are available, you will have to transport the casualty. For this reason, you must know how to transport him without increasing the seriousness of his condition.

b. Transporting a casualty by litter (FM 8-10-6) is safer and more comfortable for him than by manual means; it is also easier for you.

Manual transportation, however, may be the only feasible method because of the terrain or the combat situation; or it may be necessary to save a life. In these situations, the casualty should be transferred to a litter as soon as one can be made available or improvised.

B-8. Manual Carries

Casualties carried by manual means must be carefully and correctly handled, otherwise their injuries may become more serious or possibly fatal. Situation permitting, transport of a casualty should be organized and unhurried. Each movement should be performed as deliberately and gently as possible. Casualties should not be moved before the type and extent of injuries are evaluated and the required first aid is administered. The exception to this occurs when the situation dictates immediate movement for safety purposes (for example, it may be necessary to remove a casualty from a burning vehicle); that is, the situation dictates that the urgency of casualty movement outweighs the need to administer first aid. Manual carries are tiring for the bearers and involve the risk of increasing the severity of the casualty's injury. In some instances, however, they are essential to save the casualty's life. Although manual carries are accomplished by one or two bearers, the two-man carries are used whenever possible. They provide more comfort to the casualty, are less likely to aggravate his injuries, and are also less tiring for the bearers. The distance a casualty can be carried depends on many factors, such as—

- Nature of the casualty's injuries.
- Strength and endurance of the bearer(s).
- Weight of the casualty.
- Obstacles encountered during transport (natural or manmade).
- Type of terrain.

a. One-man Carries. These carries should be used when only one bearer is available to transport the casualty.

(1) The *fireman's carry* (Figure B-3) is one of the easiest ways for one individual to carry another. After an unconscious or disabled casualty has been properly positioned, he is raised from the ground, then supported and placed in the carrying position.

(*a*) After rolling the casualty onto his abdomen, straddle him. Extend your hands under his chest and lock them together.

(*b*) Lift the casualty to his knees as you move backward.

(*c*) Continue to move backward, thus straightening the casualty's legs and locking his knees.

(*d*) Walk forward, bringing the casualty to a standing position; tilt him slightly backward to prevent his knees from buckling.

(*e*) As you maintain constant support of the casualty with one arm, free your other arm, quickly grasp his wrist, and raise his arm high. Instantly pass your head under his raised arm, releasing it as you pass under it.

(*f*) Move swiftly to face the casualty and secure your arms around his waist. Immediately place your foot between his feet and spread them apart (approximately 6 to 8 inches).

(*g*) Grasp the casualty's wrist and raise his arm high over your head.

(*h*) Bend down and pull the casualty's arm over and down on your shoulder, bringing his body across your shoulders. At the same time, pass your arm between his legs.

(*i*) Grasp the casualty's wrist with one hand, and place your other hand on your knee for support.

(*j*) Rise with the casualty positioned correctly. Your other hand is free for use.

Figure B-3. Fireman's carry (Illustrated A—J).

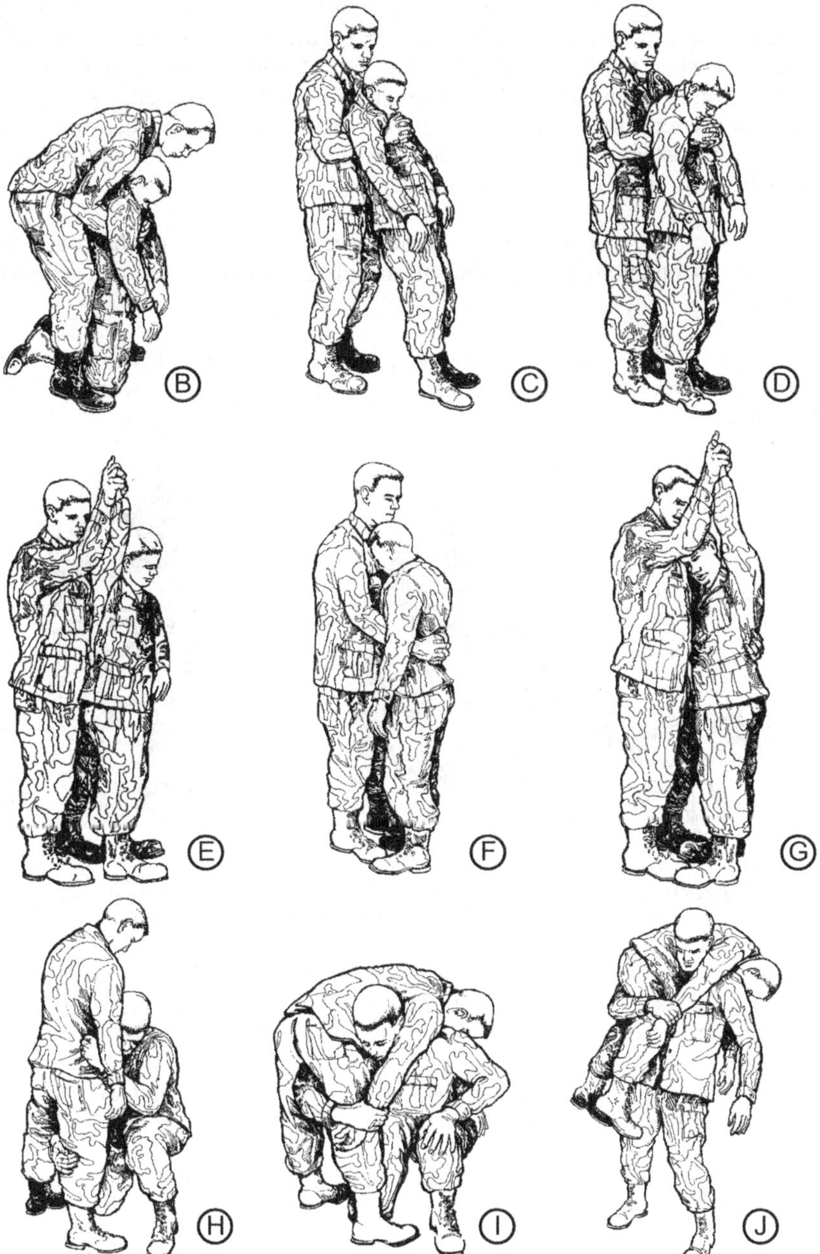

Figure B-3. Fireman's carry (Illustrated A—J) (Continued).

(2) The alternate method of the *fireman's carry* for raising a casualty from the ground is illustrated in Figure B-4; however, it should be used only when the bearer believes it to be safer for the casualty because of the location of his wounds. When the alternate method is used, care must be taken to prevent the casualty's head from snapping back and causing a neck injury. The steps for raising a casualty from the ground for the fireman's carry are also used in other one-man carries.

(*a*) Kneel on one knee at the casualty's head and face his feet. Extend your hands under his armpits, down his sides, and across his back.

(*b*) As you rise, lift the casualty to his knees. Then secure a lower hold and raise him to a standing position with his knees locked.

Figure B-4. Fireman's carry (alternate method) for lifting a casualty to a standing position (Illustrated A—B).

(3) In the *supporting carry* (Figure B-5), the casualty must be able to walk or at least hop on one leg, using the bearer as a crutch. This carry can be used to assist him as far as he is able to walk or hop.

(*a*) Raise the casualty from the ground to a standing position by using the fireman's carry.

(*b*) Grasp the casualty's wrist and draw his arm around your neck.

(*c*) Place your arm around his waist. The casualty is now able to walk or hop using you as a support.

Figure B-5. Supporting carry.

(4) The *arms carry* (Figure B-6) is useful in carrying a casualty for a short distance (up to 50 meters) and for placing him on a litter.

(*a*) Raise or lift the casualty from the ground to a standing position, as in the fireman's carry.

(*b*) Place one arm under the casualty's knees and your other arm around his back.

(*c*) Lift the casualty.

(*d*) Carry the casualty high to lessen fatigue.

Figure B-6. Arms carry.

(5) Only a conscious casualty can be transported by the *saddleback carry* (Figure B-7), because he must be able to hold onto the bearer's neck. To use this technique—

(*a*) Raise the casualty to an upright position, as in the fireman's carry.

(*b*) Support the casualty by placing an arm around his waist. Move to the casualty's side. Have the casualty put his arm around your neck and move in front of him with your back to support him.

(*c*) Have the casualty encircle his arms around your neck

(*d*) Stoop, raise him on your back and clasp your hands together beneath his thighs, if possible.

Figure B-7. Saddleback carry.

(6) In the *pack-strap carry* (Figure B-8), the casualty's weight rests high on the your back. This makes it easier for you to carry the casualty a moderate distance (50 to 300 meters). To eliminate the possibility of injury to the casualty's arms, you must hold his arms in a palms-down position.

(*a*) Lift the casualty from the ground to a standing position, as in the fireman's carry.

(*b*) Support the casualty with your arms around him and grasp his wrist closer to you.

(*c*) Place his arm over your head and across your shoulders.

(*d*) Move in front of him while still supporting his weight against your back.

(*e*) Grasp his other wrist and place this arm over your shoulder.

(*f*) Bend forward and raise or hoist the casualty as high on your back as possible so that his weight is resting on your back.

NOTE

Once the casualty is positioned on the bearer's back, the bearer remains as erect as possible to prevent straining or injuring his back.

Figure B-8. Pack strap carry.

(7) The *pistol-belt carry* (Figure B-9) is the best one-man carry for a long distance (over 300 meters). The casualty is securely supported upon your shoulders by a belt. Both your hands and the casualty's (if conscious) are free for carrying a weapon or equipment, or climbing obstacles. With your hands free and the casualty secured in place, you are also able to creep through shrubs and under low-hanging branches.

(*a*) Link two pistol belts (or three, if necessary) together to form a sling. Place the sling under the casualty's thighs and lower back so that a loop extends from each side.

NOTE

If pistol belts are not available for use, other items such as a rifle sling, two cravat bandages, two litter straps, or any other suitable material, which will not cut or bind the casualty may be used.

(*b*) Lie face up between the casualty's outstretched legs. Thrust your arms through the loops and grasp his hands and trouser leg on his injured side.

(*c*) Roll toward the casualty's uninjured side onto your abdomen, bringing him onto your back. Adjust the sling, if necessary.

(*d*) Rise to a kneeling position. The belt will hold the casualty in place.

(*e*) Place one hand on your knee for support and rise to an upright position. (The casualty is supported on your shoulders.)

(*f*) Carry the casualty with your hands free for use in rifle firing, climbing, or surmounting obstacles.

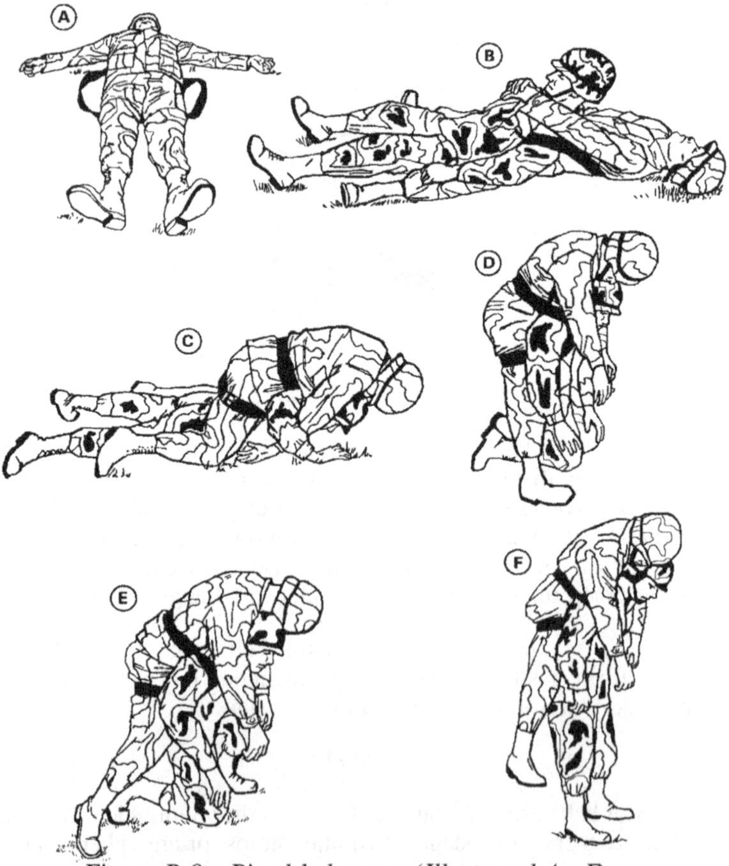

Figure B-9. Pistol-belt carry (Illustrated A—F).

(8) The *pistol-belt drag* (Figure B-10), as well as other drags, is generally used for short distances (up to 50 meters). This drag is useful in combat, since both the bearer and the casualty can remain closer to the ground than in any other drags.

(*a*) Extend two pistol belts or similar objects to their full length and join them together to make a continuous loop.

(*b*) Roll the casualty onto his back, as in the fireman's carry.

(*c*) Pass the loop over the casualty's head, and position it across his chest and under his armpits. Then cross the remaining portion of the loop, thus forming a figure eight. Keep tension on the belts so they do not come unhooked.

(*d*) Lie on your side facing the casualty.

(*e*) Slip the loop over your head and turn onto your abdomen. This enables you to drag the casualty as you crawl.

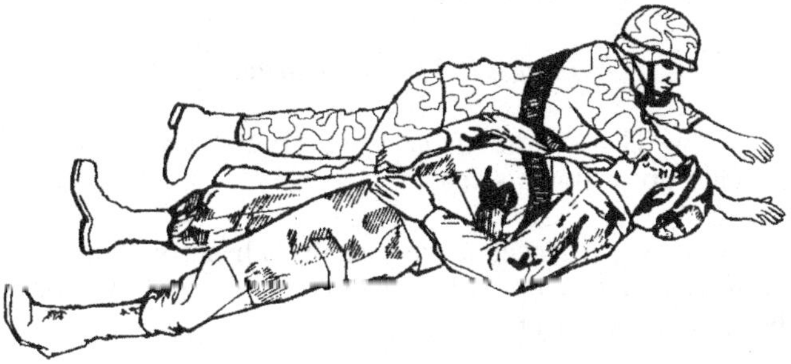

Figure B-10. Pistol-belt drag.

(9) The *neck drag* (Figure B-11) is useful in combat because the bearer can transport the casualty as he creeps behind a low wall or shrubbery, under a vehicle, or through a culvert. If the casualty is unconscious, his head must be protected from the ground. The neck drag cannot be used if the casualty has a broken arm.

NOTE

If the casualty is conscious, he may clasp his hands together around your neck.

(*a*) Tie the casualty's hands together at the wrists.

(*b*) Straddle the casualty in a kneeling face-to-face position.

(*c*) Loop the casualty's tied hands over and around your neck.

(*d*) Crawl forward dragging the casualty with you.

NOTE

If the casualty is unconscious, protect his head from the ground.

Figure B-11. Neck drag.

(10) The *cradle drop drag* (Figure B-12) is effective in moving a casualty up or down steps.

(*a*) Kneel at the casualty's head (with him lying on his back). Slide your hands, with palms up, under the casualty's shoulders and get a firm hold under his armpits.

(*b*) Rise (partially), supporting the casualty's head on one of your forearms. (You may bring your elbows together and let the casualty's head rest on both of your forearms.)

(*c*) Rise and drag the casualty backward. (The casualty is in a semisitting position.)

(*d*) Back down the steps, supporting the casualty's head and body and letting his hips and legs drop from step to step.

NOTE

If the casualty needs to be moved up the steps, you should back up the steps, using the same procedure.

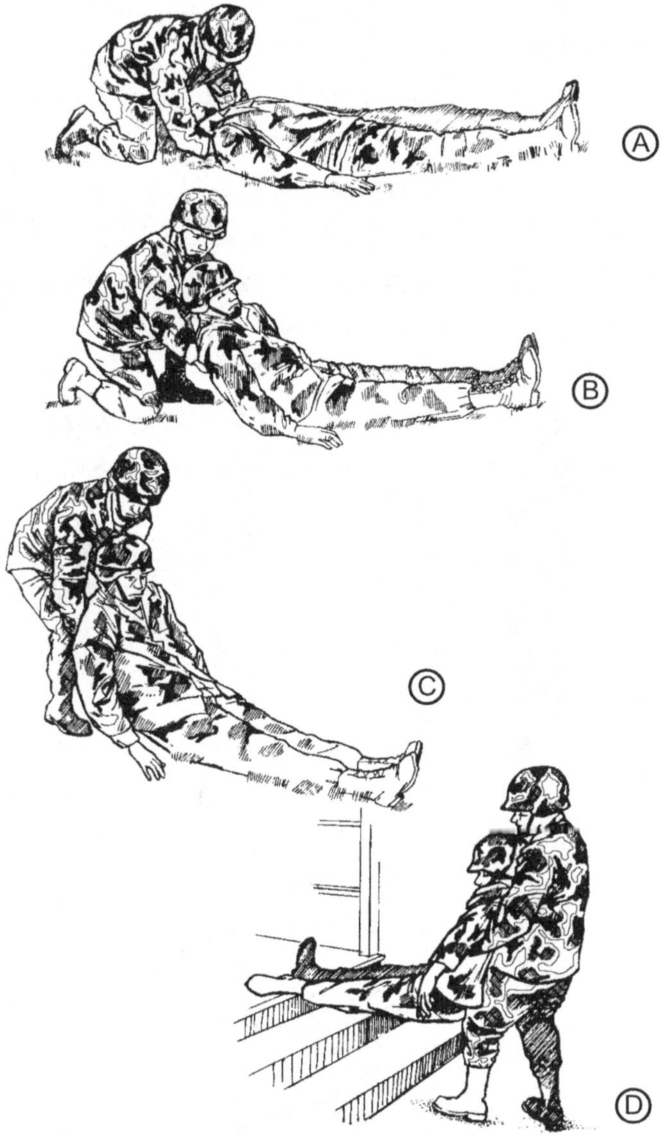

Figure B-12. Cradle-drop drag (Illustrated A—D).

(11) The *LBE carry using the bearer's LBE* can be used with a conscious casualty (Figure B-13).

(*a*) Loosen all suspenders on your LBE.

(*b*) Have the casualty place one leg into the loop formed by your suspenders and pistol belt.

(*c*) Squat in front of the standing casualty. Have him place his other leg into the loop, also.

(*d*) Have the casualty place his arms over your shoulders, lean forward onto your back, and lock his hands together.

(*e*) Stand up and lean forward into a comfortable position.

(*f*) Continue the mission.

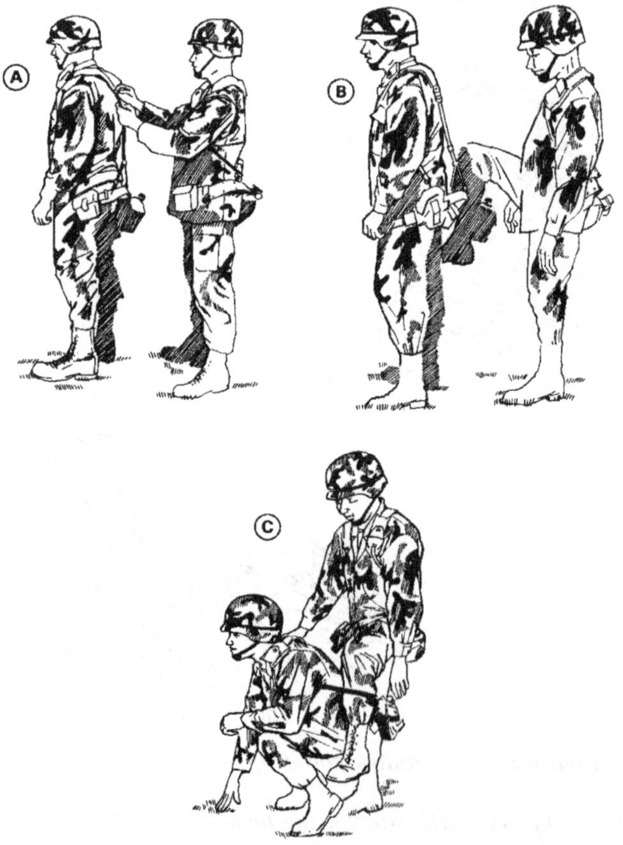

Figure B-13. Load bearing equipment carry using bearer's LBE (conscious casualty) (Illustrated A—F).

Figure B-13. Load bearing equipment carry using bearer's LBE (conscious casualty) (Illustrated A—F) (Continued).

(12) The *LBE carry using the bearer's LBE* can be used with an unconscious casualty or one who cannot stand (Figure B-14).

(*a*) Position the casualty on the flat of his back.

(*b*) Remove your LBE and loosen all suspender straps.

(*c*) Lift the casualty's leg and place it through the loop formed by your suspenders and pistol belt. Then place the other leg through the same loop. The LBE is moved up until the pistol belt is behind the casualty's thighs.

(*d*) Lay between the casualty's legs; work your arms through the LBE suspenders.

(*e*) Grasp the casualty's hand (on the injured side), and roll the casualty (on his uninjured side) onto your back.

(*f*) Rise to one knee and then push into a standing position.

(*g*) Bring the casualty's arms over your shoulders. Grasp his hands and secure them if the casualty is unconscious. If the casualty is conscious, have him lock his hands in front if he is able to do so.

(*h*) Lean forward into a comfortable position and continue the mission.

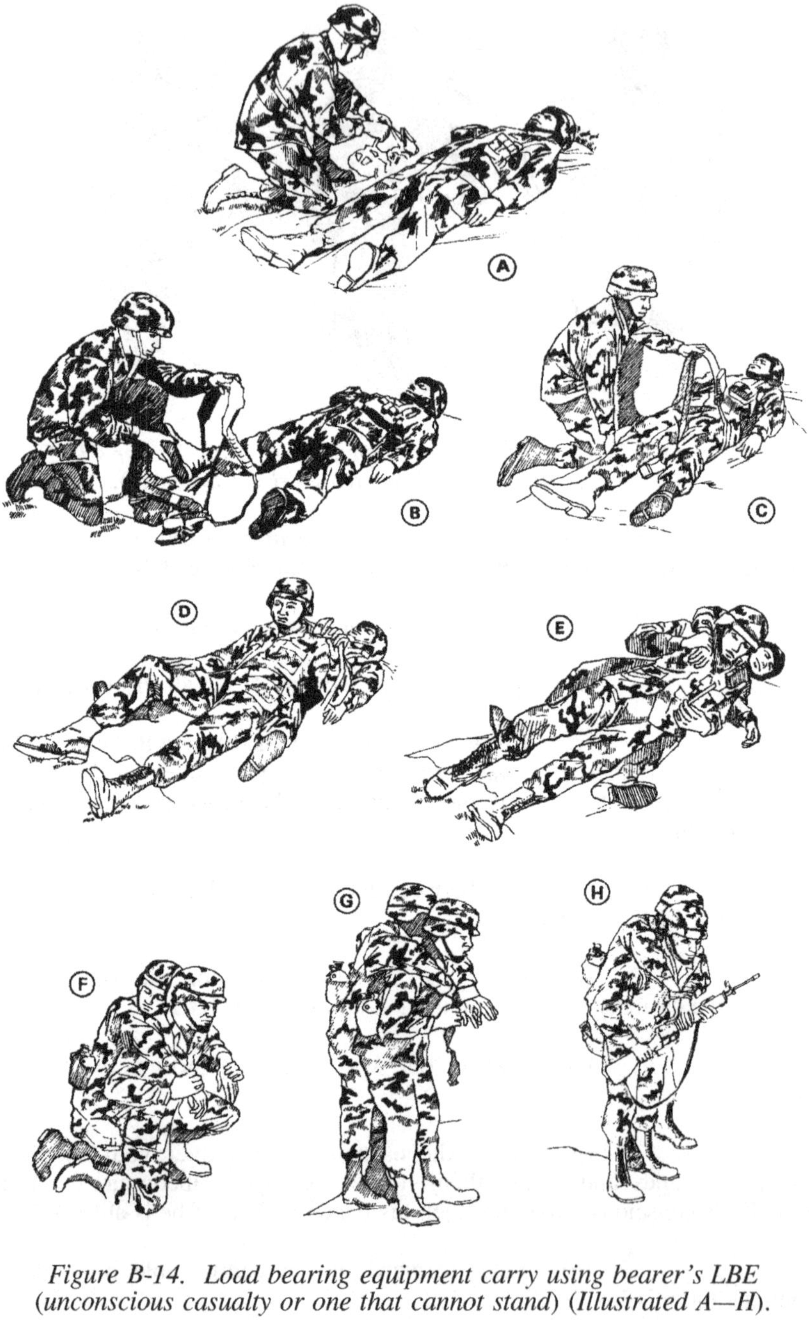

Figure B-14. Load bearing equipment carry using bearer's LBE (unconscious casualty or one that cannot stand) (Illustrated A—H).

(13) The *LBE carry using the casualty's LBE* (Figure B-15) can be used with a conscious or unconscious casualty.

(*a*) Position the casualty on his back with his LBE on.

(*b*) Loosen the casualty's two front suspenders.

(*c*) Position yourself between the casualty's legs, and slip your arms into the casualty's two front suspenders (up to your shoulders).

(*d*) Work his arms out of his LBE suspenders.

(*e*) Grasp the casualty's hand (on the injured side), and roll him (on his uninjured side) onto your back.

(*f*) Rise to one knee, then into a standing position.

(*g*) Grasp the casualty's hands and secure them, if the casualty is unconscious. Have the casualty lock his hands in front of you, if he is conscious.

(*h*) Lean forward into a comfortable position and continue the mission.

Figure B-15. Load bearing equipment carry using casualty's LBE (Illustrated A—G).

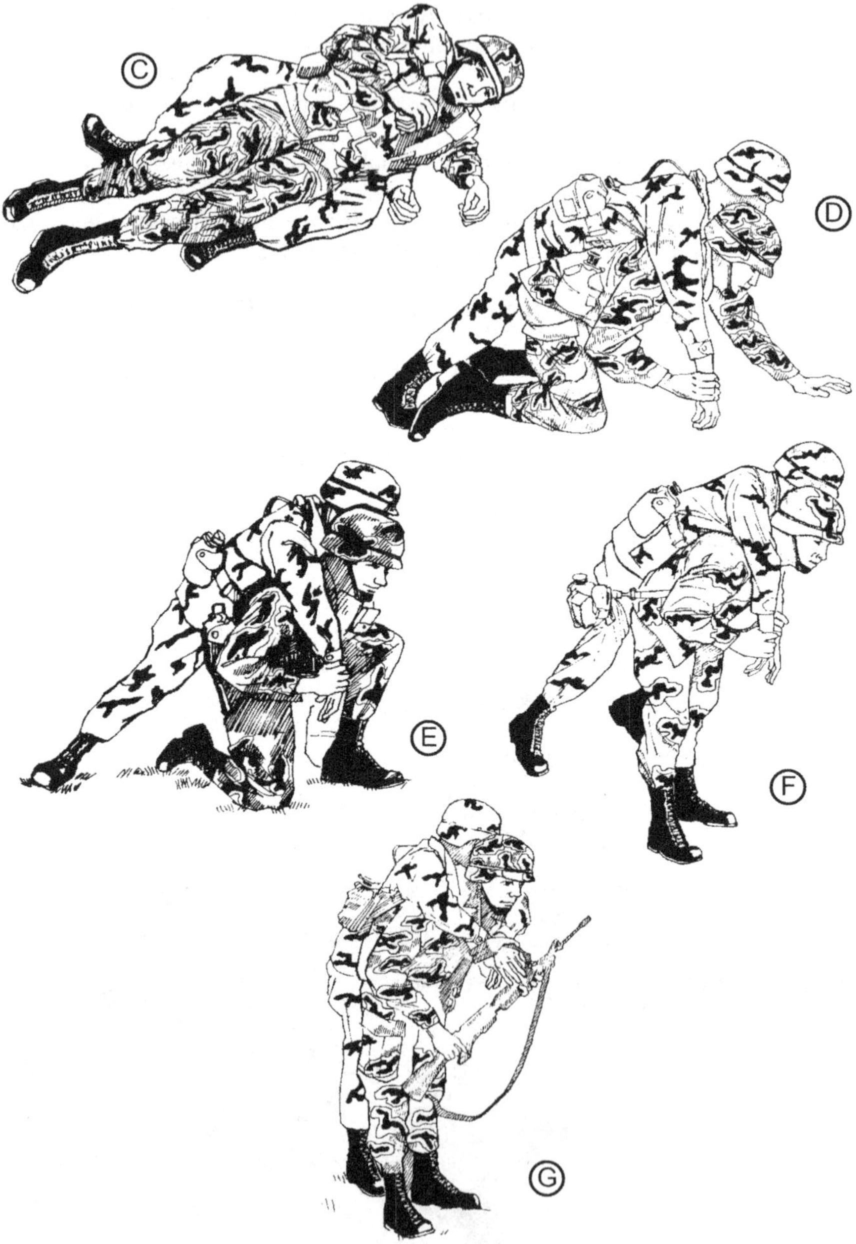

Figure B-15. Load bearing equipment carry using casualty's LBE (Illustrated A—G) (Continued).

b. Two-man Carries. These carries should be used whenever possible. They provide more casualty comfort, are less likely to aggravate injuries, and are less tiring for the bearers. Five different two-man carries can be used.

(1) The *two-man support carry* (Figure B-16) can be used in transporting either conscious or unconscious casualties. If the casualty is taller than the bearers, it may be necessary for the bearers to lift the casualty's legs and let them rest on their forearms. The bearers—

(*a*) Help the casualty to his feet and support him with their arms around his waist.

(*b*) Grasp the casualty's wrists and draw his arms around their necks.

Figure B-16. Two-man supporting carry.

(2) The *two-man arms carry* (Figure B-17) is useful in carrying a casualty for a moderate distance (50 to 300 meters) and placing him on a litter. To lessen fatigue, the bearers should carry the casualty high and as close to their chests as possible. In extreme emergencies when there is no time to obtain a spine board, this carry is the safest one for transporting a casualty with a back injury. If possible, two additional bearers should be used to keep the casualty's head and legs in alignment with his body. The bearers—

(*a*) Kneel at one side of the casualty; then they place their arms beneath the casualty's back, waist, hips, and knees.

(*b*) Lift the casualty while rising to their knees.

(*c*) Turn the casualty toward their chests, while rising to a standing position. Carry the casualty high to lessen fatigue.

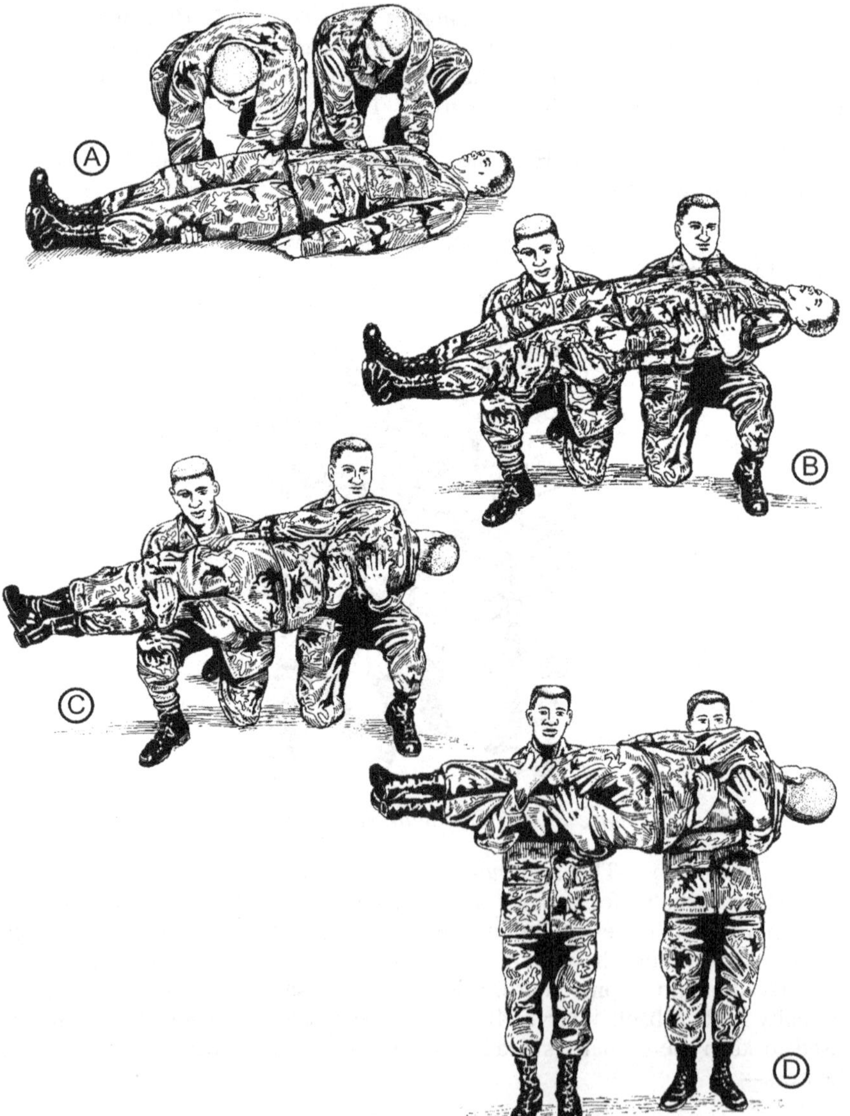

Figure B-17. Two-man arms carry (Illustrated A—D).

(3) The *two-man fore-and aft-carry* (Figure B-18) is a useful *two-man carry* for transporting a casualty for a long distance (over 300 meters). The taller of the two bearers should position himself at the casualty's head. By altering this carry so that both bearers face the casualty, it is useful for placing a casualty on a litter.

(*a*) The shorter bearer spreads the casualty's legs and kneels between them with his back to the casualty. He positions his hands behind the casualty's knees. The other bearer kneels at the casualty's head, slides his hands under the arms, across the chest, and locks his hands together.

(*b*) The two bearers rise together, lifting the casualty.

Figure B-18. Two-man fore-and-aft carry (Illustrated A—B).

(4) Only a conscious casualty can be transported with the *four-hand seat carry* (Figure B-19) because he must help support himself by placing his arms around the bearers' shoulders. This carry is especially useful in transporting a casualty with a head or foot injury for a moderate distance (50 to 300 meters). It is also useful for placing a casualty on a litter.

(*a*) Each bearer grasps one of his wrists and one of the other bearer's wrists, thus forming a packsaddle.

(*b*) The two bearers lower themselves sufficiently for the casualty to sit on the packsaddle; then, they have the casualty place his arms around their shoulders for support. The bearers then rise to an upright position.

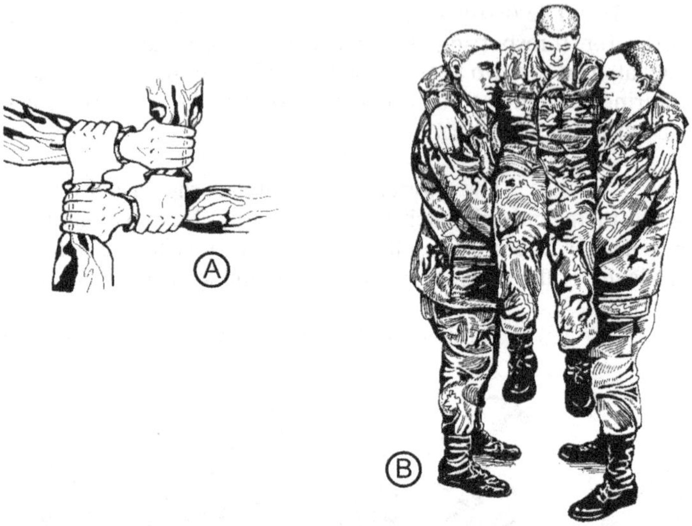

Figure B-19. Four-hand seat carry (Illustrated A—B).

(5) The *two-hand seat carry* (Figure B-20) is used when carrying a casualty for a short distance or for placing him on a litter. With the casualty lying on his back, a bearer kneels on each side of the casualty at his hips. Each bearer passes his arms under the casualty's thighs and back, and grasps the other bearer's wrists. The bearers rise lifting the casualty.

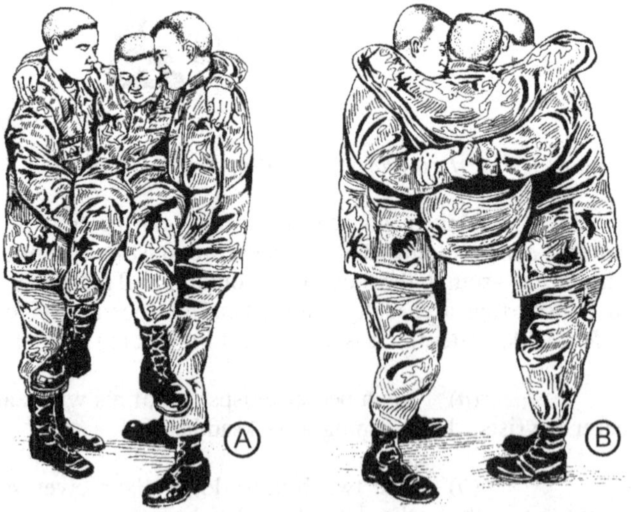

Figure B-20. Two-hand seat carry (Illustrated A—B).

B-9. Improvised Litters

Two men can support or carry a casualty without equipment for only short distances. By using available materials to improvise equipment, the casualty can be transported greater distances by two or more rescuers.

a. There are times when a casualty may have to be moved and a standard litter is not available. The distance may be too great for manual carries or the casualty may have an injury (such as a fractured neck, back, hip, or thigh) that would be aggravated by manual transportation. In these situations, litters can be improvised from materials at hand. Improvised litters must be as well constructed as possible to avoid risk of dropping or further injuring the casualty. Improvised litters are emergency measures and must be replaced by standard litters at the first opportunity.

b. Many different types of litters can be improvised, depending upon the materials available. A satisfactory litter can be made by securing poles inside such items as a blanket, poncho, shelter half, tarpaulin, mattress cover, jacket, shirt, or bed ticks, bags, and sacks (Figure B-18). Poles can be improvised from strong branches, tent supports, skis, lengths of pipe or other objects. If objects for improvising poles are not available, a blanket, poncho, or similar item can be rolled from both sides toward the center so the rolls can be gripped for carrying a patient. Most flat-surface objects of suitable size can be used as litters. Such objects include doors, boards, window shutters, benches, ladders, cots, and chairs. If possible, these objects should be padded for the casualty's comfort.

(1) To improvise a litter using a blanket and poles (Figure B-21), the following steps should be used.

Figure B-21. Litter made with blanket and poles.

(*a*) Open the blanket and lay one pole lengthwise across the center; then fold the blanket over the pole.

(*b*) Place the second pole across the center of the folded blanket.

(*c*) Fold the free edges of the blanket over the second pole and across the first pole.

(2) To improvise a litter using shirts or jackets (Figure B-22), button the shirt or jacket and turn it inside out, leaving the sleeves inside, (more than one shirt or jacket may be required), then pass the pole through the sleeves.

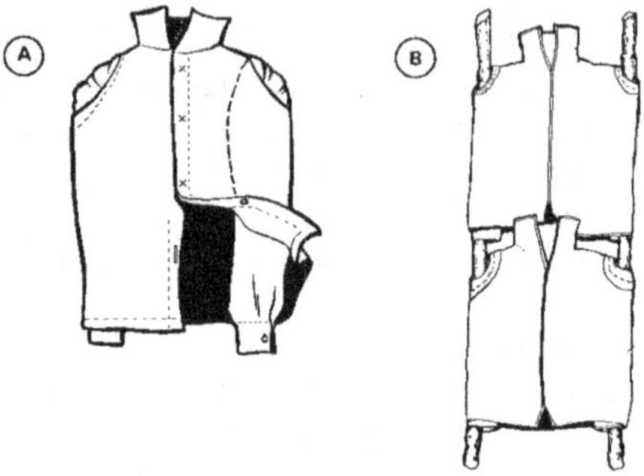

Figure B-22. Litter improvised from jackets and poles (Illustrated A—B).

(3) To improvise a litter from bed sacks and poles (Figure B-23), rip open the corners of bed ticks, bags, or sacks; then pass the poles through them.

Figure B-23. Litter improvised from bed sacks and poles.

(4) If no poles are available, roll a blanket, shelter half, tarpaulin, or similar item from both sides toward the center (Figure B-24). Grip the rolls to carry the casualty.

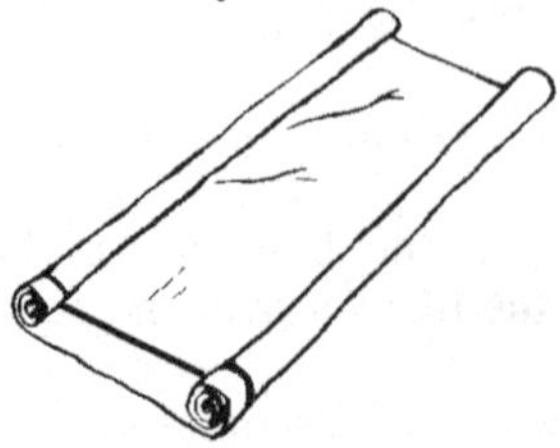

Figure B-24. Rolled blanket used as a litter.

c. Any of the appropriate carries may be used to place a casualty on a litter. These carries are:

- The one-man arms carry (Figure B-6).
- The two-man arms carry (Figure B-17).
- The two-man fore-and-aft carry (Figure B-18).
- The two-hand seat carry (Figure B-20).
- The four-hand seat carry (Figure B-19).

WARNING

Unless there is an immediate life-threatening situation (such as fire, explosion), DO NOT move a casualty with a suspected back or neck injury. Seek medical personnel for guidance on how to transport.

d. Either two or four service members (head/foot) may be used to lift a litter. To lift the litter, follow the procedure below.

(1) Raise the litter at the same time as the other carriers/bearers.

(2) Keep the casualty as level as possible.

NOTE

Use caution when transporting on a sloping incline/hill.

GLOSSARY

ACRONYMS, ABBREVIATIONS, AND DEFINITIONS

AC hydrogen cyanide
AFMAN Air Force Manual
AOC area of concentration
AR Army regulation
ATM advanced trauma management
ATNAA Antidote Treatment, Nerve Agent, Autoinjector
attn attention

BDO battle dress overgarment
BDU battle dress uniform
BZ anticholinergic drugs

C Celsius
CANA Convulsant Antidote for Nerve Agent
CASEVAC casualty evacuation
cc cubic centimeter
CG phosgene
CHS combat health support
CK cyanogen chloride
Cl chlorine
CLS Combat Lifesaver
CNS central nervous system
CO_2 carbon dioxide
COSR combat and operational stress reactions
CSR combat stress reaction
CTA common table of allowance
CX phosgene oxime

DA Department of the Army
DD Department of Defense
DM diphenylaminochloroarsine (adamsite)
DNBI disease and nonbattle injury
DOD Department of Defense
DP diphosgene
DS direct support

EMT emergency medical treatment

F Fahrenheit
FM field manual

H mustard

HD mustard
HM Hospital Corpsman
HN nitrogen mustard
HSS health service support

IPE individual protective equipment
IV intravenous

JSLIST Joint Services Light Weight Integrated Suit Technology

L lewisite
lasers laser means Light Amplification by Stimulated Emission of Radiation and sources include range finders, weapons/guidance, communication systems, and weapons simulations such as MILES [Multiple Integrated Laser Engagement System].
LBE load bearing equipment
LX lewisite and mustard

MCRP Marine Corps Reference Publication
MILES Multiple Integrated Laser Engagement System
ml milliliter
MOPP mission-oriented protective posture
MOS military occupational specialty
MTF medical treatment facility

NAPP Nerve Agent Pyridostigmine Pretreatment
NATO North Atlantic Treaty Organization
NBC nuclear, biological, and chemical
NCO noncommissioned officer
NTRP Navy Tactical Reference Publication

occlusive dressing air tight transparent dressing used to seal and cover wounds
oz ounce

PAM pamphlet
PS chloropicrin
PTSD post-traumatic stress disorder

QSTAG Quadripartite Standardization Agreement

SOP standing operating procedure
STANAG standardization agreement
STP soldier training publication

2 PAM Cl pralidoxime chloride
TB MED technical bulletin medical
TM technical manual
TSOP tactical standing operating procedure

US United States

WP white phosphorus

REFERENCES

DOCUMENTS NEEDED

These documents must be available to the intended users of this publication.

NATO STANAGs

These agreements are available on request using DD Form 1425 from Standardization Document Order Desk, 700 Robin Avenue, Building 4, Section D, Philadelphia, Pennsylvania 19111-5094.

2122. *Medical Training in First Aid, Basic Hygiene and Emergency Care.* 10 December 1975.
2126. *First Aid Kits and Emergency Medical Care Kits.* 27 September 1983.
2358. *First Aid and Hygiene Training in NBC Operations.* 3 March 1989.
2871. *First Aid Material for Chemical Injuries.* 10 March 1986.

ABCA QSTAGs

These agreements are available on request using DD Form 1425 from Standardization Document Order Desk, 700 Robin Avenue, Building 4, Section D, Philadelphia, Pennsylvania 19111-5094.

535. *Medical Training in First Aid, Basic Hygiene and Emergency Care.* 12 November 1979.

Joint and Multiservice Publications

FM 21-10. *Field Hygiene and Sanitation.* MCRP 4-11.1D. 21 June 2000.

Army Publications

AR 350-41. *Training In Units.* 19 March 1993.
DA PAM 350-59. *Army Correspondence Course Program Catalog.* 26 October 2001.
FM 3-4. *NBC Protection.* FMFM 11-9. 29 May 1992 (Reprinted with basic including Change 1, 28 October 1992; Change 2, 26 February 1996.)
FM 3-5. *NBC Decontamination.* MCWP 3-37.3. 28 July 2000. (Change 1, 31 January 2002.)

FM 3-100. *Chemical Operations Principles and Fundamentals.* MCWP 3-3.7.1. 8 May 1996.

FM 4-02.33 (8-33). *Control of Communicable Diseases Manual* (17th Edition). 3 January 2000.

FM 8-10-6 (4-02.2). *Medical Evacuation in a Theater of Operations—Tactics, Techniques, and Procedures.* 14 April 2000.

FM 4-02.7 (8-10-7). *Health Service Support in a Nuclear, Biological, and Chemical Environment.* 1 October 2002.

FM 8-284 (4-02.284). *Treatment of Biological Warfare Agent Casualties.* NAVMED P-5042; AFMAN (I) 44-156; MCRP 4-11.1C. 17 July 2000. (Change 1, 8 July 2002.)

FM 8-285 (4-02.285). *Treatment of Chemical Agent Casualties and Conventional Military Chemical Injuries.* NAVMED P-5041; AFJMAN 44-149; FMFM 11-11. 22 December 1995.

FM 22-51 (4-02.22). *Leaders' Manual for Combat Stress Control.* 29 September 1994.

CTA 8-100. *Army Medical Department Expendable/Durable Items.* 31 August 1994.

CTA 50-900. *Clothing and Individual Equipment.* 1 September 1994.

STP 21-1-SMCT. *Soldier's Manual of Common Tasks Skill Level I.* 1 October 2001.

Department of Defense Forms

DD Form 1425. *Specifications and Standards Requisition.* March 1986.

READINGS RECOMMENDED

These readings contain relevant supplemental information.

Joint and Multiservice Publications

FM 8-9 (4-02.11). *NATO Handbook on the Medical Aspects of NBC Defensive Operations AMEDP-6 (B), Part I—Nuclear, Part II—Biological, Part III—Chemical.* NAVMED P-5059; AFJMAN 44-151V1V2V3. 1 February 1996.

TB MED 81. *Cold Injury.* NAVMED P-5052-29; AFP 161-11. 30 September 1976.

TB MED 507. *Occupational and Environmental Health Prevention, Treatment and Control of Heat Injury.* NAVMED P-5052-5; AFP 160-1. 25 July 1980.

Army Publications

AR 310-25. *Dictionary of United States Army Terms (Short Title: AD)*. 15 October 1983. (Reprinted with basic including Change 1, 21 May 1986.)

AR 310-50. *Authorized Abbreviations and Brevity Codes*. 15 November 1985.

TM 3-4230-216-10. *Operator's Manual for Decontaminating Kit, Skin: M258A1 (NSN 4230-01-101-3984) and Training Aid, Skin Decontaminating: M58A1 (6910-01-101-1768)*. 17 May 1985. (Change 1, 22 January 1997.)

INDEX

References are to paragraph numbers except where specified otherwise.

FIELD MANUAL
NUMBER 4-02.4

HEADQUARTERS
DEPARTMENT OF THE ARMY
Washington, DC, 24 August 2001

MEDICAL PLATOON LEADERS' HANDBOOK
TACTICS, TECHNIQUES, AND PROCEDURES

TABLE OF CONTENTS

DISTRIBUTION RESTRICTION: Approved for public release; distribution is unlimited.

---

*This publication supersedes FM 8-10-4, 16 November 1990.

PREFACE

This field manual (FM) provides information on the structure and operation of all medical platoons and medical sections that are organic to combat and combat support (CS) battalions and squadrons. It is directed toward the medical platoon leader and medical platoon members. The tactics, techniques, and procedures (TTP) provided are not all-inclusive. They provide a way of performing a particular mission, but may require modification based on mission, enemy, terrain, troops, time available, and civilian considerations (METT-TC).

This publication provides information on the organization of the division and how medical platoons and sections organic to division units provide combat health support (CHS). It outlines the responsibilities of medical platoon/section leaders. It provides definitive information on planning, rehearsing, and conducting CHS at Echelon I. It provides TTP for directing, controlling, and managing CHS at the medical platoon/section level. It describes the troop-leading procedures for CHS operations and identifies interface and coordination requirements with other brigade medical elements. This publication provides doctrine for the Force XXI medical platoon.

This FM is not a stand-alone reference. It is a doctrine publication that speaks to the current Army of Excellence (AOE) Division and the Force XXI Digitized Division and Brigade CHS and will require the user to be familiar with FMs 4-02.21, 8-10, 8-10-1, 8-10-3, 8-10-5, 8-10-6, 8-10-7, 8-10-9, 8-10-26, 8-42, and 8-55.

This publication implements the following North Atlantic Treaty Organization (NATO) Standardization Agreements (STANAGs) and American, British, Canadian, and Australian (ABCA) Quadripartite Standardization Agreements (QSTAGs):

Title	STANAG	QSTAG
Medical and Dental Supply Procedures	2128	
Regulations and Procedures for Road Movement and Identification of Movement Control and Traffic Control Personnel and Agencies	2454	
Orders for the Camouflage of the Red Cross and the Red Crescent on Land in Tactical Operations	2931	
Medical Materiel Management during Patient Evacuation		435
Minimum Labelling Requirements for Medical Materiel		436

As the Army Medical Department (AMEDD) transitions to the 91W military occupational specialty (MOS), positions for 91B and 91C will be replaced by 91W when new unit modification table(s) of organization and equipment (MTOE) take effect.

Users of this publication are encouraged to submit comments and recommendations to improve the publication. Comments should include the page, paragraph, and line(s) of the text where the change is recommended. The proponent for this publication is the United States (US) Army Medical Department Center and School (AMEDDC&S). Comments and recommendations should be forwarded directly to **Commander, AMEDDC&S, ATTN: MCCS-FCD-L, 1400 East Grayson Street, Fort Sam Houston, Texas 78234-5052**, or by using the e-mail addresses on the Doctrine Literature website at *http://dcdd.amedd.army.mil/index1.htm* (click on Doctrine Literature).

Unless this publication states otherwise, masculine nouns and pronouns do not refer exclusively to men.

Use of trade or brand names in this publication is for illustrative purposes only and does not imply endorsement by the Department of Defense (DOD).

CHAPTER 1

COMBAT ORGANIZATIONS

Section I. THE DIVISION

1-1. Background

The division is the largest Army organization that trains and fights as a tactical team. Largely self-sustaining, it is capable of independent operations. The division is a unit of maneuver organizations with varying numbers and types of combat, CS, and combat service support (CSS) units. It may be armored, mechanized, light infantry, airborne, or air assault; each can conduct operations over a wide range of environments. The success of Army operations depends on the success of its divisions.

1-2. Role of the Division

Traditionally, divisions have operated as part of a US corps. In corps operations, divisions are normally comprised of 9 to 12 maneuver battalions, organic artillery battalions, and supporting CS and CSS units. Divisions perform a wide range of tactical missions and, for limited periods, are self-sustaining. The corps augments divisions, as missions require. All divisions must be able to deploy and conduct offensive and defensive operations, stability operations, and support operations.

Section II. TYPES OF DIVISIONS

1-3. Armored and Mechanized Divisions

a. Army of Excellence. The US Army's armored and mechanized divisions (referred to collectively as heavy divisions) provide mobile, armor-protected firepower. Armored and mechanized divisions are normally employed for their mobility, survivability, lethality, and psychological effects on the enemy. These divisions destroy enemy armored forces. They can seize and secure land areas and key terrain. During offensive operations, armored and mechanized divisions can rapidly concentrate overwhelming lethal combat power to break or envelop enemy defenses or offensive formations. These divisions then continue the attack to destroy fire support, command and control (C2), and logistics elements. Their mobility allows them to rapidly concentrate, attack, reinforce, or block enemy forces. Their collective protection systems enable them to operate in a nuclear, biological, and chemical (NBC) environment. Armored and mechanized divisions operate best in open terrain where they gain the advantage with their mobility and long-range, direct-fire weapons. Because of strategic lift requirements, armored and mechanized forces are slow to deploy from home or staging bases into an area of operations (AO). They have high consumption rates of supplies, can deploy relatively few dismounted infantry, and have limited use in restrictive terrain. See FM 71-100 for organizational structure of the division.

b. Force XXI, Digitized Division. The Army's Force XXI Division is a redesign of our current divisions that represent a leap forward into the realm of 21st Century technology. The smaller Force XXI Division possesses greater lethality and quicker mobility, as well as the CSS imperative of situational understanding. The digital technological enablers will enhance situational understanding and provide the means for information dominance by enabling friendly forces to share a complete common relevant picture (CRP). This provides the commanders a CRP while communicating and targeting in real or near real-time. Digitization permits the division to conduct operations over an extended battle space by increasing the operational areas of responsibility for all maneuver elements. Digitization will decrease decision-making time by optimizing the flow of information. This information enables Force XXI commanders to quickly mass forces allowing the division to defeat a larger, but less technologically advanced enemy. It will contribute increased lethality, survivability, and operational tempo while reducing the potential for fratricide.

1-4. Light Infantry Division

The light infantry division fights as part of a larger force in conventional conflicts and conducts missions as part of a joint force in stability operations and support operations. The light infantry division is one of our most rapidly and strategically deployable divisions. Its C2 structure readily accepts any augmentation forces, permitting task organizing for any situation. The factors of METT-TC largely determine the augmentations required for the division. The optimum use of light forces is as a division under corps control, its mission capitalizing on its capabilities. The division exploits the advantages of restricted terrain and limited visibility. It achieves mass through the combined effects of synchronized small-unit operations and fires, rather than through the physical concentration of forces on the battlefield. Light division forces physically mass only when risk to the force is low and the payoff is high. The division deploys as an entity; widely dispersed to conduct synchronized, but decentralized, operations primarily at night or during periods of limited visibility. Light force limitations include their austere CS and CSS systems, and their requirement for support from the corps or joint force headquarters, based on METT-TC. For organizational structure of the light infantry division, see FM 71-100.

1-5. Airborne Division

The airborne division can rapidly deploy anywhere in the world to seize and secure vital objectives. The airborne division must be able to conduct forced entry operations. It conducts parachute assaults to capture initial lodgments, execute large-scale tactical raids, secure intermediate staging bases or forward operating bases for ground and air operations, or rescue US nationals besieged overseas. It also can serve as a strategic or theater reserve as well as reinforcement for forward presence forces. The airborne division can assault deep into the enemy's rear areas to secure terrain or interdict enemy supply and withdrawal routes. It can seize and repair airfields to provide a forward operating base and airheads for follow-on air-landed forces. It is capable of all other missions assigned to light infantry divisions. The airborne division uses its strategic and operational mobility to achieve surprise on the battlefield. The US Air Force (USAF) can accurately deliver the airborne division into virtually any objective area under almost any weather condition. All equipment is air transportable; most are air-droppable. All personnel are trained for parachute assaults and airborne operations. Engagements with enemy armored or motorized formations require special consideration. The division does not have sufficient armored protection to defeat heavier armored formations at close range. Antitank weapons in the division compensate for, but do not completely offset, this deficit. For division organizational structure, see FM 71-100.

1-6. Air Assault Division

The air assault division combines strategic deployability with tactical mobility within its AO. It attacks the enemy deep, fast, and often over extended distances and terrain obstacles. The air assault division must be able to conduct forced entry operations. The airmobile division of the Vietnam era provided the US Army the operational foundation, experience, and tactics for today's air assault operations. Air assault operations have evolved into combat, CS, and CSS elements (aircraft and troops) deliberately task-organized for tactical operations. Helicopters are completely integrated into ground force operations. Air assault operations generally involve insertions and extractions under hostile conditions, as opposed to mere air movement of troops to and from secure locations about the battlefield. Once deployed on the ground, air assault infantry battalions fight like battalions in other infantry divisions; however, normal task organization of organic aviation results in greater combat power and permits rapid aerial redeployment. The rapid tempo of operations over extended ranges enables the division commander to rapidly seize and maintain the tactical initiative. For division organizational structure, see FM 71-100.

1-7. Medium Division

The medium division consists of one armored brigade, one mechanized brigade, and one air assault brigade, and traditional heavy division aviation, CS, and CSS units. The Army designed this division to provide commanders operational flexibility with armor lethality and light infantry strength in restrictive terrain.

Section III. THE DIVISION STAFF

1-8. The Division Commander

The division commander is responsible for everything the division does. He assigns missions, delegates authority, and provides guidance, resources, and support to accomplish the mission.

1-9. Assistant Division Commanders

Within a division there are two assistant division commanders (ADC). The division commander prescribes their duties, responsibilities, and relationships with the staff and subordinate units. Normally, the responsibilities are broken down as operations and training (or maneuver) and support. Light divisions have an ADC for operations and training (ADC-OT), while in the heavy division, he is referred to as the ADC for maneuver (ADC-M) and, in both heavy and light divisions, there is an ADC for support (ADC-S).

1-10. Chief of Staff

The chief of staff directs the efforts of both the coordinating and special staffs. His authority usually amounts to command of the staff.

1-11. Staff Section

The command sergeant major (CSM), Assistant Chief of Staff, G1 (Personnel), Assistant Chief of Staff, G2 (Intelligence), Assistant Chief of Staff, G3 (Operations and Plans); and Assistant Chief of Staff, G4 (Logistics) function at division level in much the same way their counterpart staffs function at battalion and brigade level. The Assistant Chief of Staff, G5 (Civil Affairs) is the civil-military operations officer. This position is normally authorized only at division level and higher. For a detailed discussion of staff organization and functions, see FM 101-5. Provided in paragraph 1-33 is a detailed discussion of the coordinating staff and, in paragraph 1-34, a detailed discussion of the special staff.

Section IV. ORGANIZATIONS OF DIVISIONS

1-12. Division Organizations

All divisions are generally organized with a similar basic design. This design comprises a division headquarters and headquarters company (HHC), three ground maneuver brigades, an aviation brigade, an artillery brigade (referred to as division artillery [DIVARTY]), a support command, a cavalry squadron, an air defense artillery (ADA) battalion, an engineer battalion or brigade, a signal battalion, a military intelligence battalion, a military police (MP) company, and in most cases, a chemical company. The division headquarters provides C2 for the division's organic, attached, or supporting units. The HHC provides logistics support and personnel for the division headquarters and staff sections. Ministry teams in each division unit provide religious support to soldiers and their families. These teams provide worship opportunities, pastoral care, religious education, and spiritual fitness. The HHC is normally located near the division's main command post (CP). See FM 71-100 for definitive information on division organizations.

1-13. Ground Maneuver Brigade

The maneuver brigade headquarters provides the C2 facilities necessary to employ maneuver and fires. The only unit permanently assigned to the brigade is the brigade HHC. The necessary combat, CS, and CSS units to accomplish the brigade mission are attached, under operational control (OPCON), or placed in support of the brigade. The brigade's HHC furnishes logistics support (including equipment and personnel) and security for the brigade headquarters staff sections. The brigade normally controls from two to five attached maneuver battalions. It can be employed in autonomous or semiautonomous operations when properly organized for combat. Field Manuals 7-30 and 71-3 contain details on employment of the brigade. Ground maneuver battalions and additional units are placed in a command relationship to the brigade headquarters. This allows the division to accomplish missions in any environment. As units are added to brigades and the division, the division support command (DISCOM) is modified to meet changes in the division's supply, maintenance, and medical requirements.

1-14. Aviation Brigade

The aviation brigade is a maneuver force of organic, attached, and supporting Army aviation units. They include attack, air assault, reconnaissance (RECON), electronic warfare (EW), and general support

(GS) units. The division and aviation brigade commanders can tailor the brigade for virtually any combat, CS, and CSS operation to accomplish division missions. The brigade is most effective when its aerial forces concentrate at critical times or places to destroy units and exploit enemy vulnerabilities. The brigade extends the division capability to simultaneously strike the enemy throughout his depth and from multiple directions. The aviation brigade commander may be required to operate over great distances with his forces spread throughout the division's AO. This makes timely and accurate coordination difficult. Coordination is one the most important functions of the aviation brigade staff. Corps aviation brigades may augment or support the division. Aviation units in these organizations include attack helicopter and assault helicopter battalions, medium helicopter and theater aviation companies, and command aviation battalions. Field Manual 1-111 is the doctrinal base for aviation brigade operations.

1-15. Division Artillery

The DIVARTY is the division's primary organic indirect fire support organization. It normally comprises cannon and rocket artillery. Fire support systems neutralize, suppress, or destroy enemy forces. The DIVARTY provides close support, interdiction, and counterfire fire support to division operations. The division normally receives additional field artillery (FA) support from the corps or joint force commander. This support could include cannon and multiple launch rocket system (MLRS) battalions or brigades. Corps 155-millimeter cannons and MLRS battalions are organized similar to DIVARTY battalions. Corps artillery battalions have a liaison section to coordinate with supported units. Field Manual 6-20-2 provides further detail concerning DIVARTY operations.

1-16. Engineer Brigade

Divisional engineers accomplish mobility, countermobility, survivability, general engineering, and geospatial missions. Additionally, they may perform infantry combat missions and tasks when required. Armored and mechanized divisions have an organic engineer brigade; light, airborne, and air assault divisions have only an engineer battalion organic to the division. A division, when fully committed, normally requires a corps combat engineer battalion and a CS equipment company to augment its organic engineer units. Corps provides additional engineer units based on an METT-TC analysis. The airborne, light, and air assault divisions have limited hauling and earth-moving capabilities. This reduces their capacity for obstacle creation and reduction, protective shelter construction, and combat route missions (mobility, counter-mobility, and survivability). The engineer brigade commander (the engineer battalion commander in light divisions) serves as the division engineer. He coordinates the efforts of all engineers working within the division sector. He requests support from higher headquarters based on his engineer estimate and the commander's concept for the operation. Additional assets may be attached, under OPCON, or in direct support (DS) of the division. Corps engineer units often reinforce a division. Field Manual 5-71-100 details division engineer operations.

1-17. Division Support Command

The DISCOM provides division-level CSS to all organic assigned and attached elements of the division. It furnishes limited CSS to nondivisional units in the division area. The DISCOM routinely performs the functions of arming, fueling, fixing, moving, and sustaining soldiers and their systems. The DISCOM commander's role is complex. He is a brigade-level commander and the division's principal CSS operator. He exercises full command authority over organic units in the support command. He also has a close relationship with the division G4 and the ADC-S because of their overlapping interests. Although the division G4 has coordinating staff responsibility for logistics planning and develops division-level plans, policies, and priorities, the DISCOM commander advises the division staff during the formulation of plans, estimates, policies, and priorities. The ADC-S, on the other hand, commands and supervises all rear area operations. The G3, with the G4, and the DISCOM commander normally locate the CSS elements in the division rear area. The forward support battalions (FSB) provide DS to brigades and are positioned in the brigade support areas (BSA). Assigned to the FSB is the forward support medical company (FSMC) that provides Echelon II medical care to those battalions with organic medical platoons. The company provides both Echelon I and Echelon II medical treatment on an area support basis to units operating in the BSA and brigade AO. The FSMC establishes treatment facilities (division clearing stations) in the BSA. The FSMC performs the following functions:

- Treatment of patients with disease and nonbattle injuries (DNBI), triage of mass casualties, advanced trauma management (ATM), initial resuscitation and stabilization, and evacuation of patients incapable of returning to duty from battalion aid stations (BAS) to the FSMC.

- Ground evacuation for patients from BAS and designated collecting points to the FSMC.

- Sick call services for the BSA and brigade rear area.

- Maintenance of field health records for personnel receiving their primary care from the FSMC according to Army Regulation (AR) 40-66.

- Operational dental care (includes emergency and essential dental care).

- Emergency Class VIII resupply to unit operating in the brigade AO.

- Unit-level medical maintenance.

- Diagnostic medical laboratory and radiology services commensurate with Echelon II medical treatment facilities (MTF).

- Outpatient consultation services for patients referred from Echelon I MTF.

- Limited reinforcement and augmentation to supported medical platoons.

- Patient holding for up to 40 patients (20 patients in light infantry divisions) able to return to duty (RTD) within 72 hours.

- Reconstitution/regeneration support for maneuver medical platoons

The main support battalion (MSB) is located in the division support area (DSA) to provide area support to divisional units in the DSA and backup support to the FSB. Assigned to the MSB is the main support medical company (MSMC). The MSMC provides Echelons I and II medical care to units without organic medical treatment elements operating in the DSA. The MSMC establishes its MTF and base of operations in the DSA and—

- Provides advice and guidance to the MSB commander and his staff on the health of the command and CHS activities.

- Performs triage, initial resuscitation and stabilization, and preparation for evacuation of sick, injured, and wounded personnel.

- Provides medical evacuation (ground) support on an area support basis in the DSA. (Corps ambulances are normally used to evacuate patients from the BSA; however, this mission could be assigned to the MSMC.)

- Provides treatment squads that may operate independently of the division clearing station for limited time periods.

- Provides reconstitution support for forward deployed medical companies/elements.

- Provides sick call services for the BSA and brigade rear area.

- Maintains the field health records of personnel receiving their primary care from the FSMC according to AR 40-66.

- Provides operational dental care (includes emergency and essential dental care).

- Provides Class VIII resupply through the division medical supply office (DMSO) to division and nondivisional units via supply point distribution, logistical packages, redirected ambulance backhaul, and emergency deliveries.

- Performs unit-level medical maintenance.

- Provides diagnostic medical laboratory and radiology services commensurate with Echelon II MTF.

- Provides outpatient consultation services for patients referred from Echelon I MTF.

- Provides patient holding for up to 40 patients (20 patients in light infantry divisions) able to RTD within 72 hours.

- Provides mental health (MH) and neuropsychiatric consultation services and combat stress control (CSC) for the division.

- Provides preventive medicine (PVNTMED) and environmental health surveillance, inspections and consultation services for division units.

• Provides optometry support limited to eye examinations, spectacle assembly using presurfaced single-vision lenses, and repair services.

Medical elements from the BSA and DSA may be deployed forward with a forward logistics element (FLE) and temporarily provide support for forward areas. Additionally, corps support organizations may use FLE for special support requirements and to rapidly resupply as far forward as possible. Although the division has its own organic CSS units, it relies on corps and corps support command (COSCOM) units to sustain the division for continuous operations. (It may also coordinate through the civil affairs [CA] staff for available host-nation [HN] support.) The medical brigade supporting the corps provides CHS that includes hospitals in GS of the division. In DS, it provides a forward surgical team (FST), air and ground medical evacuation (MEDEVAC), CSC, and veterinary, PVNTMED, and combat health logistics (CHL) services that include Class VIII resupply and blood management. See FMs 63-2, 63-2-2, 63-20, 63-20-1, 63-21, and 63-21-1 for information on DISCOM CHS operations. See FMs 4-02.1, 4-02.21, 8-10-1, 8-10-3, 8-10-5, 8-10-9, 8-10-25, 8-10-26, and 8-51 for definitive information on CHS for the division.

1-18. Division Cavalry Squadron

The division cavalry squadron performs RECON and security for division operations. This helps the division commander to maneuver his brigades and battalions and attack the enemy at the most critical points. It fills gaps between units and establishes physical contact with divisional units and adjacent units. The division cavalry squadron also—

• Facilitates the division's movement with RECON, establishing contact points and passage points, and coordinating with higher and adjacent headquarters.

• Performs RECON and security operations in the division's rear area.

• Performs damage control and combat operations in the division's rear area when tasked as, or as part of, a tactical combat force (TCF).

See FM 17-95 for definitive information on cavalry squadron operations.

1-19. Air Defense Battalion

The division ADA battalion retains the division's freedom of maneuver, protects critical division assets, destroys enemy aircraft before they release their ordnance, and denies the enemy aerial RECON. The division's ADA battalion commander is the air defense coordinator. He integrates the division's air defense plan with both corps and echelons above corps (EAC) air defenses and integrates any OPCON air defense units. See FM 44-100 for definitive information on air defense operations and procedures.

1-20. Military Intelligence Battalion

The military intelligence (MI) battalion provides the division dedicated intelligence, counterintelligence (CI), and EW support, including communications intercept, direction finding capability, and electronic attack.

The MI battalion responds to the G2- and G3-assigned intelligence, CI, and EW requirements and missions. The MI battalion receives, analyzes, and disseminates intelligence information from echelons above division (EAD) as well as from assets within the division. The division MI battalion conducts G2-tasked intelligence collection operations according to the division commander's priority intelligence requirements and information requirements through the analysis and control element (ACE). The ACE is organic to the MI battalion and operates under the direction of the G2. Additionally, the MI battalion conducts G3-tasked EW operations missions per the commander's EW priorities. The MI battalion coordinates and directs corps and EAC intelligence and electronic warfare (IEW) assets supporting the division. A MI brigade performs IEW support at the corps level. It provides GS to the corps and, on occasion, may reinforce the division's intelligence efforts. When required, intelligence, imagery, EW, operations security (OPSEC) support, and interrogation assets directly support the divisions. See the FM 34-series (MI) manuals that provide the foundations for MI operations.

1-21. Signal Battalion

Signal support to the division is a collective and integrated application of communications, automation, and information services and systems. The signal battalion uses three distribution systems to cover the division's AO. They are the area common user system, the combat net radio (CNR) system, and the automatic data distribution system. Their integrated operation forms the division communications system. Signal units attached to or supporting the division will be under OPCON of the division signal battalion commander unless otherwise assigned or attached. The corps signal brigade links the division communications system to corps and higher echelons. The resulting architecture provides deployed forces with secure global, voice, and digital data communications. If required for special missions, corps and EAC signal assets can augment the division's existing signal capabilities. Specific responsibilities for the signal staff and the signal battalion are contained in FMs 11-50 and 101-5.

1-22. Military Police Company

The division MP company performs five functions in support of division operations:

- Maneuver and mobility support, to include support to river crossing operations, breaching operations, and passage of lines: straggler and dislocated civilian control; route RECON and surveillance; and main supply route (MSR) enforcement.

- Area security, to include RECON operations, area damage control, base/air base defense, response force operations and critical site, assets, or high-risk personnel security.

- Internment and resettlement operations, to include enemy prisoner of war (EPW) and civilian internee operations, confinement of US military prisoners and populace, and resource control.

- Law-and-order operations, to include law enforcement, criminal investigation, and US military prisoner confinement.

- Police intelligence operations, to include criminal information used to support the intelligence preparation of the battlefield (IPB), police assessment process, and joint information.

The MP company commander normally collocates with the division rear CP in the DSA to direct the use of his MP unit. Finally, the division normally receives one corps MP company to support sustained operations. Like the divisional MP company, this attached corps MP company works under the staff supervision of the provost marshal. Field Manual 3-19.1 details MP operations.

1-23. Division Chemical Company

The division chemical company is found in all divisions except light infantry. It reduces the effects of enemy NBC weapons and counters enemy sensor systems by using smoke and obscurants on division combat operations. Its primary focus is NBC RECON, decontamination (less patient decontamination), and smoke generation. Several NBC organizations and detachments provide added NBC defensive capabilities to the division. Corps chemical units may support corps CS and CSS units operating in the division area and may reinforce the divisional chemical company. Other sources of chemical support for divisions are mechanized or motorized smoke units, NBC RECON units, NBC decontamination units, and dual-purpose (light divisions only) smoke and decontamination companies. When additional companies are attached to or under OPCON of a division, a separate chemical battalion headquarters and headquarters detachment may be needed for command, control, and communications (C3). It will report directly to the division chemical officer for operational taskings. Field Manual 3-100 provides the doctrinal base for chemical company operations.

1-24. Division Band

The mission of the Army bands is to provide music to enhance unit cohesion and morale and to musically support military operations. Bands provide important support to information operations and should be integrated into public affairs, CA, and psychological operations (PSYOP) plans. Army bands are assigned secondary missions (except the US Army Band, the US Army Field Band, and the US Army Military Academy Band which have unique missions). These principal secondary missions are augmentation of security operations for CPs, or augmentation of perimeter security for EPW/civilian internee operations.

Section V. MANEUVER BRIGADES AND REGIMENTS

1-25. Organization of the Armored or Mechanized Infantry Brigade

The armored or mechanized infantry brigade is a combination of armored and mechanized infantry battalion task forces (TF) and other supporting units grouped under the command of a brigade headquarters. It participates in division or corps operations according to the principles and concepts set forth in FMs 71-100 and 100-5.

a. Divisional Brigades.

(1) Close combat-heavy brigades are the major subordinate maneuver commands of armored and mechanized infantry divisions. The only permanent unit assigned to a brigade is its HHC. The HHC provides direction and control over units assigned to, attached to, or supporting the brigade. The new Force XXI digitized brigade has gone through some organizational changes; for example, the maneuver battalion now has only three companies. There was also a redesign of its CSS in the maneuver battalions and the supporting DISCOM FSB. The Force XXI maneuver battalion with its enhanced computers, communications, and digitalization has real-time situational understanding that provides personnel a CRP.

(2) Divisional infantry, armored, and mechanized battalions are attached to brigades to destroy the enemy and to seize and hold terrain. Normally, each brigade can control three or four maneuver battalions with their CS and CSS units. When it is necessary to concentrate forces, control of more battalions may be necessary. However, the battalions assigned to a brigade must be limited to a number that can be controlled in a very complex battle situation.

(3) Light infantry battalions may be attached to the heavy brigade for specific missions and for a short duration. Use of light forces requires careful consideration of key employment and logistics support.

(4) The divisional brigade does not have support assets assigned to it; however, a habitual support relationship does exist between a designated FSB and the maneuver brigade. Normally, brigade support is also provided by a DS FA battalion; an ADA battery; an engineer company; a forward area signal platoon; an MP platoon; combat IEW elements; a tactical air control party (TACP); and a DISCOM FSB. Attack helicopter units may also operate with the brigade. When sorties are allocated for planning, USAF tactical air operations support the brigade.

b. Separate Mechanized Infantry and Armored Brigades.

(1) Since separate brigades conduct operations under corps command, they are organized to provide their own support. Units organic to the separate brigade include—

- A brigade HHC to provide C2 and limited CS assets, to include MP, chemical, and air defense elements.
- Tank and mechanized battalions to fight battles, destroy or disrupt enemy forces, and seize and hold terrain.
- An armored cavalry troop for RECON, security, and economy-of-force operations.
- A DS FA battalion to provide fire support.
- An engineer company for combat engineer support.

- An MI company to assist in collecting, processing, and disseminating intelligence and to support EW operations.

- A support battalion organized to provide CSS in the same way as the DISCOM's FSB provides CSS to divisional brigades, but with the added ability to link directly with the COSCOM for augmentation.

(2) Additional combat, CS, and CSS units may be attached to a separate brigade as required by the brigade's mission and operating circumstances. The separate brigade may be attached to a division (less support) but is usually controlled by a corps.

1-26. Organization of Infantry Brigades

a. Divisional Brigades.

(1) Light infantry, airborne, or air assault brigades are the major subordinate maneuver commands of infantry, airborne, or air assault divisions.

(2) There may be three or four brigades assigned to an infantry division, depending on operational requirements but in most cases it is three.

(3) Combat support and CSS are provided to the *brigade* by the *division*. Normally, FA support is provided by a light FA DS battalion. An engineer company, a forward area signal center platoon, combat IEW elements, and DISCOM forward support elements also routinely support a brigade. From time to time, attack helicopter units and USAF bombers may operate in support of the brigade.

b. Separate Light Infantry Brigades.

(1) Since separate brigades sometimes conduct independent operations, they are organized to provide their own support. Each is generally organized with—

- A brigade HHC to provide C2.
- Infantry battalions to destroy the enemy and to seize and hold terrain.
- A support battalion with several support units to provide CSS.
- A combat IEW company to assist in collecting, processing, and disseminating intelligence and to support EW operations.
- A light FA battalion to provide fire support.
- An engineer company for combat engineer support.
- An armored cavalry troop for RECON, security, and economy of force operations.

(2) Additional combat, CS, and CSS units may be attached to the separate brigade, as required. The separate brigade may be attached to a division or placed under the control of a higher command such as a corps.

1-27. Armored Cavalry Regiments

The armored cavalry regiments (ACR) are designed as either an ACR (armored) or an ACR (light) (L).

a. Armored Cavalry Regiment (Armored). The ACR is a self-contained combined arms organization composed of armored cavalry squadrons (ACS), an aviation squadron, a support squadron, and separate CS companies and batteries. The ACR is a separate unit that supports the corps or a joint task force (JTF). It is often reinforced by corps CS units and divisional maneuver battalions. The ACR operates independently over a wide area and at extended distances from other units. The ACR is a highly mobile, armored force capable of fighting the fully mechanized threat in the environmental states of war or conflict. The ACR may be rapidly deployed to a theater of operations by sealift. When supporting a light corps, limitations may exist in corps support capabilities, strategic mobility, and terrain restrictions. The regimental ACS, a highly mobile, armor-protected force, is discussed below. The regimental aviation squadron (RAS) provides the regiment with combat aviation assets. It is organized with air cavalry/RECON troops, attack helicopter troops, and an assault helicopter troop. The squadron adds a very responsive, terrain-independent combat capability to the regiment. The maneuverability and flexibility of the RAS enhance the combat flexibility of the regiment. The RAS may operate independently of or in close coordination with the ACS, or it may provide troops to the ACS.

b. Armored Cavalry Regiment (L). The ACR (L) is a self-contained combined arms organization capable of being packaged and rapidly deployed by air or sealift as part of a force projection Army responding rapidly to worldwide contingencies. The role of the ACR (L) may be traditional, initial entry, or follow-on. The traditional role would support a US corps or TF through a RECON, security, and economy-of-force capability. As an initial entry force, the ACR (L) would support Army or JTF operations with credible force as a demonstration of US resolve. In the follow-on role, the ACR (L) will follow an opposed entry force (division-ready brigade type) to expand the point of entry, to provide RECON and security, and to serve as the initial combat-capable maneuver force. Because of the C2 structure and support base within the regiment, it is a very modular organization capable of rapidly integrating armored forces into its task organization. This factor supports the Army with a force package that can be tailored for the situation and expanded once it is deployed to a theater if the situation dictates the need for armored forces. The ACR (L) is composed of ACS, an aviation squadron, a support squadron, and separate CS companies and batteries. The ACR (L) is a separate unit that supports the contingency corps. It is often reinforced by corps CS units and divisional maneuver battalions. The ACR (L) operates independently over a wide area and at extended distances from other units. The ACR (L) is a highly mobile force capable of executing the full range of doctrinal cavalry missions against a comparable threat in the environmental states of war or conflict. It is also capable of conducting stability operations and support operations. When supporting a light corps, limitations may exist in corps support capabilities, strategic mobility, and terrain restrictions. The organizational structure for the ACR (L) is similar to the ACR (armored) with some significant exceptions. These differences may require modification of the TTP prescribed for the ACR according to FM 17-95. The following assets or capabilities are not organic to the ACR (L):

- Neither the ACR (armored) nor the ACR (L) is authorized a tactical CP.
- The chemical company of the ACR (L) is not organized with a smoke platoon.
- The engineer company of the ACR (L) is not organized with bridging assets.
- Organic assets do not include digital terrain database development.
- The MI company of the ACR (L) does not have ground surveillance radar.
- The aviation squadron of the ACR (L) does not have attack helicopter troops.

Limited ballistic protection offered by the high-mobility multipurpose wheeled vehicle (HMMWV) and lack of organic tank assets require judicious application of standard cavalry doctrine. The ACR (L) is ideally suited for force packaging and employment by the contingency corps against a comparably equipped threat, but must be reinforced to defeat a modern mechanized or armored force. For definitive information on both the ACR (armored) and the ACR (L), see FM 17-95.

Section VI. THE BATTALION

1-28. Organization of the Infantry Battalion

a. Organization. An infantry battalion is organized and equipped to give it the capabilities needed to accomplish its missions. It is large enough to engage enemy regiments using a full range of organic and nonorganic weapons and support. However, it is small enough that the battalion commander can personally lead and immediately influence the action of his units in battle.

(1) To understand the organizational structure of the battalion, one must understand the organizational roles of echelons above and below the battalion and how the battalion serves as the interface for these echelons.

(2) Within the context of organizational roles, platoons normally fight as part of a company. Companies fight using their subordinate platoons as fire or maneuver elements. Battalions provide support to the companies; ensure the battlefield has depth; and synchronize the various arms and services to achieve the maximum effect from the available forces. The brigade task-organizes the battalion, fitting the forces to the ground, mission, and enemy situation. Divisions provide CS and CSS force multipliers. Corps conducts operational-level warfare, providing additional CS and CSS assets in accordance with the corps main effort.

(3) To execute doctrine, the infantry battalions require adequate troop strength; an organic antiarmor capability; supporting arms; optimized task organization based on the mission; and adequate support. These requirements are met through the organization of the infantry battalions and through augmentation and task organization where required.

b. Types of Battalions. There are six basic types of infantry battalions: mechanized infantry, air assault, airborne, ranger, light, and mountain. The fundamental combat mission of the infantry battalion, regardless of type, is to destroy or capture the enemy by means of fire and maneuver; or to repel his assault by fire, close combat, and counterattack. To accomplish specific missions, the battalion is normally augmented with combat, CS, and CSS assets.

c. Task Organization. Normally, infantry battalions operate as table of organization and equipment (TOE) units only in garrison. For training and for combat, they are task organized for the mission at hand. Task organizing tailors the unit to get the most from its capabilities and to minimize its limitations. It is a temporary grouping of forces designed to accomplish a particular mission. Task organization involves the distribution of available assets to subordinate control headquarters by attaching or placing assets under OPCON to the subordinate. Task organization is made after analysis of the mission and all of the other METT-TC factors. When developing the task organization, the commander must clearly understand the capabilities and limitations of his organic and supporting units; he must consider the existing C2 relationships.

1-29. Organization of the Mechanized Infantry and Armored Battalions

Mechanized infantry battalions and armored battalions are organized, equipped, and trained to accomplish specific missions; each type battalion has unique capabilities and limitations.

a. Missions.

(1) The mission of the mechanized infantry battalion is to destroy or capture the enemy by means of fire and maneuver, or to repel his assault by fire, close combat, and counterattack.

(2) The mission of the armored battalion is to close with and destroy enemy forces using fire, maneuver, and shock effect, or to repel his assault by fire and counterattack.

(3) Battalion TF accomplish missions and tasks as part of a brigade's operation. Occasionally, TF will conduct operations directly under a division's or an ACR's control; such as participating in the higher headquarters covering force; acting as a reserve; or forming a TCF in rear area operations.

b. Capabilities.

(1) The capability of the mechanized infantry battalion and the armored battalion is increased through task organization. Based on situational estimates, the brigade commander task-organizes armored and mechanized infantry battalions by cross-attaching companies between these units. As a rule, cross-attachment is done at battalion level because it has the necessary C2 and support capabilities to employ combined arms formations. The brigade commander determines the mix of companies in a TF. Similarly, the TF commander may cross-attach platoons to form company teams for specific missions.

(2) Tank and mechanized infantry battalion TF apply their mobility, firepower, and shock effect to—

- Conduct sustained combat operations in all environments.
- Accomplish rapid movement and limited penetrations.
- Exploit success and pursue a defeated enemy as part of a larger formation.
- Conduct security operations (advance, flank, or rear guard) for a larger force.
- Conduct defensive, retrograde, or other operations over assigned areas.
- Conduct offensive operations.

c. *Limitations.*

(1) Because of the high density of tracked vehicles, the battalion has the following limitations:

- Mobility and firepower are restricted by urban areas, dense jungles and forests, very steep and rugged terrain, and significant water obstacles.
- Strategic mobility is limited by substantial quantities of heavy equipment.
- Consumption of supply items is high, especially Classes III, V, and IX.

(2) Battalions are task-organized according to mission; they are routinely augmented to improve engineer, fire support, air defense, intelligence, and CSS capabilities.

1-30. Battalion Task Force on the AirLand Battlefield

a. The foundation of AirLand Battle doctrine at the TF level is classical maneuver warfare. In its simplest form, maneuver warfare involves using a part of the force to find, then contain the enemy, while the remainder of the force moves to a position of advantage and attacks his weakest point—usually a flank or the rear. The goal is to mass enough combat power at the critical place and time to destroy or threaten the enemy with destruction, while preserving freedom for future action.

b. The TF commander must understand the intent of the brigade and division commander to properly employ his force. The TF commander develops his intent and concept and accepts risks to achieve decisive results. He seizes the initiative early and conducts offensive action aimed at imposing his will on the enemy. The objective of his maneuver is to position strength against weakness, throw the enemy off balance, and aggressively follow-up to defeat and destroy the enemy.

Section VII. THE BATTALION STAFF

1-31. Command and Control Responsibilities of the Battalion

The commander establishes a standard C2 system by defining the functions of key individuals, organizations, and facilities. He organizes his staff in a manner to accomplish the mission. He will develop a basic organization flexible enough to be modified to meet changing situations. This section discusses the individual and staff functions and responsibilities and how they are organized to facilitate C2.

1-32. Staff

a. Commander. The commander commands and controls subordinate combat, CS, and CSS elements that are organic or attached to his unit or under its OPCON. The commander's main concerns are to accomplish his unit's mission and to ensure the welfare of his soldiers.

(1) The commander cannot win the battle alone. He must rely on his staff and subordinate commanders for advice and assistance in planning and supervising operations. He must completely understand their limits and capabilities. He must train subordinate commanders to execute his concept in his absence. Also, he must cross-train his staff to continue unit operation when staff elements suffer combat losses.

(2) The staff reduces the demands on the commander's time; they assist him by—

- Providing information.
- Making estimates and recommendations.
- Preparing plans and orders.
- Supervising the execution of orders issued by, or in the name of, the commander.

The commander assigns clear-cut responsibility for functions to unit staff officers to ensure that conflicts do not arise. As a rule, staff officers are delegated the authority to say "yes" to requests by subordinate unit commanders. They defer to the commander when the answer is "no." The staff must be responsive to subordinate unit commanders.

b. Executive Officer. The executive officer (XO) is second in command and is the principal assistant to the battalion commander. The XO is prepared to assume the duties of the commander. He formulates and announces staff operating policies and ensures the commander and staff are informed on matters affecting the command. The XO ensures that—

- The required liaison is established.

- All staff officers, unless otherwise instructed by the commander, inform him (the XO) of any recommendations or information they gave directly to the commander, or of any instructions they received directly from the commander.

- The XO functions as the *Chief of Staff* and directs and coordinates staff activities.

He represents the commander, when required, and exercises supervision of the tactical operations center (TOC) and its operations.

c. Command Sergeant Major. The CSM is the senior noncommissioned officer (NCO) in the unit. He acts in the name of the commander when dealing with other NCO in the unit; he is the commander's primary advisor concerning the enlisted ranks. He should understand the administrative, logistical, and operational functions of the unit to which he is assigned. Since he is the senior enlisted soldier in the unit, his attention should be focused on operations, training, and how well the commander's decisions and policies are being carried out. He is the senior enlisted soldier who is responsible to the commander for training in the organization. He coaches and trains first sergeants (1SGT) and platoon sergeants (SGT); he works very closely with company commanders in this regard. He maintains close contact with subordinate and attached unit NCOs. The CSM may act as the commander's representative in supervising critical aspects of an operation. The CSM can lead the advance/quartering party during a major movement. He may also help in the CSS effort during the battle.

1-33. Coordinating Staff

a. Adjutant (US Army).

(1) The Adjutant (US Army) (S1) has primary responsibility to provide all activities and functions associated to sustain personnel manning of the unit, provide personnel support, and other human resource support to service members. The S1 section is principally structured to provide critical wartime functions of personnel strength accounting and casualty reporting. Battalion S1 sections can provide limited personnel services and support within the battalion, but require augmentation support to provide the full spectrum of human resource support and services. The S1 participates in the planning process, personnel estimates, loss estimates, and recommends replacement priorities. Within the battalion, the S1 coordinates EPW support, limited legal support, CHS planning (in conjunction with the battalion surgeon), and public affairs. The battalion S1 also coordinates legal actions with the brigade S1.

(2) Information linkages and split-base operating procedures should minimize the distinction between wartime/peacetime and forward/rear in regards to personnel support. Personnel (S1) sections provide additional services at the home station. When deployed, the S1 performs replacement operations, postal operations, strength management, personnel actions (awards, promotions, evaluations, reassignments, and military pay), morale welfare and recreation support, and Red Cross coordination. The commander must leverage his reach capability with these functions.

(3) During deployment, the S1 operates from the combat trains and is normally collocated with the Logistics Officer (US Army) (S4). While collocated, functional responsibilities between the S1 and S4 are not intended to be interchangeable.

b. Intelligence Officer (US Army). The Intelligence Officer (US Army) (S2) exercises overall staff responsibility for intelligence. He conducts the IPB with the commander and Operations and Training Officer (US Army) (S3) using—

- Higher collection sources.
- Ground and aerial RECON.
- Observation posts.
- Ground surveillance radar.
- Target acquisition.
- Electronic warfare assets.
- Organic scout platoon.

In conjunction with the IPB process, he prepares and disseminates intelligence estimates. The tactical intelligence officer assists the S2. The tactical intelligence officer is part of a two-man battalion information coordination center (BICC). The BICC's primary responsibility is to manage the unit's intelligence collecting, processing, and disseminating effort for the S2. The BICC develops and initiates the reconnaissance and surveillance (R&S) plan; identifies requirements that cannot be met by the battalion's assets; and notifies the brigade S2.

c. Operations and Training Officer (US Army). The S3, as the operations officer, is the commander's principal assistant for coordinating and planning the battle. The S3—

- Monitors the battle.
- Makes sure that CS assets are provided when and where required.
- Anticipates developing situations.

He advises the commander on—

- Courses of action (COA).
- Combat and CS matters.
- Organization and training.
- Operational matters during the battle.

He prepares the operations estimate and conducts planning and coordination with other staff sections resulting in published operation orders (OPORD), operation plans (OPLAN), and training programs.

In conjunction with his planning duties, he is responsible for PSYOP; EW activities, OPSEC; deception; and, in conjunction with the S4, tactical troop movement. He establishes priorities for communications to support tactical operations and coordinates with the XO and the battalion signal officer on the location of the main CP.

(1) Operations and Training Officer (US Army) (Air). The S3 (Air), the principal assistant to the S3, is normally in the TOC. He assumes the duties of the S3 in his absence. He coordinates the employment of close air support with the fire support element (FSE) and the TACP, as well as the air defense section leader.

(2) Nuclear, biological, and chemical personnel. The assistant S3/chemical officer is assigned to the S3 section of combat battalions with a chemical NCO as his assistant. A decontamination specialist is assigned to the HHC of airborne and air assault battalions. The chemical officer and NCO train and supervise the battalion decontamination crew. During combat operations, chemical personnel provide a 24-hour capability within the S3 section to receive, correlate, and disseminate information on NBC attacks. They consolidate subordinate units' operational exposure guide (OEG) radiation status and report to higher headquarters as required. They provide recommendations concerning mission-oriented protective posture (MOPP) levels and employment of supporting NBC RECON and smoke units. If the unit comes under NBC attack, battalion NBC personnel organize and establish a battalion NBC center. They supervise activities of the radiological survey and monitoring teams and the chemical detection teams. The NBC personnel also coordinate and supervise decontamination missions (less patient decontamination) conducted with or without support-level decontamination assets.

d. Logistics Officer (US Army). The S4 has primary staff responsibility for determining CSS requirements and priorities. The S4 is responsible for CSS planning in the military decision-making process. His section is responsible for the procurement, receipt, storage, and distribution of supplies; and for transportation of units, personnel, and CSS items to their required locations. He designates lines of communications (LOC) and supply routes and locations of CSS elements; and prepares and develops CSS plans in concert with the current tactical plan. The S4 is responsible for the preparation, authentication, and distribution of CSS support plans and orders when published separately. The S4 establishes the requirements for civilian labor and the collection and disposal of excess property, salvage, and captured material.

e. Command and Control, Communications, and Computer (C4) Operations Officer (US Army). The Assistant Chief of Staff, (Signal) (G6)/Communications Staff Officer (US Army) (S6), C4 operations officer, is the principal staff officer for all matters concerning C4. A G6/S6 is located at all echelons of command from battalion through corps. He is responsible for advising the commander, staff, and subordinate commanders on C4 operational matters. Command and control, communications, and computer operations are inclusive of network operations and information management. Network operations include network management, information dissemination management, and information assurance. Information management includes relevant information and information systems functions.

f. Battalion Maintenance Officer. The battalion maintenance officer (BMO) plans, coordinates, and supervises the maintenance and recovery efforts of the maintenance platoon and ensures that adequate maintenance support is provided to the TF. Although he is a staff officer in the battalion headquarters, he is also the maintenance platoon leader. The maintenance warrant officer assists the BMO by providing technical assistance and supervision of the maintenance platoon. The BMO supervises the unit maintenance collection point in the armored and mechanized infantry battalions only.

1-34. Special Staff

a. Battalion Surgeon. In this role the battalion surgeon is a special staff officer and advisor to the battalion commander on employment of the medical platoon and on the health of the battalion. He is also the supervising physician (medical officer/field surgeon) of the medical platoon's treatment squad. This officer is responsible for all CHS provided by the platoon. His responsibilities include—

- Planning and directing CHS for the maneuver battalion. He does this in conjunction with the battalion S1, who is the coordinating staff officer responsible to the commander for health and welfare of the troops.
- Advising the maneuver battalion commander and his staff on CHS operations and the medical threat.
- Supervising the health, welfare, organizational training, administration, discipline, maintenance of equipment, supply functions, and employment of medical platoon personnel.
- Examining, diagnosing, treating, and prescribing courses of treatment for patients, to include ATM.
- Training combat lifesavers (CLS).
- Supervising the battalion CSC program.
- Supporting humanitarian assistance programs, when directed.
- Providing PVNTMED support for the battalion. Requesting PVNTMED support from the brigade for PVNTMED requirements beyond his (battalion surgeon) capabilities.
- Planning and overseeing PVNTMED training for battalion personnel.
- Advising the commander on the health of the battalion.
- Supervising the training of unit field sanitation teams.
- Ensuring the field health records are maintained according to AR 40-66.
- Advising the commander on the effects of the Geneva Conventions on CHS. See Appendix A, The Geneva Conventions, for additional information.

b. Chaplain. The chaplain advises the commander and staff on religious matters and of the influence of indigenous religious groups and customs on the commander's courses of actions. The chaplain is a member of the battalion's unit ministry team (UMT). This team is the staff section that provides religious support to the battalion. The team advises the commander on unit morale and ethical issues and assists in meeting the religious and spiritual needs of the soldiers. The team consists of a chaplain and a chaplain assistant. The chaplain provides the clergy-related support to the unit. These include worship and prayer services, funeral and memorial services, and in-depth grief counseling. The chaplain assistant provides the administrative and logistical management for the team as well as the team's security. The chaplain exercises the necessary staff authority for developing, coordinating, and executing the religious support plan. Additionally, the chaplain facilitates soldiers' free exercise of their religious rights, beliefs, and worship practices and makes recommendations for ethical decision-making and moral leadership programs. See FM 16-1 for definitive information on religious support for the battalion.

1-35. Other Staff Assets

a. Headquarters and Headquarters Company Commander. The HHC commander has the responsibility of ensuring that the command facilities are provided logistical support. Normally, he places his XO with the main CP to supervise support, security, and movement. The HHC commander locates himself at the field trains CP to monitor and coordinate all battalion activities there. He uses landlines and messengers to control all elements in the field trains and communicates with the combat trains using the administration/logistics net (a frequency modulated [FM] radio net). The HHC commander is available for other tactical missions as dictated by the estimate of the situation. These roles normally come into play during operations other than sustained ground combat. They may include coordination and control of the RECON/counter-RECON effort; combat patrols; or any other task designated by the battalion commander.

b. Physician Assistant. The physician assistant (PA), normally a Captain, Army Medical Specialist Corps, performs general technical health care and administrative duties. The PA is ATM-qualified and works under the clinical supervision of the medical officer. This officer serves as the medical platoon leader in the absence of an assigned physician.

c. Field Medical Assistant. The field medical assistant, a Medical Service Corps (MS) officer, normally a Lieutenant, is the operations/readiness officer for the medical platoon. He is the principal assistant to the battalion surgeon for operations, administration, and logistics. The field medical assistant coordinates CHS operations with the battalion S3 and S4 and coordinates patient evacuation with the FSMC.

d. Fire Support Officer. The integration of fire support into the maneuver operation is a decisive factor in the success of battle. The maneuver commander is responsible for the whole of his operation including the fire support plan. The fire support officer (FSO) is responsible for advising the commander on the best available fire support resources; for developing the fire support plan; for issuing the necessary orders in the name of the commander; and for implementing the approved fire support plan. The FSO normally locates with the commander, but it may be necessary to locate where he can communicate best.

e. Air Defense Artillery Officer. The senior leader of any supporting ADA unit(s) advises the commander on the employment of ADA assets. During the planning process, he is at the TOC to ensure the integration of air defense into the concept of operation. During the execution of the plan, he positions himself in a location that will enable him to best C2 the air defense assets. He monitors the command net to remain responsive to the needs of the commander. He also monitors the early warning net to assist in the acquisition and dissemination of early warning information as a member of the Army airspace C2 system.

f. Engineers. The leader of the supporting engineer unit advises the commander on the employment of engineer assets. During the initial planning, he is at the TOC to advise the commander on employment of his unit. During the battle, the engineer unit provides a representative with a radio at the TOC, if possible, to coordinate the engineer effort. If no representative is available, the TOC periodically monitors the engineer net. Regardless of the system used, the engineer leader is responsible for maintaining constant communications with the battalion.

g. Antiarmor Company Commander/Platoon Leader (Light Battalion). This leader advises the commander on the tactical employment of his weapon systems. He may serve as a fourth maneuver element or as an alternate battalion CP when properly task-organized. The first alternate CP for the battalion is the combat trains CP (CTCP).

h. Scout Platoon Leader. He advises the commander and the S2 on the employment of his element. He is responsible for conducting tactical RECON in support of the battalion. He assists the S2 in developing the R&S plan.

i. Battalion Mortar Platoon Leader. He advises the battalion commander and the FSO on tactical employment of the battalion mortar platoon; he may assist the FSO with his fire support coordinator responsibilities. His platoon headquarters may also serve as an alternate CP.

CHAPTER 2

DIVISION COMBAT HEALTH SUPPORT

Section I. OVERVIEW OF DIVISION AND CORPS MEDICAL ASSETS

2-1. Tables of Organization and Equipment

The TOE provides a model for fielding a unit at full capability, or at a reduced capability if resource constraints so mandate. The TOE also specifies the capabilities that the unit has to accomplish its mission. Tables of organizations and equipment are scheduled for revision when changes in doctrine occur, upon introduction of new or improved equipment, or to incorporate more effective organizational design. New TOEs are developed to accommodate the requirements of new organizations. If the TOE is not scheduled for revision or replacement by a new TOE, it will be scheduled for cyclic review every 3 years. There are different editions of TOE that are identified as under the TOE modernization system. For example, A-Edition TOE were developed to identify restructuring initiatives for existing L-Edition TOE. Example of TOE editions based on modernization of the Army include—

- Division 86, H-Edition TOE.
- Army of Excellence (AirLand Battle/Medical Force 2000), L-Edition TOE.
- Medical Reengineering Initiative (MRI), A-Edition TOE.
- Force XXI (Digitized Division), F-Edition TOE.

The MTOE is the document that is seen at the unit level. The MTOE is a modified version of a Headquarters, Department of the Army (DA)-approved TOE that prescribes the unit organization, personnel, and equipment necessary to perform a mission in a specific geographical or operational environment. At unit level, the MTOE is the base document for requesting personnel and equipment; distributing personnel and equipment resources; unit status reporting; and reporting supply and maintenance status. In addition to the TOE, some organizations have a table of distribution and allowances (TDA). The TDA prescribes the organizational structure for a unit having a support mission for which a TOE does not exist and which may include civilian positions. For the remainder of this chapter only the Army of Excellence/AirLand Battle and the Force XXI/Digitized Division TOE (with some information pertaining to MRI units) will be discussed.

2-2. Division Medical Assets, L-Edition TOE (Army of Excellence/AirLand Battle)

The Army of Excellence/L-Edition TOEs were developed to support the AirLand Battle/Medical Force 2000 doctrine. They will be in the Army's inventory until the Army completes its Army XXI and other modernization initiatives. Medical assets that are organic to the division with units under the L-Edition TOE include—

- Division surgeon's section (DSS) that has a surgeon and three additional medical personnel.
- Division support command, division medical operations center (DMOC), which could have up to 18 personnel assigned, depending on type of communications equipment the DMOC is using.

- Main support battalion, health service support officer (HSSO).
- Main support medical company.
    - Headquarters section.
    - Treatment platoon.
    - Ambulance platoon.
    - Division medical supply office.
    - Mental health section.
    - Preventive medicine section.
    - Optometry section.
- Forward support battalion, HSSO.
- Forward support medical company.
    - Headquarters section.
    - Treatment platoon.
    - Ambulance platoon.
- Aviation brigade flight surgeon's section

In addition, medical platoons/section are assigned to the following combat arms battalions or squadrons:

- Armored.
- Mechanized infantry.
- Airborne infantry.
- Air assault.
- Light infantry.
- Armored cavalry.
- Artillery.

- Reconnaissance, surveillance, and target acquisition (RSTA) squadron.
- Air defense artillery battalion.

Also, medical personnel may be assigned to the following units.

- Engineer battalion.
- Signal battalion.
- Antitank company.
- Military police company.

2-3. Division Medical Assets, F-Edition TOE (Force XXI/Digitized Division)

The F-Edition TOEs were developed to support the Force XXI/Division Redesign Initiative and Force XXI doctrine. Digitization of the new Force XXI Division is a high priority.

NOTE

Digitization is defined as the application of information to acquire, exchange, and employ timely battlefield information. It will *enhance situational understanding* and provide the means for information dominance by enabling friendly forces (ten divisions, Reserve Components, and joint/combined forces) to *share a common picture of the battlefield* while communicating and targeting in real or near real-time. Digitization will reduce the "fog of war" and *decrease decision-making time* by *optimizing the flow of information*. It will allow the "orchestration" of combat power at critical times and places faster than an adversary can. It will contribute *increased lethality, survivability, and operational tempo* while *reducing the potential for fratricide* to ensure *seamless digital communications* from the sustaining base to the tactical and strategic levels.

Medical assets that are organic to the division with units under the F-Edition TOE include—

- Division surgeon's section (a surgeon and 12 additional medical personnel).
- Division support command medical operations branch, including an officer and an NCO.
- Division support command medical materiel management branch (MMMB) (one officer and one NCO).

- Division support battalion HSSO.
- Division support medical company (DSMC).
  - Headquarters section.
  - Treatment platoon.
  - Ambulance platoon.
  - Mental health section (a division psychiatrist, an NCO, and one MH specialist are assigned to this section.
  - Preventive medicine section (one environmental science officer and a PVNTMED NCO are assigned to this section).
  - Optometry section.
- Forward support battalion, CHS cell/HSSO (an officer and an NCO are assigned to this section).
- Forward support medical company.
  - Headquarters section.
  - Treatment platoon (one less treatment squad than the L-Edition TOE).
  - Ambulance platoon.
  - Mental health section (a division psychiatrist, an NCO, and one MH specialist are assigned this section).
  - Preventive medicine section (one environmental science officer and a PVNTMED NCO are assigned to this section).
- Maneuver brigade surgeon's section (BSS) (six personnel assigned).
- Medical platoons/sections are assigned to the following combat arms and CS battalions:
  - Armored.
  - Mechanized infantry.
  - Artillery.

- Engineer.
- Signal.

- Medical personnel may be assigned to the following units:
  - Engineer company.
  - Antitank company.

2-4. Division Surgeon, L-Edition TOE (Army of Excellence/AirLand Battle)

The division surgeon is an Medical Corps (MC) officer, area of concentration (AOC) 60A. He is a special staff officer and normally coordinates his CHS activities through the G1. Generally, the surgeon's duties are administrative; the division commander charges him with full responsibility for the technical control of all medical activities in the command. The division surgeon's staff is assigned to the DSS of the division HHC. Personnel assigned to this section include an operations NCO (MOS 91W40), a clerk typist (MOS 71L10), and a patient administration specialist (MOS 71G10). These personnel along with the DMOC staff located in the DISCOM, assist the division surgeon in the performance of his duties. The division surgeon's responsibilities include—

- Advising on the health status of the command and of the occupied or friendly territory within the commander's area of responsibility.

- Briefing the division commander and/or his representative during all routine and emergency division briefings on CHS operations.

- Participating in the preparation of division OPLAN and contingency plans and identifying potential medical hazards associated with geographical locations and climatic conditions.

- Determining reporting requirements and frequencies.

- Advising on the health effects of the environment.

- Advising on the health effects of NBC devices/weapons, to include OEG.

- Exercising technical supervision of subordinate brigade surgeons, physicians, and PAs.

- Providing consultation and mentoring to subordinate brigade surgeons, physicians, and PAs.

- Advising on the health effects of directed-energy devices/weapons.

- Determining requirements for the requisition, procurement, storage, maintenance, distribution management, and documentation of Class VIII supplies within the division.

- Monitoring critical Class VIII items and keeping the division commander informed.

- Determining requirements for medical personnel and making recommendations concerning their assignments.

- Coordinating with medical unit commanders, to include leaders of medical platoons and sections, for continuous CHS.

- Submitting to higher headquarters those recommendations on professional medical problems that require research and development.

- Recommending use of captured medical supplies in support of EPW and other recipients.

- Advising on medical intelligence requirements (including the examination and processing of captured medical supplies as directed by the corps surgeon).

- Providing recommendations on allocation and redistribution of AMEDD personnel, CHL, and CHS during the reconstitution process.

- Advising commanders about the PVNTMED aspects of reconstitution and availability and use of CSC teams.

- Forwarding the Command Health Report (RCS MED-3 [R7]) according to Chapter 3, AR 40-5.

- Advising commanders on the effects of accumulated fatigue, radiation exposure, possible delayed effects from exposure to chemical warfare (CW) or biological warfare (BW) agents, and use of countermeasures and pretreatments.

- Advising commanders on disposition of personnel exposed to lethal, but not immediately life-threatening, doses of radiation or CW and BW agents.

- Ensuring the division's CHS annex is developed for all contingency plans. For CHS planning factors, see FM 8-55.

- Initiating, through the division commander and the G3, medical training, first aid training, and CLS training programs for the division.

- Overseeing the continuing health education program for the division and ensuring compliance with AR 351-3.

- Ensuring that all physicians and nonphysician health care providers have gone through a credentialing committee according to AR 40-68 to validate clinical privileges.

- Ensuring that clear and accurate patient records are maintained of all clinical encounters for supported/deployed personnel through the use of appropriate forms as directed by AR 40-66. See Appendix B for management of individual health records in the field. For additional information on the division surgeon, see FMs 8-10-3 and 8-10-5.

2-5. Division Surgeon, F-Edition TOE (Force XXI/Digitized Division)

The division surgeon, an MC officer (Lieutenant Colonel [LTC], AOC 60A00), is a division-level special staff officer. He normally works under the staff supervision of the division Chief of Staff. The division surgeon is responsible for the technical control of all medical activities in the command. He oversees and coordinates CHS activities through the DSS. The division surgeon advises the division commander on all medical or medical-related issues. The division surgeon's responsibilities are the same as those identified above except for—

- Briefing the division commander and/or his representative during all routine and emergency division briefings on CHS operations. This is normally accomplished using the Combat Service Support Control System (CSSCS).

- Monitoring the status of critical Class VIII items list and providing the G4 a list of medical items that should be a part of the commander's tracked items list.

- Developing the division's CHS annex for all contingency plans. For CHS planning factors, see FM 8-55.

- Ensuring that clear and accurate patient records are maintained of all clinical encounters for supported/deployed personnel through the use of appropriate forms as directed by AR 40-66. See Appendix B for management of individual health records in the field. Also, digital patient records at the division and brigade level will be available through the fielding of Medical Communications for Combat Casualty Care (MC4) and the Theater Medical Information Program (TMIP); see Chapter 3. For additional information on the division surgeon, see FM 4-02.21.

2-6. Division Surgeon's Section, F-Edition TOE (Force XXI/Digitized Division)

The DSS is normally located with the division main and consists of a medical plans and operations cell, a CHL cell, a patient disposition and reports cell, and a PVNTMED cell. Figure 2-1 shows the typical organization and staffing of the DSS.

a. Medical Plans and Operations Cell. The medical plans and operations cell is responsible for coordinating, planning, synchronizing, rehearsing, and conducting CHS for the division. For definitive information on the organization, functions, and operations of this section, see FM 4-02.21.

b. Combat Health Logistics Cell. The CHL cell is responsible for planning, coordinating, and prioritizing CHL and medical equipment maintenance programs for the division. The CHL cell is staffed with a health service materiel officer (HSMO). The HSMO (Major, AOC 70K67) works closely with the DISCOM MMMB and medical logistics (MEDLOG) company. The HSMO coordinates and oversees the CHL support for the division.

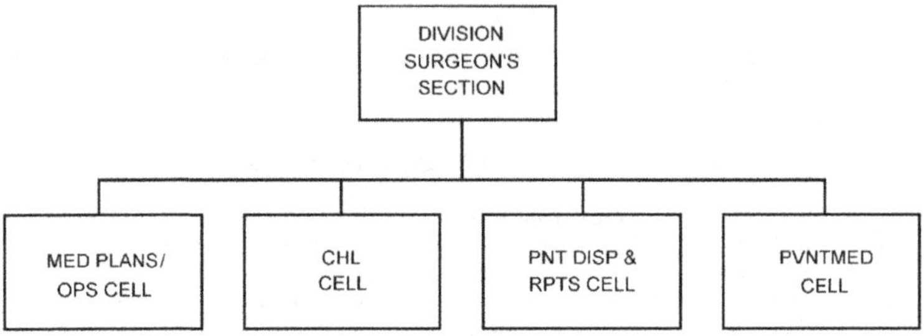

Figure 2-1. Division surgeon's section.

c. Patient Disposition and Reports Cell. The patient disposition and reports cell is responsible for coordinating patient disposition throughout the division. The branch obtains and coordinates disposition of patients with the medical plans and operations cell and the corps medical regulating office(r) (MRO). It prepares and forwards appropriate medical statistical reports as required. The patient disposition and reports cell is staffed with a patient administration NCO and two patient administration specialists.

d. Preventive Medicine Cell. The division PVNTMED cell is responsible for—

- Supervising the command PVNTMED program, to include health assessment and medical surveillance; see AR 40-5 and FM 4-02.17.

- Ensuring PVNTMED measures are implemented that protect division personnel against food-, water-, and vectorborne diseases, as well as environmental injuries (for example, heat and cold injuries).

- Monitor disease trends within the division.

The PVNTMED missions are accomplished according to the division CHS plan and coordinated by the PVNTMED officer through the medical plans and operations cell with the DSMC and FSMCs. Division PVNTMED personnel provide advice and consultation in the areas of environmental sanitation, epidemiology, and entomology, as well as limited sanitary engineering services and pest management. Additional information pertaining to the PVNTMED personnel and their specific functions is discussed in FMs 4-02.17, 8-10, 8-10-1, and 8-10-3. The PVNTMED cell is staffed with a PVNTMED officer. The PVNTMED officer (Major, AOC 60C00) is responsible for the implementation of the command PVNTMED program.

2-7. Corps Medical Assets in Support of Divisions

Corps medical units in GS and DS of the division are normally assigned to the corps medical command (MEDCOM) or medical brigade. The MEDCOM/brigade will provide subordinate units to support the division by establishing a command relationship of OPCON or attachment. The MEDCOM/brigade could also choose to maintain only a support relationship of DS or GS to support the division. The division surgeon and DSS (Force XXI) and the DMOC under Army of Excellence interface with corps medical units according to the MEDCOM/brigade tactical standing operating procedures (TSOP). The DSS or DMOC and other division staff elements must be prepared to integrate corps-level medical units/elements into the medical, as well as the logistical, support structure. The MEDCOM/brigade will normally deploy a liaison officer to the division to coordinate and synchronize corps CHS. Information concerning the organization, functions, and responsibilities of the corps MEDCOM/brigade is found in FM 8-10.

a. Corps Medical Command and Medical Brigade. The corps MEDCOM and medical brigade provide C2, including—

- Staff planning.
- Supervision of operations.
- Administration of the assigned and attached units.

b. Medical Logistics Battalion. The MEDLOG battalion is organic to the corps medical brigade. The MEDLOG battalion provides C2 for assigned MEDLOG companies and the blood support detachment. The MEDLOG battalion is responsible for receiving, storing, and distributing medical materiel; single and multivision optical fabrication and repair; medical maintenance; blood and blood product collection, manufacturing, and distribution; medical gas production and distribution; and building of medical assemblages/push packages. The MEDLOG battalion will employ standard state-of-the-art MEDLOG information management and communications systems, to include satellite links.

(1) *Medical logistics company.* The MEDLOG company provides Class VIII supplies, DS/GS medical maintenance, and optical support. The MEDLOG company will use line-item requisitioning to support customers and will have the capabilities of building and maintaining preconfigured push packages in support of forward deployed medical units.

(2) *Blood support detachment.* The MEDLOG battalion's blood support detachment serves as the Army's blood supply unit (BSU). Blood and blood products will be stored and distributed under rigid specifications and managed by standard automated systems. Air movement will be the mode of choice for transporting blood and blood products. Army blood support in the AO will be the responsibility of the supporting MEDLOG battalion. The MEDLOG battalion's blood support detachment will collect, manufacture, receive, store, and distribute blood and blood products on an area basis.

c. Medical Evacuation Battalion. The headquarters and headquarters detachment, MEDEVAC battalion serves as the central manager of ground and air evacuation assets in the corps. Its mission is to provide C2 of ground and air MEDEVAC units within its AO. Information pertaining to the organization, functions, and capabilities of this unit is discussed in FM 8-10-6; air ambulance operations in support of the division are discussed in FMs 4-02.21, 8-10-3, and 8-10-26. A ground ambulance and one air ambulance company are normally placed in support of a division. The air ambulance company provides aeromedical evacuation on a DS basis. This company may be attached for support (less OPCON) to the division aviation brigade. Air ambulances will operate from the DSA and BSA, providing 24-hour immediate response MEDEVAC capability.

d. Medical Detachment, Combat Stress Control. The CSC detachment provides DS to a division's maneuver brigades and GS/reinforcing support to the DSA, including corps units in those areas. The detachment must function with its elements widely dispersed, some working in and for the supported division and others working in the corps. For definitive guidance on the medical detachment, CSC, see FM 8-51.

e. Veterinary Elements. The division will have corps veterinary team/personnel positioned at Class I supply points. They may also be deployed to the division to monitor and evaluate zoonotic diseases and environmental data, to include food exposed to NBC agents. Veterinary personnel will provide animal care for military working dogs and may perform investigations and postmortem examination of reported animal deaths. For additional information see FM 8-10-18.

f. Preventive Medicine Elements. Corps PVNTMED team/personnel may be deployed into the division when requested to augment division PVNTMED sections. Corps PVNTMED personnel provide advice and consultation in the areas of environmental sanitation, epidemiology, and entomology, as well as sanitary engineering services and pest management. Additional information pertaining to corps PVNTMED units and their specific functions is discussed in FM 4-02.17.

g. Dental Services. In planning the concept of operations, unit-level dental support is dependent upon corps-level area dental support assets in numbers sufficient enough to support the manpower requirement criteria for operational dental care. Unit dental support relies on corps-level area dental support units for assistance in providing operational care. Modules of area dental support units also augment or reconstitute unit dental elements when required. Corps-level dental units provide dental modules to reinforce or reconstitute the division dental modules when necessary and to operate field dental clinics. See FM 4-02.19 for definitive information on dental services.

Section II. DIVISION SUPPORT COMMAND MEDICAL UNITS AND ELEMENTS

2-8. Division Support Command, L-Edition TOE (Army of Excellence/AirLand Battle)

The DISCOM is organized to provide the maximum amount of CSS within prescribed strength limitations while providing the most effective and responsive support to tactical units in a combat environment. In order to provide responsive support to the tactical commander, logistics, medical, and personnel services support must be effectively organized and positioned as far forward as necessary to support the tactical plan. Division-level CHS for the Army of Excellence division is coordinated and provided by the DISCOM medical elements listed below:

- Division medical operations center, DISCOM HHC, located in the DSA.

- Main support battalion.
- Main support medical company, MSB, located in the DSA.
- Forward support battalion.
- Forward support medical company, FSB, located in the BSA.

a. Division Medical Operations Center. The DMOC's mission is to plan, coordinate, and synchronize the division's CHS with technical medical advice from the division surgeon. The division surgeon and the DMOC chief have joint responsibilities for CHS operations in the division. Their staff positions in the division and DISCOM require a close working relationship and coordination of their CHS activities. The DMOC staff is responsible to the DISCOM commander for staff supervision of CHS within the DISCOM. The division surgeon and DMOC chief will develop operating procedures that will enhance the flow of information and facilitate the synchronization of CHS operations within the division. It is imperative that the division surgeon and the DMOC chief work as a team. Both share equal responsibility for planning and overseeing CHS operations. The DMOC is responsible for monitoring CHS activities within the division area and keeping the DISCOM commander informed of the status of CHS. The division surgeon is informed of the DISCOM's CHS status through reports prescribed by the TSOP. For definitive information on the DMOC, see FMs 8-10-3 and 63-2.

b. Main Support Battalion. Information pertaining to the structure and operations of the MSB is provided in FM 63-21. The DMOC will interface with elements of the MSB, as required and approved by the DISCOM commander. The DMOC may interface with elements of the MSB through the DISCOM support operations section. The interactions and coordination between the DMOC and the MSB are driven by CHS requirements of the division and changes with the tactical situation. The chief, DMOC, and the MSB commander must develop policies and procedures that clearly delineate responsibilities and coordination requirements for an effective working relationship. Tasking of the MSMC elements by the DISCOM will be through command channels.

c. Main Support Medical Company. The MSMC provides division- and unit-level CHS and medical staff advice and assistance on an area basis to units operating in the DSA. Combat health support operations are coordinated by the DISCOM DMOC medical operations branch through technical channels. The DISCOM will task elements of the MSMC through command channels to provide division-level CHS. The interface between the MSMC and the DMOC is essential for providing required division CHS. For definitive information of the MSMC, see FMs 8-10-1, 8-10-3, and 63-21.

d. Forward Support Battalion. The DMOC will interface with elements of the FSB as required and approved by the DISCOM commander. The DMOC may interface with elements of the FSB through the DISCOM support operations section. This interface between the DMOC and elements of the FSB is driven by CHS requirements in the forward areas. This information will assist the DMOC in planning, coordinating, and managing division medical elements and resources in support of the battle. Communications and coordination between elements of the DMOC and the FSB are essential for successful accomplishment of the DMOC's and FSB's CHS mission. For additional information, see FM 63-20.

e. Forward Support Medical Company. The FSMC provides CHS for the brigade as well as area medical support for the brigade rear. Combat health support operations are coordinated by the FSMC commander and the FSB HSSO. The DISCOM tasks elements of the FSMC through command channels to provide division-level CHS. The FSMC commander positions are documented 05A, AMEDD immaterial, meaning any qualified AMEDD officer can assume command. When the FSMC commander is not a physician, medical decisions and technical supervision of physicians is performed by the senior physician assigned to the FSMC. The FSMC commander keeps the FSB commander informed on the CHS aspect of FSB operations and the health of the command. He regularly attends FSB staff meetings to obtain information to facilitate the execution of medical operations. He provides staff estimates and assists the FSB and brigade staffs with development of the CHS plan. For additional information on the FSMC commander, see FMs 8-10-1 and 63-20.

2-9. Division Support Command, F-Edition TOE (Force XXI/Digitized Division)

The DISCOM is a multifunctional organization capable of providing, coordinating, and synchronizing logistical support to the division. The DISCOM's mission of sustaining the divisions combat power is more critical than ever. The DISCOM consists of an FSB, a division support battalion (DSB), a division aviation support battalion, and the HHC. The DISCOM provides CSS for the division. It provides arming through its Class V operations, fueling through Class III operations, repairing through its maintenance operations, transportation through the truck company and the supply and transportation sections in the FSB, and sustaining, through the provision of rations, individual equipment and medical support. The manning function is provided by the personnel sections throughout the division. The DISCOM organization shown in Figure 2-2 identifies DISCOM units in support of maneuver brigades and to the division.

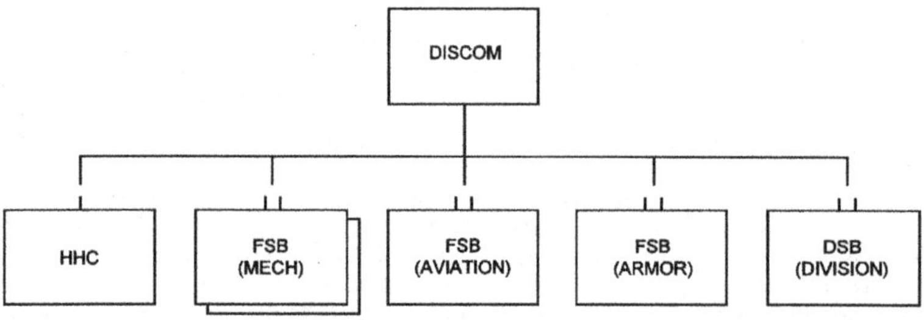

Figure 2-2. Division support command organization, Force XXI.

Division support command medical assets are organic to all of the subordinate battalions except for the division aviation support battalion (DASB). These medical assets include—

- Medical operations branch, DISCOM headquarters.
- Medical materiel management branch, DISCOM headquarters.
- Division support battalion, CHS cell.
- Division support medical company.
- Forward support battalion, CHS cell.
- Forward support medical company.

a. Medical Operations Branch (Division Support Command Headquarters). The medical operations branch is assigned to the distribution management center (DMC). The DMC has four branches—plans branch, operations branch, procurement branch, and medical operations branch. The DMC provides the division support operations the overall total asset visibility and the in-transit visibility of all commodities, movements, and units within, assigned, or inbound to the division AO. The primary responsibility of the medical operations branch is to assist with synchronization of the division CHS plan. The medical operations branch works with the DSS and the DISCOM staff in assisting with the development of the division CHS plan. This branch briefs the DISCOM commander and staff on CHS initiatives, as required. The medical operations branch and the DSS plan and coordinate for the employment of division medical assets and relocation of DISCOM CHS elements. The medical operations branch coordinates the CHS plan with synchronization of the division CHS plan. This branch collects medical information of intelligence value from reporting medical assets and forwards it to the appropriate division and DISCOM staff elements. The medical operations branch coordinates the placement of DS corps medical assets with supported DISCOM units, either in the DSA or DSA. The branch is responsible for—

- Briefing the DISCOM commander on the CHS planning and operations required.
- Providing current information that will assist the DSS with development of staff estimates and the division CHS plan.
- Coordinating the attachment of corps medical units/elements with DISCOM units.
- Providing information to the DSS on the DISCOM commander's intent for logistics and CHS operations.
- Coordinating the division CHS plan with all DISCOM staff elements.
- Coordinating operations information with the DSS and making recommendations to ensure synchronization of CHS activities in support of the division.

- Ensuring CHS information from the DSS is staffed to all DISCOM elements in a timely manner.

- Coordinating with the BSS, as required, for synchronization of division CHS.

b. Medical Materiel Management Branch. The MMMB is assigned to the general supply office (GSO). The GSO coordinates and supervises the supply management for water and Classes I, II, III(B), and III(P) supplies and recommends priorities for the allocation and other controls of supplies. The MMMB provides advice on the receipt, storage, and distribution of supplies within its area of responsibility. This office consists of a Class I branch, Classes II/III(P)/IV branch, Class III and water supply branch, Class V branch, Class VIII branch (MMMB), and Class IX branch. The MMMB manages the Class VIII supply system in the division. The branch coordinates and recommends the prioritization of medical supplies and blood products. It also coordinates for the disposition of captured enemy medical materiel. Under the technical control of the HSMO of the DSS, the MMMB monitors and coordinates Class VIII resupply for division medical units/elements. Using the CHL functional module of the TMIP/MC4 system, the Theater Army Medical Management Information System (TAMMIS), Joint Total Asset Visibility, Transportation Coordinators' Automated Information for Movement System II, and/or other automated logistics management systems, the MMMB manages all Class VIII requisitions submitted from the division to the supporting MEDLOG company. The MMMB maintains a record of the requisition until it is filled. The MMMB coordinates shortfalls in throughput distribution with the DSS and division support operation section. The MMMB may update priorities with the MEDLOG company to correct deficiencies in the delivery system. The MMMB provides Class VIII situational understanding to the DISCOM staff and the DSS according to the TSOP. For definitive information on division Class VIII resupply operations, see FMs 4-02.1, 4-02.21, 8-10-9, and 63-2-2. The MMMB, in coordination with the CHL cell of the DSS, manages the distribution of blood and blood products for division medical units. It also coordinates through the DSS with the G5 for disposition of captured enemy medical material.

c. Division Support Battalion Health Services Support Officer. The HSSO is assigned to the command section and is the medical plans and operations officer. The HSSO coordinates internal medical support. He coordinates the schedules, locations, and capabilities of medical support with the DSMC. He prepares and provides an area medical plan to the subordinate units. The HSSO is also responsible for coordinating the placement of supporting corps medical elements attached to the DSB within the battalion's assigned area of the DSA. For additional information on the DSB HSSO, see FM 63-23-2.

d. Division Support Medical Company. The DSMC has the overall mission of providing Echelons I and II CHS to units located in the DSA and division rear areas. It provides C2 for organic elements and attached medical units. The DSMC is dependent on appropriate elements of the corps and division for patient evacuation (including air ambulance), CHS operations planning, guidance, legal, finance, and personnel and administrative services. It is also dependent on the headquarters and headquarters detachment of the DSB for food service and religious support. The DSMC is organized into a company headquarters, a treatment platoon, an ambulance platoon, an optometry section, a PVNTMED section, and a MH section.

e. Forward Support Battalion Combat Health Support Cell. The CHS cell is assigned to the support operations section. It is staffed with a medical planner/HSSO who is the FSB commander's special staff officer for CHS and a member of the FSB battle staff and a medical operations NCO who is the primary assistant to the HSSO. This cell is responsible for—

- Providing the CHS input for logistics preparation of the battlefield for the FSB.

- Providing the CHS estimates and medical threat input for inclusion in the FSB commander's estimate.

- Coordinating and synchronizing FSB medical operations for the supported brigade.

- Coordinating the delivery of Class VIII supplies via logistics packages (LOGPAC).

- Overseeing all FSB CHS planning activities to ensure such planning is synchronized laterally and vertically.

- Developing the CHS portion of the FSB's OPLAN in coordination with the FSB staff, the FSMC commander, and the DISCOM medical operations branch.

- Coordinating the placement of supporting corps medical elements attached to the FSB within the BSA. Identifying CHS support requirements for the BSA including space requirements for the FST (tents, equipment, vehicles, and trailers), the forward support MEDEVAC team (FSMT) (includes fuel truck and forward area refueling equipment [FARE], aircraft parking, tents for billeting), and the supporting corps ground ambulance teams and ambulances.

- Coordinating communications access for supporting corps elements as required.

- Coordinating through the BSS with the brigade S3 (Air) for current Army airspace command and control (A2C2) information that is provided to the FSMT crews. Also provides a copy of the brigade OPORD/OPLAN A2C2 annex that provides the air corridors for medical evacuation.

- Coordinating CHS taskings from the DISCOM medical operations branch with the FSB staff and the FSMC commander. Tasking may include area medical/dental, PVNTMED, CSC, and CHS reinforcement, or reconstitution support.

- Coordinating for the training and use of nonmedical personnel for patient decontamination in the event of an NBC or weapons of mass destruction attack. (See FM 8-10-7.)

- Coordinating and synchronizing CHS requirements with the BSS and the DISCOM medical operations branch.

- Monitoring the status of the FSB and brigade medical elements via the medical situational reporting on Force XXI Battle Command Brigade and Below (FBCB2).

- Monitoring the status of division medical units/elements via medical reporting on CSSCS.

- Advising the FSB commander on CHS operations in the BSA and brigade rear.

- Maintaining situational understanding of lateral and supporting medical units.

- Submitting and forwarding status reports in accordance with DISCOM and brigade TSOP.

f. Forward Support Medical Company. The FSMC has the overall mission of providing Echelon I and Echelon II CHS on a DS basis for the supported maneuver brigade. It provides C2 for organic elements and attached medical units. The FSMC is dependent on appropriate elements of the corps, division, brigade, and FSB for patient evacuation (including air ambulance), CHS operations planning and guidance, and for legal, finance, and personnel and administrative services. It is also dependent on the headquarters and distribution company of the FSB for food service and religious support and the base support company for maintenance. The FSMC is organized into a company headquarters, a treatment platoon, an ambulance platoon, a PVNTMED section, and a MH section (see Figure 2-3). For more detailed information on the operations and functions of the medical company, see FM 8-10-1. The FSMC performs these functions:

- Treatment of patients with DNBI, battle fatigue (BF), and trauma injuries. It provides routine sick call, triage of mass casualties, ATM, surgical resuscitation/stabilization (when the FST from the corps is deployed/collocated with the FSMC), and preparation of patients incapable of returning to duty for further evacuation.

- Ground ambulance evacuation for patients from BAS and designated casualty collection points (CCP).

- Operational dental care (emergency and essential dental care).

- Class VIII resupply and medical equipment maintenance for supported units.

- Medical laboratory and radiology services commensurate with Echelon II/division-level treatment.

- Outpatient consultation services for patients referred from unit-level MTF.

- Patient holding for up to 40 patients able to RTD within 72 hours.

- Limited reinforcement and augmentation to supported maneuver battalion medical platoons.

- Coordination with the UMT for required religious support.

- Preventive medicine consultation and support.

- Combat stress control, to include management of BF and stress-related casualties.

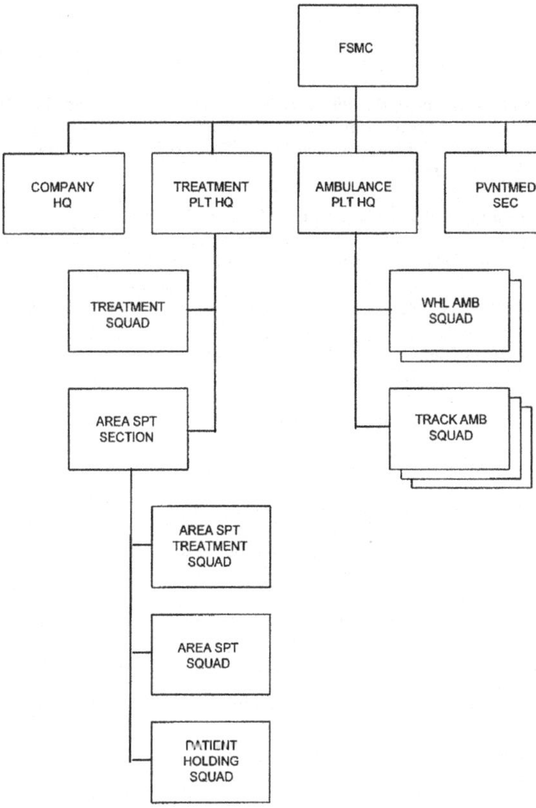

Figure 2-3. Forward support medical company.

Section III. BRIGADE HEADQUARTERS MEDICAL ASSETS

2-10. Brigade Surgeon, F-Edition TOE (Force XXI Digitized Division)

The brigade surgeon is an MC officer (Major, AOC 62B00). He is a special staff officer who plans and coordinates brigade CHS activities with the brigade staff. The brigade surgeon is assigned to the HHC of the maneuver brigade. The surgeon is responsible for the technical control of all medical activities in the command. The brigade surgeon oversees and coordinates CHS activities through the BSS and the brigade S3. The brigade surgeon keeps the brigade commander informed on the status of CHS for brigade operations and the health of the command. He provides input and obtains information to facilitate medical planning. His specific duties in this area include—

- Ensuring implementation of the CHS section of the brigade TSOP.
- Determining the allocation of medical resources within the brigade.
- Supervising technical training of medical personnel and the CLS program within the brigade.
- Determining procedures, techniques, and limitations in the conduct of routine medical care, emergency medical treatment (EMT), and ATM.
- Monitoring aeromedical and ground ambulance evacuation.
- Monitoring the implementation of automated medical systems.
- Informing the division surgeon on the brigade's CHS situation.
- Monitoring the health of the command and advising the commander on measures to counter disease and injury threats.
- Exercising technical supervision of subordinate battalion surgeons and PAs.
- Providing consultation and mentoring for subordinate battalion surgeons, physicians, and PAs.
- Providing the medical estimate and medical threat for inclusion in the commander's estimate.
- Monitoring the command PVNTMED program (includes health assessment and medical surveillance); see AR 40-5 and FM 4-02.17.
- Ensuring field health records are maintained by primary care providers according to AR 40-66.

2-11. Brigade Surgeon's Section, F-Edition TOE (Force XXI/Digitized Division)

The BSS is assigned to the HHC of the brigade and operates out of the brigade TOC. The section, in coordination with the HSSO of the FSB support operations section and the FSMC commander, is responsible for the development of the medical portion of the brigade OPLAN/OPORD and takes part in the brigade planning process. The BSS staff is responsible to the brigade commander for staff supervision of CHS within the brigade. The BSS is also responsible for coordinating GS and DS relationships of organic medical units and medical units/elements whether under OPCON or attached to the brigade. The brigade commander is updated as required on the status of CHS in the brigade. Figure 2-4 shows the typical organization and staffing of the BSS. It consists of a medical plans and operations cell and a patient disposition and reports cell. The staff of the BSS assists the brigade surgeon in planning and conducting brigade CHS operations.

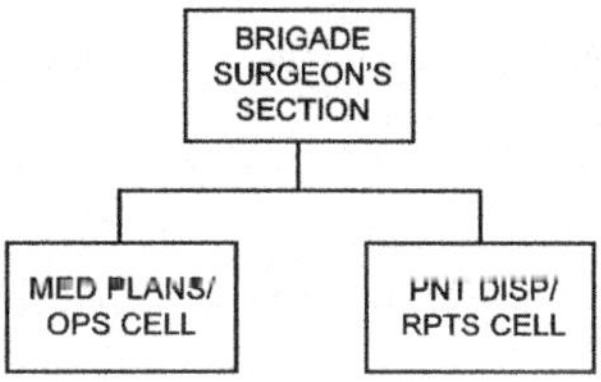

Figure 2-4. Brigade surgeon's section.

2-12. Brigade Surgeons, L-Edition TOE (Army of Excellence/AirLand Battle)

The FSMC commander positions are documented 05A, AMEDD immaterial, meaning any qualified AMEDD officer can assume command. When the FSMC commander is not a physician, medical decisions and technical supervision of physicians is performed by the senior physician/treatment platoon leader. When a brigade surgeon is not assigned to the brigade headquarters, the treatment platoon leader, who is always a physician, will perform the brigade surgeon's duties. His duties and responsibilities as the brigade surgeon require that he work closely with the FSMC commander and include, but are not limited to—

- Ensuring the implementation of the CHS section of the division TSOP.
- Determining the allocation of CHS resources within the brigade.
- Supervising the technical training of medical personnel and the CLS program within the brigade.
- Developing and monitoring the MEDEVAC plan (ground and air) which supports the brigade's maneuver plan.
- Writing the CHS portion of brigade TSOP, OPLAN, and OPORD.
- Monitoring requests for aeromedical evacuation from supported units.
- Monitoring the health of the command and advising the commander on measures to counter the medical threat.
- Monitoring and advising units on their mild to moderate BF cases and determining the capability to restore BF casualties within the brigade's AO.

- Informing the division surgeon and the DMOC of the brigade's CHS situation.

- Supervising corps medical elements within the brigade's AO when directed.

- Exercising technical control over subordinate battalion surgeons.

- Assuming technical supervision of PAs organic to subordinate units in the absence of their assigned physicians.

- Advising PAs assigned to artillery and engineer battalions, as required.

2-13. Armored Cavalry Regiment and Separate Brigade Surgeons (Army of Excellence)

The duties of the regimental surgeon and the separate brigade surgeon are the same as those identified in paragraph 2-10.

2-14. Armored Cavalry Regiment Medical Troop

The mission of the ACR medical troop is to provide Echelons I and II medical care within the ACR. The capabilities of this unit are to—

- Provide C2 of attached medical elements (including CHS planning; policies and procedures; support operations; and MEDEVAC coordination for movement of patients within and out of the regiment AO).

- Advise the regiment commander and support squadron commander on the health of the command and other CHS activities affecting the regiment.

- Develop, prepare, and coordinate the CHS portion of OPLAN and OPORD.

- Allocate medical resources (personnel and equipment) to all assigned and attached units of the regiment.

- Perform triage, initial resuscitation and stabilization, and preparation for further evacuation of patients generated in the regiment rear area.

- Provide ground evacuation for patients from Echelon I MTFs.

- Employ treatment squads to perform reinforcement/augmentation to maneuver squadrons' medical platoons. (These squads/teams are routinely placed OPCON to supported maneuver squadrons.

They are normally attached to the squadron medical platoon under technical control of the squadron surgeon.)

- Provide CHL and medical equipment maintenance repair parts and support to the regiment on an area support basis. (The regiment medical supply section maintains a 5-day stock of emergency *push* packages and individual medical items. Emergency supply requests are sent to the supporting MEDLOG battalion or the nearest medical unit.)

- Provide dental support (including treatment of maxillofacial injuries; operational dental care that includes emergency and essential dental treatment).

- Provide laboratory service commensurate with the regiment's Echelon II facility.

- Perform patient holding for up to 40 patients awaiting evacuation or RTD within 72 hours.

- Provide outpatient consultation services for patients referred from Echelon I MTFs.

Section IV. MEDICAL PLATOONS (ARMY OF EXCELLENCE/AIRLAND BATTLE)

2-15. Assignment

A medical platoon is organic to each combat battalion HHC. Under the Army of Excellence TOE, the platoon is organized with a headquarters section, a treatment squad (two treatment teams), an ambulance squad, and a combat medic section. The medical platoon is organized as shown in Figures 2-5 and 2-6.

NOTE

> Mechanized infantry and armor units have four ambulance squads consisting of two ambulance teams each assigned to their medical platoon ambulance squads. Airborne and air assault medical platoons have 12 trauma specialists assigned to their combat medic section, while light infantry has 9 trauma specialists assigned to their combat medic sections. The armor medical platoons have 3 trauma specialists assigned to their combat medic section and the mechanized infantry medical platoons have 3 health care SGTs (91W30) and 9 trauma specialists (91W10) assigned to their combat medic section.

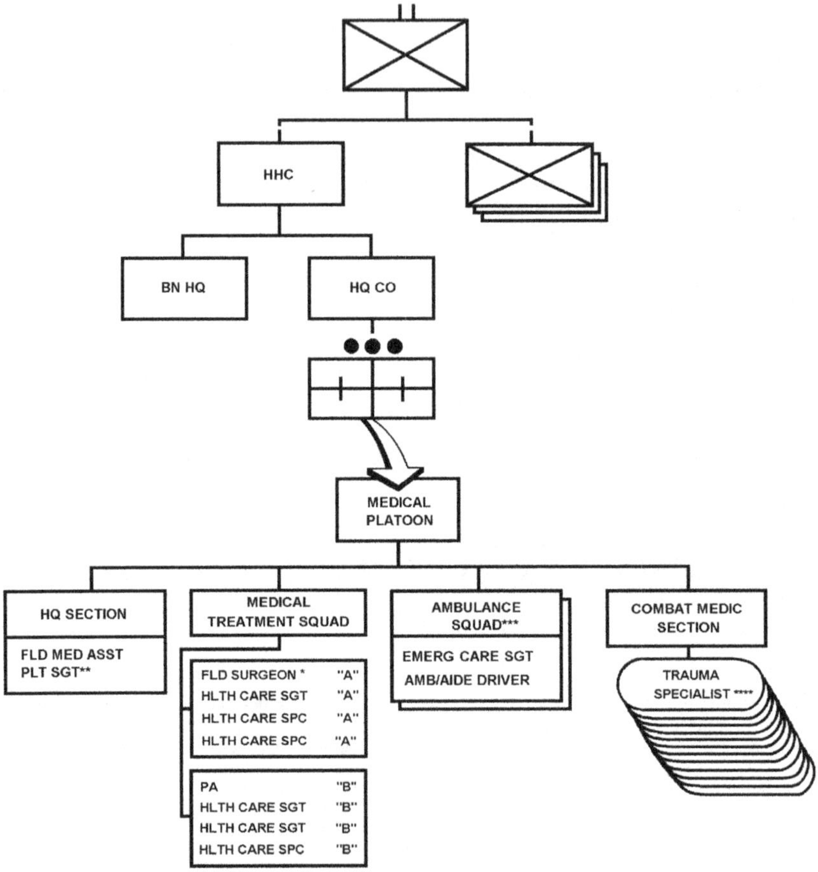

* ALSO BATTALION SURGEON
** ALSO SERVES AS AMBULANCE SECTION SGT
*** 4 AMBULANCE TEAMS
**** AIRBORNE BATTALIONS HAVE 12 TRAUMA SPECIALISTS ASSIGNED TO THE COMBAT MEDIC SECTION AND AIR ASSAULT BATTALIONS HAVE 10 TRAUMA SPECIALISTS ASSIGNED TO THE COMBAT MEDIC SECTION
"A" ALFA TREATMENT TEAM
"B" BRAVO TREATMENT TEAM

Figure 2-5. Medical platoon, headquarters and headquarters company light infantry battalion.

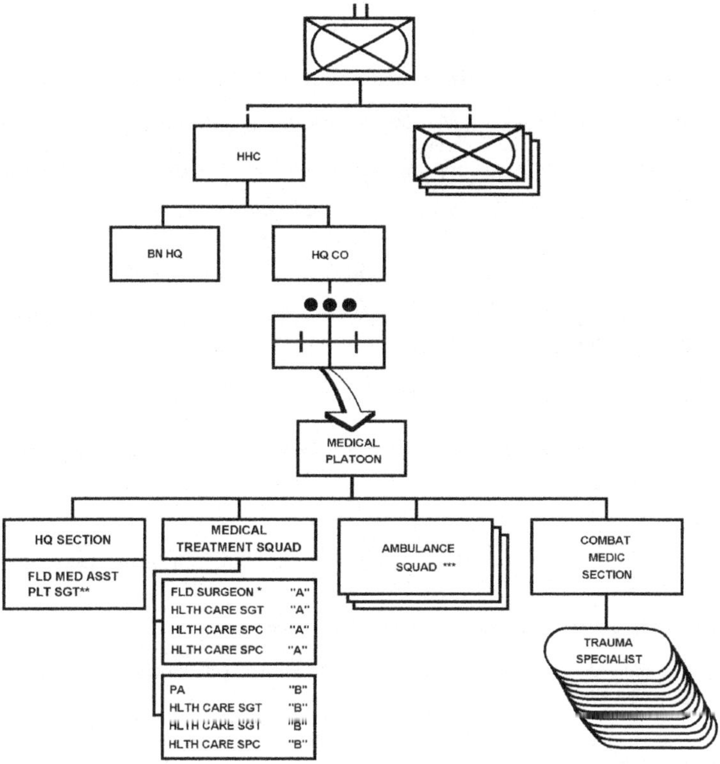

* ALSO BATTALION SURGEON
** ALSO SERVES AS AMBULANCE SECTION SGT
*** MECHANIZED INFANTRY BATTALIONS HAVE 3 AMBULANCE SQUADS AND ARMOR BATTALIONS HAVE 4 AMBULANCE SQUADS
"A" ALFA TREATMENT TEAM
"B" BRAVO TREATMENT TEAM

Figure 2-6. Medical platoon, mechanized infantry battalion.

2-16. Battalion Surgeon/Medical Platoon Leader, L-Edition TOE

The battalion surgeon/medical platoon leader (MC, AOC 62B00) is the medical advisor to the battalion commander and his staff. He is the supervising physician (operational medicine officer) of the medical platoon treatment squad. This officer is responsible for all medical treatment provided by the platoon. His responsibilities include—

- Planning and directing Echelon I CHS for the battalion.
- Advising the battalion commander and his staff on the status of the health of the command.
- Supervising the administration, discipline, maintenance of equipment, supply functions, organizational training, and employment of assigned or attached personnel.
- Examining, diagnosing, treating, and prescribing courses of treatment for patients, to include ATM.
- Coordinating the establishment and training of nonmedical personnel for patient decontamination teams.
- Training CLS.
- Supervising the battalion MH/CSC program, to include training troop leaders in the preventive aspect of stress on soldiers.
- Supporting humanitarian assistance programs when directed.
- Overseeing the common task training, continuing medical education, and clinical training of subordinate medical personnel.
- Monitoring the command PVNTMED program, to include health assessment and medical surveillance; see AR 40-5 and FM 4-02.17.
- Ensuring field health records are maintained by primary care providers according to AR 40-66.

2-17. Platoon Headquarters, L-Edition TOE

a. The headquarters section, under the direction of the battalion surgeon/medical platoon leader, provides for the C3 and resupply for the platoon. The platoon headquarters is manned by the field medical assistant and the platoon SGT. It is normally collocated with the treatment squad to form the BAS. The CP includes the plans and operations functions performed by the field medical assistant. The platoon has access to the battalion wire communication network for communications with all major elements of the battalion and with supporting units. Wireless communications for this section consists of a tactical FM radio mounted in the platoon headquarters vehicle. The medical platoon employs an FM radio network for CHS operations (Figure 2-7). The headquarters section serves as the net control station (NCS) for the platoon.

b. The field medical assistant, an MS Corps officer, is the operations/readiness officer for the platoon. He is the principal assistant to the battalion surgeon/medical platoon leader for operations, administration, and logistics. The field medical assistant coordinates CHS operations with the battalion S3 and S4 and coordinates patient evacuation with the FSMC.

c. The platoon SGT assists the platoon leader and supervises the operations of the platoon. He also serves as the ambulance section SGT. This NCO prepares reports; requests general supplies as well as medical supplies; advises on supply economy procedures; and maintains authorized stockage levels (ASL) of expendable supplies. He supervises the activities and functions of the ambulance section, to include operator maintenance of ambulances and equipment; OPSEC; and EMT.

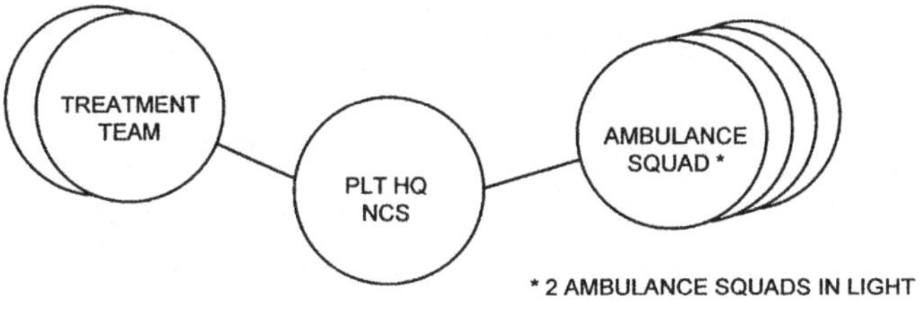

Figure 2-7. Medical platoon operations net.

d. The PA performs general technical health care and administrative duties. The PA is ATM-qualified and works under the clinical supervision of the medical officer. This officer serves as the medical platoon leader in the absence of an assigned physician. He performs the following duties:

- Establishes and operates a BAS or BAS minus (one treatment team).

- Treats, within his ability, sick or injured patients. He refers those patients requiring treatment beyond his capability to the supervising physician.

- Provides initial resuscitation to wounded personnel, as required.

- Conducts training for battalion personnel in first aid procedures (self-aid/buddy aid), CLS, field sanitation, evacuation of the sick, injured, and wounded, and the medical aspects of injury prevention. For additional information on the CLS Program, see Appendix C.

- Assists in the conduct of the battalion preventive psychiatry program, to include training troop leaders in the preventive aspects of stress on soldiers.

- Trains medical personnel in emergency medical procedures and, in the absence of a physician, ensures common task training, continuing medical education, and clinical training of subordinate medical personnel.

2-18. Treatment Squad, L-Edition TOE

The treatment squad is the basic medical treatment element of the BAS. It provides routine sick call services, emergency medical care, triage, and ATM. This squad is staffed with an operational medicine officer (primary care physician/battalion surgeon), a PA, two health care SGTs, and four health care specialists. The squad's physician and PA are trained in ATM procedures.

2-19. Battalion Aid Station/Treatment Squad Operations, L-Edition TOE

Battalion aid station is the generic term used in designating the unit-level/Echelon I MTF.

a. The treatment squad can split into two treatment teams and operate as two separate aid stations (BAS minus), normally not to exceed 24 hours. In continuous operations, when operating for longer periods, personnel efficiency and unit capability will tend to deteriorate. Each team employs treatment vehicle(s) with two medical equipment sets (MES)—one trauma field MES, and one sick call field MES.

b. For communications, each treatment team uses an FM tactical radio and is deployed in the medical platoon's operations net. However, under certain tactical conditions, the battalion S4 may require BAS elements to use the S4 net.

c. The BAS is under the tactical control of the battalion S4 and is normally deployed in the vicinity of combat trains (see Figures 2-8 and 2-9 for suggested layout of a BAS). To reduce ambulance turnaround time in providing ATM to patients within 30 minutes of wounding, the BAS may split and place its treatment teams as close to maneuvering companies as tactically feasible. The battalion S4 closely coordinates locations for forward positioning CSS elements (including medical treatment elements) with the battalion S3. This is to ensure that the location of these elements is known by commanders of maneuvering forces. Coordination ensures that CSS elements are not placed in the way of friendly maneuvering forces, in line of fires, or in areas subject to be overrun by rapidly advancing enemy forces. Treatment teams situated close to (within 1,000 meters of) maneuvering companies in contact must be prepared to withdraw to preplanned, alternate positions on short notice.

d. When maneuvering companies anticipate large numbers of casualties, augmentation of the medical platoon with one or more treatment teams from the FSMC should be made. Augmenting treatment teams are under the tactical control of the battalion S4; but are under the OPCON of the battalion surgeon/ medical platoon leader. A suggested scheme of employment is to place a team in close support of each maneuvering company while locating one treatment team in the combat trains. Medical treatment facilities should not be placed near targets of opportunity such as ammunition, petroleum, oils and lubricants (POL), distribution points, or other targets that may be considered lucrative by the opposing force. Considerations for the location of the BAS should include—

- Tactical situation/commander's plan.
- Expected areas of high casualty density.
- Security.
- Protection afforded by defilade.
- Convergence of lines of drift.
- Evacuation time and distance.
- Accessible evacuation routes.

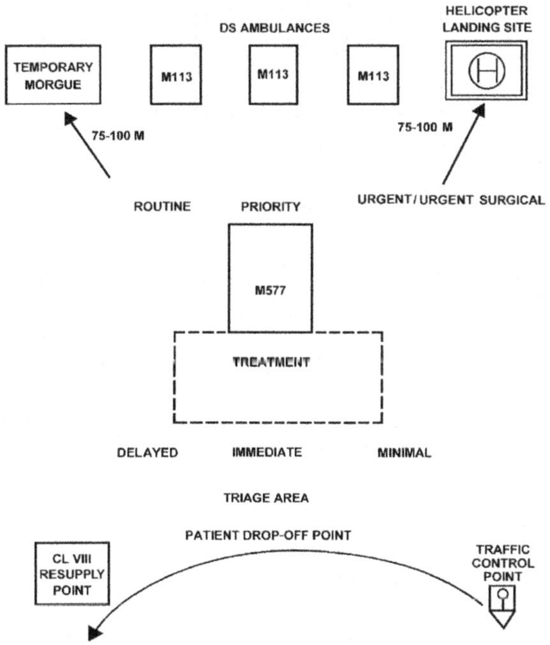

NOTE: A PATIENT DECONTAMINATION SITE MUST BE IDENTIFIED. SEE FM 8-10-7.

Figure 2-8. Layout of a battalion aid station (heavy).

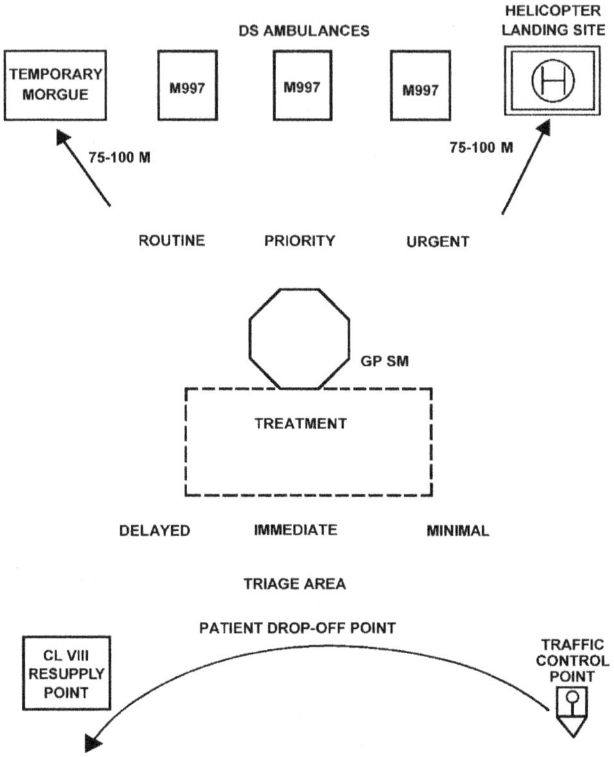

Figure 2-9. Layout of a battalion aid station (light).

- Avoidance of likely target areas such as bridges, fording locations, road junctions, and firing positions.
- Solid ground with good drainage.
- Near an open area suitable for helicopter landing.
- Available communication means.
- Additional space near this site for establishing a patient decontamination site if required.

e. At the BAS, patients requiring further evacuation to the rear are stabilized for movement. Constant efforts are made to prevent unnecessary evacuation; patients with minor wounds or illnesses are treated and RTD as soon as possible. Other functions of the BAS include—

- Receiving and recording patients.

- Notifying the S1 of all patients processed through the BAS, giving identification and disposition of patients.

- Preparing Field Medical Cards (FMC) (Department of Defense [DD] Form 1380), as required. See FM 8-10-6 for information on completion and disposition of the form.

- Verifying information contained on the FMC of each patient evacuated to the BAS.

- Requesting and monitoring MEDEVAC of patients.

- Monitoring personnel, when necessary, for NBC contamination prior to medical treatment.

- Supervising patient decontamination and treating NBC patients (refer to FMs 8-10-7, 4-02.283, 8-284, and 8-285).

NOTE

Patient decontamination is performed by a pretrained team. This team is composed of eight nonmedical personnel from supported units working under the supervision of medical personnel. Patient decontamination teams perform best when they train and exercise their skills with the supporting BAS.

f. Medical evacuation from the BAS is performed by ground ambulances from the FSMC and by corps air ambulance teams.

g. Patient holding and food service is not available at the BAS. Therefore, only procedures necessary to preserve life or limb, or enable a patient to be moved safely, are performed at the BAS.

h. Ammunition and individual weapons belonging to patients evacuated from the BAS are disposed of as directed by command standing operating procedures (SOP)/policy. All excess equipment collected at the BAS is disposed of by the battalion S4 or as directed by command SOP.

NOTE

Patients will always retain their protective mask when evacuated to the next echelon of care, as long as they are in the combat zone. Based on the threat, they may retain the protective mask until evacuated out of the theater.

i. Patients requiring dental treatment are provided relief for dental pain, if required, then evacuated to the supporting medical company where operational dental care (emergency and essential dental treatment) is provided.

j. Patients requiring optometry services initially report to the BAS. For those patients requiring only routine replacement of spectacles, necessary information is obtained from the individual and forwarded to the division optometry section. The required spectacles are fabricated and forwarded to the BAS for issue to the patient. For optometry services other than routine repair or replacement of spectacles, patients are transported to the optometry section, MSMC, located in the DSA.

2-20. Combat Medic Section, L-Edition TOE

To foster good interpersonal relations and morale of combat troops, every effort should be made to attach the same trauma specialists to the same unit they habitually support each time the unit deploys. However, during lulls in combat operations, they should return to the medical platoon for consultation and proficiency training. Functions of trauma specialists are as follows:

- Performs triage and EMT for the sick and wounded.
- Arranges MEDEVAC for litter patients and directs ambulatory patients to CCP or to the BAS.
- Initiates a FMC for the sick and wounded and, as time permits, prepares a FMC on deceased personnel.
- Screens, evaluates, and treats, within his capabilities, those patients suffering minor illnesses and injuries.
- Keeps the company commander and the battalion surgeon/medical platoon leader informed on matters pertaining to the health and welfare of the troops.
- Manages Class VIII resupply for the unit's CLS.
- Maintains sufficient quantities of medical supplies to support the tactical situation.
- Serves as a member of the unit field sanitation team. In this capacity, he advises the commander and supervises unit personnel on matters of personal hygiene and field sanitation (FM 21-10-1).

2-21. Ambulance Squads, L-Edition TOE

a. Medical platoon ambulance squads provide evacuation within the battalion. Ambulance teams provide medical evacuation and en route care from the soldier's point of injury or a CCP to the BAS. In mass casualty situations, nonmedical vehicles may be used to assist in casualty evacuation (CASEVAC) as directed by the commander. Plans for the use of nonmedical vehicles to perform CASEVAC should be included in the battalion's TSOP.

b. Under the modular medical system, the ambulance squad consists of two ambulance teams.

(1) The emergency care SGT—

- Performs triage and EMT procedures in the care and management of trauma patients.
- Assists in the care and management of BF patients.
- Prepares patient for movement.
- Provides patient care en route.
- Maintains contact with supported units.
- Collects casualties.
- Performs NBC detection procedures.

(2) The ambulance/aide driver is trained in EMT procedures. He operates and maintains the ambulance and all onboard equipment. He assists the aide/evacuation NCO in the care and handling of patients.

c. Specific duties of the ambulance team are to—

- Maintain contact with supported elements.
- Find and collect the wounded.
- Administer EMT as required.
- Initiate or complete the FMC.
- Evacuate litter patients to the BAS.
- Direct or guide ambulatory patients to the BAS.
- Perform triage when necessary.

- Provide Class VIII resupply to trauma specialists.
- Serve as messengers within medical channels.
- Maintain operational readiness of assigned vehicle.

d. The number of ambulance squads in a section varies and is based on the type of parent organization. The infantry, airborne, and air assault maneuver battalions' ambulance sections have two ambulance squads; each is equipped with HMMWV ambulances. The mechanized infantry and the armored battalions ambulance sections have four ambulance squads equipped with M-113 tracked ambulances.

2-22. Employment and Functions of the Ambulance Team, L-Edition TOE

a. The ambulance team is a mobile trauma specialist team. Its function is to collect, treat, and evacuate the sick and wounded to the nearest treatment station or ambulance exchange point (AXP). For communications, the ambulance team employs an FM tactical radio mounted on its assigned ambulance. The team uses the medical platoon's internal operations net; however, in certain circumstances it may operate in the battalion administration/logistics net or as established by the battalion signal operating instructions (SOI).

b. In the heavy maneuver battalions, the track ambulance team routinely deploys with the maneuver company trains; however, it operates as far forward as the tactical situation permits and evacuates patients from the point of injury, if possible. In the light maneuver battalion, the wheeled ambulance team is either dispatched from the BAS, pre-positioned as close to the supported units as the tactical situation permits, or positioned with the maneuver company trains. Ambulance teams operating in a company's AO are normally under the tactical control of the company XO or 1SG, but remain under the technical and OPCON of the medical platoon. An ambulance team from the BAS will habitually support the same company. To become familiar with the specific terrain and battlefield situation, the team maintains contact with the company during most combat operations.

c. During static situations where the company is not in enemy contact or is in reserve, the team returns to the BAS to serve as backup support for other elements in contact. However, during movement to contact, the ambulance team immediately deploys to its regularly supported company. During combat operations, the team may dismount (leaving the ambulance in the company trains area), find, treat, and move patients to safety, and later evacuate them to the BAS. When moving patients to the ambulance location, CCP, or company aid post, the team is normally assisted by nonmedical personnel.

2-23. Medical Evacuation, L-Edition TOE

a. Optimum patient care and treatment is dependent upon an evacuation system that provides a continuous movement of patients. Medical evacuation is the process of moving patients from the point of injury or illness to an MTF, while providing en route medical care, or between MTFs. Each stop in the process is to provide medical treatment to enhance the patient's early RTD or to stabilize him for further evacuation. The responsibility for patient evacuation rests with the echelon of CHS to which the patient is to be evacuated (see Figure 2-10). Ambulances go forward, pick up patients, and move them to the supporting MTFs.

(1) Ambulance teams of the medical platoon evacuate patients from the company aid post or CCP to the BAS.

(2) Ambulance squads of the FSMC evacuate patients from the BAS to the division clearing station.

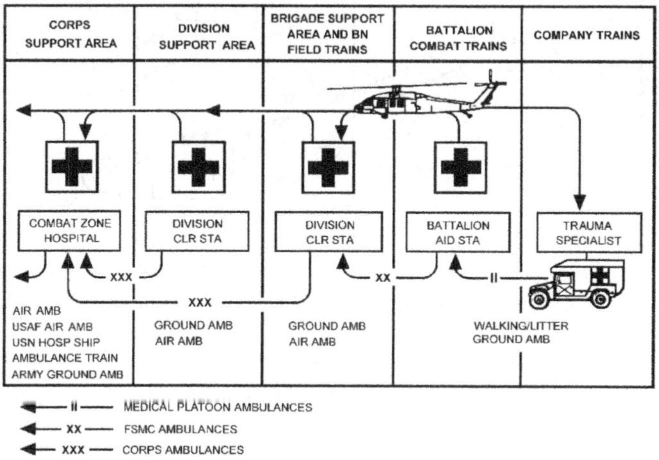

Figure 2-10. Patient evacuation flow.

b. An ambulance shuttle system may be set up between the FSMC division clearing station and the BAS. An AXP is established (Figure 2-11) so that ambulances are moving forward as others move rearward; thus enabling a continuous rearward evacuation flow, while decreasing ambulance turnaround time. Patients are evacuated no further to the rear than their conditions require.

c. Aeromedical evacuation in the combat zone should be used to the maximum extent possible for critically ill or wounded patients. See FMs 8-10-6 and 8-10-26 for additional information on aeromedical evacuation. Refer to FM 8-10-6 for MEDEVAC request procedures. Normally, ground ambulances are used to evacuate the minimally ill or wounded and those patients who cannot be evacuated by air. The specific mode of evacuation is determined by the patient's condition, aircraft/vehicle availability, the tactical situation, and weather conditions (METT-TC factors). When both air and ground ambulances are used, specific factors are considered in determining which patients are to be evacuated by air and which are to be evacuated by ground ambulances (see FM 8-10-6). Normally, the physician or PA treating the patient (or the senior trauma specialist in their absence) makes this determination; it is based on the medical condition of the patient. However, the goal is to get the trauma patient to the initial treatment/ATM element within 30 minutes of wounding.

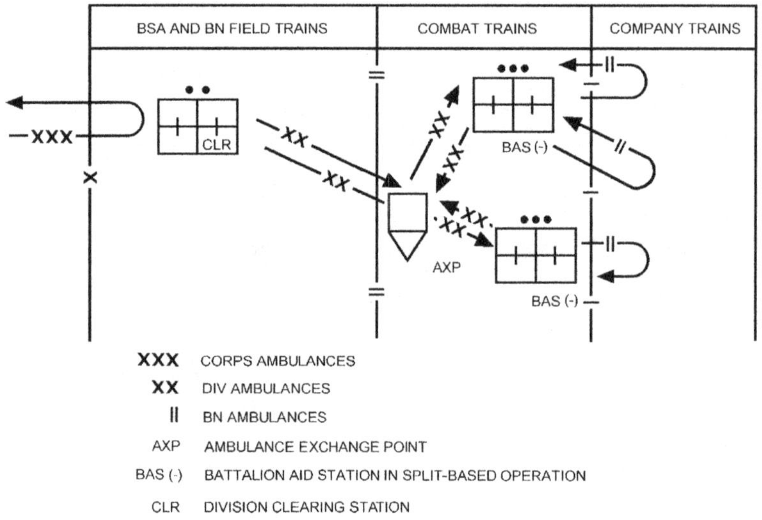

Figure 2-11. Ground ambulance shuttle system.

2-24. Evacuation and Disposition of Remains

a. The transportation and disposition of remains is a Quartermaster function. Air and ground ambulance personnel do not clear the battlefield of remains nor do they carry remains in their dedicated medical vehicles or aircraft. Medical units do not accept remains or provide temporary morgues in which to hold remains for other units. Other units are responsible for evacuation of remains to mortuary affairs collection points.

b. The only remains that medical units/elements handle are those of its own unit members or of patients who are dead on arrival (DOA) or who died of wounds (DOW) while in their care. Whenever a medical unit/element establishes a temporary morgue, it should be out of sight of the triage and treatment areas. The temporary morgue/holding area can be established behind a natural barrier, such as a stand of trees or it can be shielded from the view of others by using either tents or tarpaulins.

2-25. Class VIII Resupply, L-Edition TOE

a. The medical platoon maintains a 2-day (48-hour) stockage of Class VIII supplies within its MES. The following MES are authorized for the medical platoon treatment section and they include—

- Chemical Agent Patient Decontamination, National Stock Number (NSN) 6545-01-176-4612 (1).
- Chemical Agent Patient Treatment, NSN 6545-01-141-9469 (2).
- Sick Call Field, NSN 6545-01-228-1886 *(2).
- Trauma Field, NSN 6545-01-228-1667 *(2).

* Indicates the numbers of MES authorized for each treatment team.

Normal medical resupply of the platoon is performed by the DMSO through LOGPACs, backhaul, or in coordination with the movement control office(r) (MCO). Medical resupply may also be by preconfigured Class VIII packages (*push* packages) throughput from the MEDLOG battalion located in the corps support area (Figure 2-12).

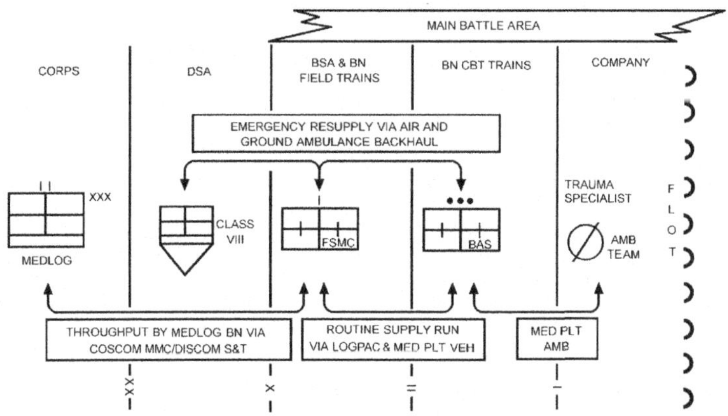

Figure 2-12. Flow of Class VIII supplies.

b. In a tactical environment, the emergency medical resupply (ambulance backhaul) system is used. In this environment, medical supplies are obtained informally and as rapidly as possible, using any available medical transportation assets. The medical platoon submits supply requests to the supporting FSMC, who in turn fills requests and ships supplies forward. Request for items not available at the FSMC are forwarded to the DMSO; the request is filled from division stocks and shipped to the requestor by the most expedient means available. Air ambulances from corps and ground ambulances from the DISCOM transport medical supplies directly to BAS. Class VIII resupply of trauma specialists are performed by ambulances of the medical platoon. The trauma specialist can also be resupplied from the ambulance crew from supplies onboard the ambulance.

Section V. MEDICAL PLATOONS, FORCE XXI/DIGITIZED DIVISION

2-26. Medical Platoon Assignment, F-Edition TOE

The medical platoon is organic to all maneuver battalions. In the armored battalions and mechanized infantry battalions, the platoon is organized with a headquarters section, a treatment section, ambulance squads, and a combat medic section.

NOTE

1. One 91W10 per armor company and one 91W30 and three 91W10s per mechanized infantry company.
2. One ambulance team per maneuver company supported.

The medical platoon receives Echelon II CHS from the supporting FSMC.

2-27. Platoon Headquarters, F-Edition TOE

The headquarters section, under the direction of the platoon leader, provides the C3 and logistics for the platoon. The platoon headquarters is manned by the field medical assistant and the platoon SGT. It is normally collocated with a treatment team/squad to form the BAS. The CP includes the plans and operations functions performed by the field medical assistant. The platoon has access to the HHC and the maneuver battalion wire communication network for communications with all major elements of supported and supporting units. Wireless communications for this section consists of a tactical FM radio mounted in the platoon headquarters vehicle. The medical platoon employs an FM radio network for CHS operations, to include telemedicine and teleconsultation procedures. The headquarters section serves as the NCS for the platoon. Each of the medical platoon vehicles have CSS functions for the FBCB2 system. The FBCB2 is a hardware/software suite that digitizes C2 at brigade level and below. The FBCB2 concept provides a seamless battle command capability for performance of missions throughout the operational continuum at the tactical level. The FBCB2 is the implementation of information age technology to provide increased battlefield operational capabilities. The system, positioned on the ambulance and treatment vehicles, will perform combat, CS, and CSS functions for the planning and execution of operations. The FBCB2 represents a major paradigm shift for the CSS and CHS communities. For the first time, the CSS organizations are digitally linked to the platforms and organizations that they support and the CHS elements are digitally linked to brigade and FSB medical elements. The FBCB2 provides a common operations picture enabling CHS and CSS providers to maintain the operational tempo set by maneuver commanders and to have near-time situational understanding of what is taking place on the battlefield.

2-28. Battalion Surgeon/Medical Platoon Leader, F-Edition TOE

The battalion surgeon/medical platoon leader (MC, AOC 62B) is a working physician on Treatment Team "Alpha." He is the medical advisor to the supported battalion commander and his staff. He is also the supervising physician (field surgeon) of the medical platoon's treatment teams. This officer is responsible for all medical treatment provided by the platoon. His responsibilities include—

- Planning and directing CHS for the supported maneuver battalion.
- Advising the supported maneuver battalion commander and his staff on CHS operations and the medical threat.
- Supervising the administration, discipline, maintenance of equipment, supply functions, organizational training, and employment of assigned or attached personnel.
- Examining, diagnosing, and treating (or prescribing courses of treatment) for patients, to include telementoring (TMEN) and ATM.
- Training CLS.
- Supervising the battalion CSC program, to include individual and leader training on the prevention of BF and other stress-related conditions.
- Planning and conducting humanitarian assistance programs when directed.
- Coordinating the medical evacuation of patients, as required.

The field medical assistant, an MS officer, is the operations/readiness officer for the platoon. He is the principal assistant to the platoon leader for operations, administration, and logistics. The field medical assistant coordinates CHS operations with the forward support company (FSC) support operations, the supported TF S1 and S4, and MEDEVAC with the FSMC. The platoon SGT assists in supervising the operations of the platoon. He also serves as the ambulance section SGT. Physician assistants are assigned to the Bravo and Charlie treatment teams. The PA (AOC 65D) performs general technical health care and administrative duties. He is ATM-qualified and works under the clinical supervision of the medical officer. The PA performs the following duties:

- Establishes and conducts treatment team operations when deployed to other locations away from the BAS.

- Treats, within his ability, sick or injured patients. He refers those patients requiring treatment beyond his capability to the supervising physician.

- Provides ATM for wounded and injured patients.

- Provides medical treatment for DNBI patients.

- Conducts training for battalion personnel in first-aid procedures (self-aid, buddy aid, and CLS), field sanitation, evacuation of the sick and wounded, and the medical aspects of injury prevention.

- Assists in the conduct of the battalion CSC program, to include individual and leader training on the prevention of BF and other stress-related conditions.

- Trains medical personnel in EMT procedures.

2-29. Treatment Section, F-Edition TOE

The three treatment teams (Teams Alpha, Bravo, and Charlie) are the basic medical treatment elements of the BAS. They provide Echelon I medical care and treatment. This includes sick call, EMT, ATM, and triage for the management of mass casualty situations. Each treatment team is staffed with a primary care physician or a PA, one health care SGT (E-5 or E-6) and two health care specialists (E-4 or E-3). The physician, PA, and health care personnel are all trained in ATM procedures, commensurate with their positions and skill levels.

2-30. Combat Medic Section, F-Edition TOE

Trauma specialists are allocated to mechanized infantry companies on the basis of one trauma specialist per platoon and a senior health care SGT for each company. In armored units, the allocation is one health care SGT and, normally, one ambulance team per company.

a. Trauma Specialist Location. The mechanized infantry platoon trauma specialist normally locates with, or near, the element leader. When the platoon is moving on foot in the platoon column formation, he positions himself near the element leader trailing the base squad forward of the second team. This formation is the platoon's primary movement formation. When the platoon is mounted, the trauma specialist will normally ride in the same vehicle as the platoon SGT.

b. Health Care Sergeant. The company health care SGT or specialist with the armor company normally collocates with the 1SG. When the company is engaged, he remains with the 1SG and provides medical advice, as necessary. As the tactical situation allows, he will provide medical treatment and prepare patients for MEDEVAC. The ambulance team supporting the company works in coordination with the trauma specialists supporting the platoons. When a casualty occurs in a tank or an armored fighting vehicle, the ambulance team will move as close to the vehicle as possible, making full use of cover, concealment, and defilade. Assisted, if possible, by the vehicle's crew, they will extract the casualty from the vehicle and administer EMT. They move the patient to the treatment team or to a CCP to await further MEDEVAC. The company health care SGT normally remains with the company CP, but may be used anywhere in the company, assisting the ambulance teams in some situations.

2-31. Ambulance Squads, F-Edition TOE

There are three ambulance squads assigned to the medical platoon. Each squad has two ambulance teams. Armored ambulance teams have three emergency care personnel while wheeled ambulances have two emergency care personnel. The platoon ambulances provide medical evacuation within the supported maneuver battalion/TF. Ambulance teams provide medical evacuation and en route care from the soldier's point of injury to the BAS. In mass casualty situations, nonmedical vehicles may be used to assist in CASEVAC as directed by the supported commander. Plans for the use of nonmedical vehicles to perform CASEVAC should be included in the maneuver battalion's TSOP and OPORD.

2-32. Property Exchange

a. United States Army Medical Evacuation Operations. Whenever a patient is evacuated from one MTF to another or is transferred from one ambulance to another, medical items of equipment (casualty evacuation bags [cold weather-type bags], blankets, litters, and splints) remain with the patient. To prevent rapid and unnecessary depletion of supplies and equipment, the receiving Army element exchanges like property with the transferring element. This reciprocal procedure will be practiced to the fullest extent possible through all phases of evacuation from the most forward element through the most rearward hospital.

This subparagraph implements STANAG 2128 and QSTAGs 435 and 436.

b. Medical Property of Allied Nations (NATO and ABCA Armies). Medical property accompanying patients of allied nations will be returned at once, if possible. If it is not possible, like items will be exchanged as in paragraph *a* above.

c. Medical Property of Coalition Forces or Allied Nations Without Ratified Standardization Agreements. Absence of a formal agreement, such as an Acquisition and Cross-Servicing Agreement, medical property accompanying patients of coalition and allied forces without ratified STANAGs will be returned to the parent nation as soon as practicable. Commanders should consult with their Staff Judge Advocate early in the planning process to ensure appropriate policy and procedures are developed and disseminated.

Section VI. TREATMENT TEAMS, MEDICAL SECTIONS, AND SPECIAL PURPOSE MEDICAL PLATOONS (ARMY OF EXCELLENCE/ AIRLAND BATTLE)

2-33. Combat Support Unit and Division Headquarters Treatment Team, L-Edition TOE

Treatment teams are organic to CS units and the division headquarters. With the exception of the combat engineer battalion, a medical support element in the light division normally consists of one treatment team. This treatment team is designed to provide Echelon I CHS for personnel of supported units. A treatment team normally with two ambulance teams is relatively small in comparison to a medical platoon; therefore, it will require augmentation from a supporting medical company in mass casualty situations.

2-34. Medical Section, Headquarters and Headquarters Battery, Division Artillery, L-Edition TOE

a. Organizations and Functions. The DIVARTY medical element includes a treatment team and an ambulance team. It is organized as shown in Figure 2-13. Personnel staffing of this section includes a PA, health care SGT, two health care specialists and an ambulance team.

(1) *Division artillery physician assistant.* The PA is the medical advisor to the DIVARTY commander and his staff. The PA works under the supervision of a physician, normally the division surgeon or treatment platoon leader of the MSMC. He coordinates with the division and brigade surgeons to ensure that all PAs/medical element leaders in the subordinate FA battalions, are working under the clinical supervision of a physician. Certain situations may require that the clinical supervision of PAs in FA units be passed to the physician in charge of the nearest supporting MTF. Such requirements, however, are coordinated through the division surgeon. The PA is responsible to the supervising physician for the medical treatment provided by DIVARTY medical personnel (inclusive of medical personnel assigned to FA battalions). His duties include—

- Operating the DIVARTY aid station.
- Planning and directing unit Echelon I CHS for members of the DIVARTY headquarters and FA battalions.
- Arranging for Echelon II CHS.
- Arranging for patient evacuation to the supporting medical company.
- Supervising the administration and maintenance of equipment, the supply function, technical training, and the employment of medical personnel.
- Examining, diagnosing, and treating (or prescribing courses of treatment for) patients, to include ATM for the trauma patient under the clinical supervision of a physician.
- Coordinating patient evacuation.

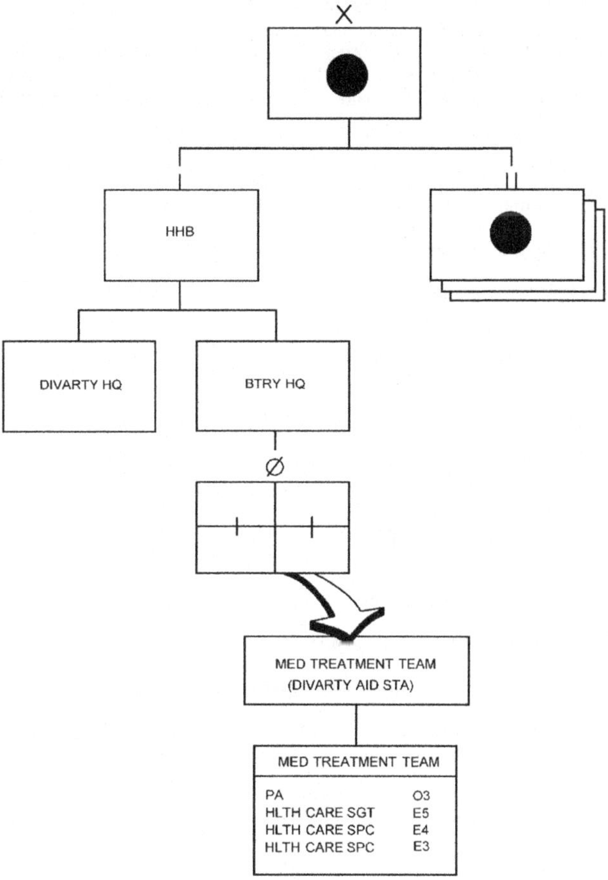

NOTE: THE LIGHT INFANTRY AND AIRBORNE DIVISION ARTILLERY DO NOT HAVE AN AMBULANCE TEAM.

Figure 2-13. Medical section, headquarters and headquarters battery, division artillery.

(2) *Health care sergeant*. The health care SGT assists the PA in accomplishing his duties; he supervises personnel on the treatment and ambulance teams. He prepares reports, requests general and medical supplies, maintains supply economy procedures, and maintains the ASL of expendable supplies. This NCO also performs triage and assists with ATM procedures in the care of trauma and NBC-insulted patients, and care and management of BF patients. He also performs routine patient care and NBC detection procedures. His duties also include—

- Establishing and assisting with operating the DIVARTY aid station.
- Maintaining the patient accountability/casualty reporting system.
- Maintaining MES.
- Conducting tactical and technical proficiency training for subordinate members of the section.
- Conducting sanitation inspections of troop living areas, food service areas, waste disposal areas, and potable water distribution points and equipment.
- Maintaining field health records of all patients seen according to AR 40-66.

(3) *Health care specialists*. These specialists assist the health care SGT in accomplishing his duties. They perform triage and EMT. Their specific duties include—

- Erecting and breaking down field medical shelter systems, to include chemical/ biological protective shelters.
- Performing patient care.
- Initiating patient records (FMC).
- Maintaining the patient daily disposition log.
- Operating and maintaining assigned vehicle, tactical radio, and power generation equipment. (Also may serve as a member on the battery field sanitation team.)

b. *Employment*. The medical section establishes an aid station near the DIVARTY headquarters and provides Echelon I CHS for members of the DIVARTY headquarters and headquarters battery (HHB).

(1) The section employs a treatment HMMWV, a cargo trailer, and two MESs (one trauma treatment set and one general sick call set).

(2) For communications, the section employs a telephone set (TA 312/PT) and is deployed in the HHB wire net. It employs an FM tactical radio and is deployed as designated by the DIVARTY SOI. This section also has access to the supporting medical company's tactical operations net to request Echelon II CHS.

c. Operations. The preceding paragraphs describe BAS operations; these are equally applicable to the DIVARTY BAS. Figures 2-8 and 2-9 show suggested layouts of a BAS.

d. Medical Evacuation. The DIVARTY HHB, depending on the type, may have one ambulance team to provide medical evacuation support. Those units assigned to the DIVARTY without an ambulance team are dependent on the supporting medical company. Evacuation of patients to and from the DIVARTY aid station is provided by the MSMC in the DSA.

e. Class VIII Supplies. The medical section maintains a 2-day (48-hour) stock level of Class VIII supplies for the HHB. Routine requests for medical supplies are submitted through command channels to the DMSO that is assigned to the MSMC. Class VIII supplies may be picked up by the requesting unit or forwarded to the DIVARTY BAS during routine ambulance runs. For emergency resupply procedures, see paragraph 2-25.

f. Property Exchange. See paragraph 2-32.

2-35. Treatment Team, Headquarters and Headquarters Battery, Field Artillery Battalion, L-Edition TOE

Organic to the HHB of the FA battalion is a treatment team, an ambulance team, and a combat medic section. The treatment team operates the BAS and the ambulance team provides limited ground ambulance medical evacuation support for the battalion. Trauma specialist from the combat medic section, are deployed with each firing platoon and with the service battery. Medical elements of the HHB, FA battalion are organized as shown in Figure 2-14. Personnel staffing for the treatment team include a PA, a health care SGT, two health care specialists, MOS 91W20, and one health care specialist, MOS 91W10.

a. Physician Assistant. In the absence of a physician, the PA is the principal advisor to the battalion commander and his staff in the areas of health and medical readiness. Working under the clinical supervision of a physician, he is the primary medical care provider for the battalion and supervises all activities of battalion medical personnel. The PA is trained in ATM procedures and as stated, works under the clinical supervision of a medical officer. He is responsible to the supervising physician for all treatment provided by medical personnel of the section. His specific duties include—

- Establishing and operating the BAS.
- Planning and supervising Echelon I CHS and coordinating with the supporting medical company for Echelon II CHS for the battalion.
- Treating, within his ability, patients reporting to him.
- Referring patients who require treatment beyond his capability to the supervising physician.
- Providing initial resuscitation (ATM) for the wounded.
- Training medical personnel and CLSs in emergency medical procedures.

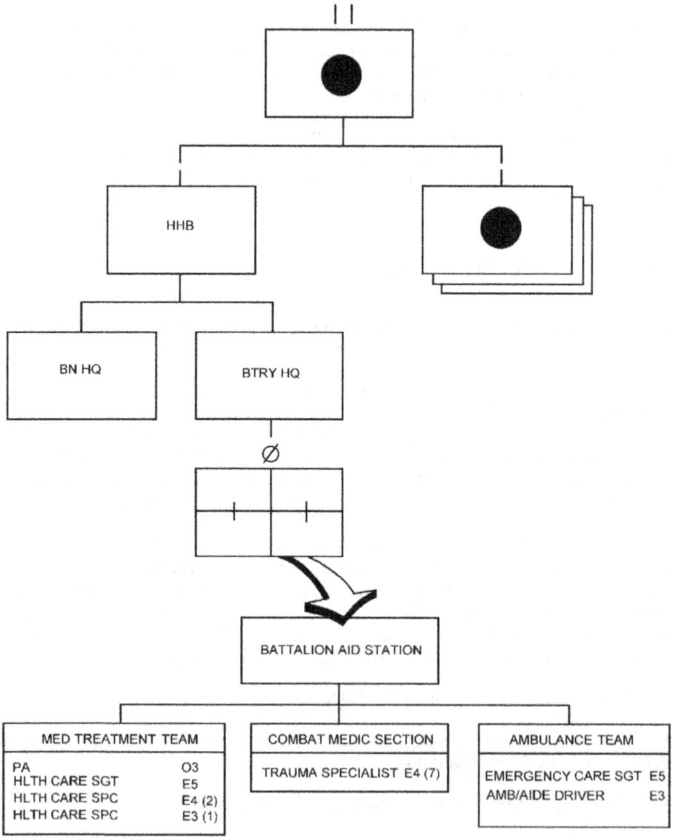

Figure 2-14. Medical elements, headquarters and headquarters battery, field artillery battalion.

b. Health Care Sergeant. This NCO assists the PA in accomplishing his duties. The specific duties of this NCO are the same as those described for the health care SGT in the DIVARTY HHB (refer to paragraph 2-34*a*[2]).

c. Health Care Specialists. The duties and functions of these specialists are the same as those discussed in paragraph 2-34*a*(3).

d. Trauma Specialists. Trauma specialists are allocated to a DS FA battalion on the basis of one to each firing platoon and the service battery. The duties and functions of trauma specialists are described in paragraph 2-20.

e. Employment. The treatment team establishes a BAS near the battalion headquarters and provides Echelon I CHS.

(1) The section employs a treatment HMMWV, a cargo trailer, and two MESs (one trauma treatment set and one general sick call set).

(2) For communications, the section employs a telephone set (TA 312/PT) and is deployed in the HHB wire communications net. It also employs an FM tactical radio and is deployed in the net designated by the battalion SOI. This section also has access to the supporting medical company's tactical operations net to request Echelon II CHS.

f. Operations. Paragraphs 2-19 describes a BAS operation; these are equally applicable to the FA BAS. Figures 2-8 and 2-9 show suggested layouts of a BAS.

g. Medical Evacuation. The HHB ambulance team evacuates patients to the BAS and to the supporting medical company if Echelon II ground ambulance support is not available.

h. Property Exchange. See paragraph 2-32

2-36. Medical Treatment Team, Headquarters and Headquarters Company, Division Aviation Brigade/Combat Aviation Squadron, L-Edition TOE

a. Organization and Functions. The division aviation brigade/combat aviation squadron medical treatment team is organized as shown in Figure 2-15. Personnel staffing for this section include a flight surgeon, a health care SGT, and two health care specialists. The flight surgeon (brigade surgeon) is the medical advisor to the aviation brigade commander and his staff. He is the primary care physician of the brigade. The flight surgeon is responsible for treatment provided by the medical treatment team (brigade aid station). His duties include—

- Operating the brigade aid station.

- Examining and determining the medical qualification for flying status of aviators within the brigade headquarters, or aviators referred to him by units without a flight surgeon.

- Planning and directing Echelon I CHS for members of the brigade headquarters.
- Coordinating for evacuation of patients to the division clearing station.
- Coordinating for division CHS augmentation, as required..
- Supervising the administration and maintenance of equipment, the supply function, technical training, and the employment of medical personnel.
- Examining, diagnosing, treating, and prescribing courses of treatment for patients, to include ATM for trauma patients.

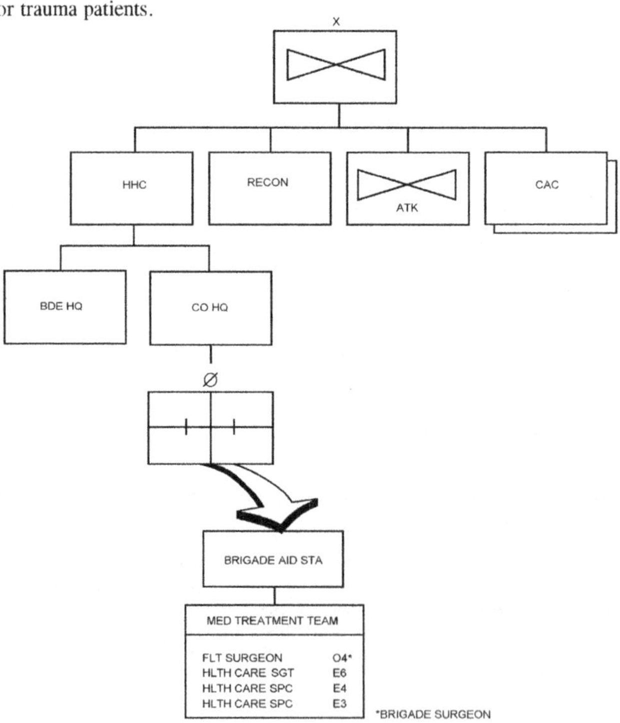

Figure 2-15. Medical treatment team, headquarters and headquarters company, division aviation brigade.

b. Operations. Paragraph 2-19 describes aid station operations; these are equally applicable to the DIVARTY BAS. Figures 2-8 and 2-9 show suggested layouts of a BAS.

c. Medical Evacuation. The brigade HHC medical section has no MEDEVAC assets. Evacuation of patients is provided by the supporting medical company.

d. Class VIII Resupply. See paragraph 2-25.

e. Property Exchange. See paragraph 2-32.

2-37. Medical Section, Headquarters and Headquarters Company, Attack Helicopter Battalion, Division Aviation Brigade, L-Edition TOE

a. Organization and Functions. The attack helicopter battalion medical section is organized as shown in Figure 2-16. Personnel staffing this section include a flight surgeon, a health care SGT, and two health care specialists. For further explanation, see paragraph 2-36*a*.

b. Property Exchange. See paragraph 2-32.

2-38. Medical Platoon, Headquarters and Headquarters Troop, Reconnaissance Squadron, Division Aviation Brigade, L-Edition TOE

a. Organization and Functions. The headquarters and headquarters troop (HHT) RECON squadron, division aviation brigade medical section is organized as shown in Figure 2-17. The medical section has a medical treatment squad, an ambulance squad, and a combat medic element.

(1) The patient treatment squad includes a flight surgeon (AOC 61N00), a PA, a section SGT/health care SGT (MOS 91W30), two health care SGTs (MOS 91W20), and three health care specialists (MOS 91W10).

(*a*) For flight surgeon responsibilities, see paragraph 2-36.

(*b*) The PA assists the flight surgeon in performance of his duties. He serves as the aviation brigade flight surgeon in the absence of the flight surgeon. His duties include—

- Examining and determining the medical qualification for flying status of aviators within the brigade headquarters; or aviators referred to his treatment section by units without a flight surgeon.

- Examining, diagnosing, treating, and prescribing courses of treatment for patients, to include ATM for trauma patients.

- Performing general technical health care and administrative duties (refer to paragraph 2-17*d*).

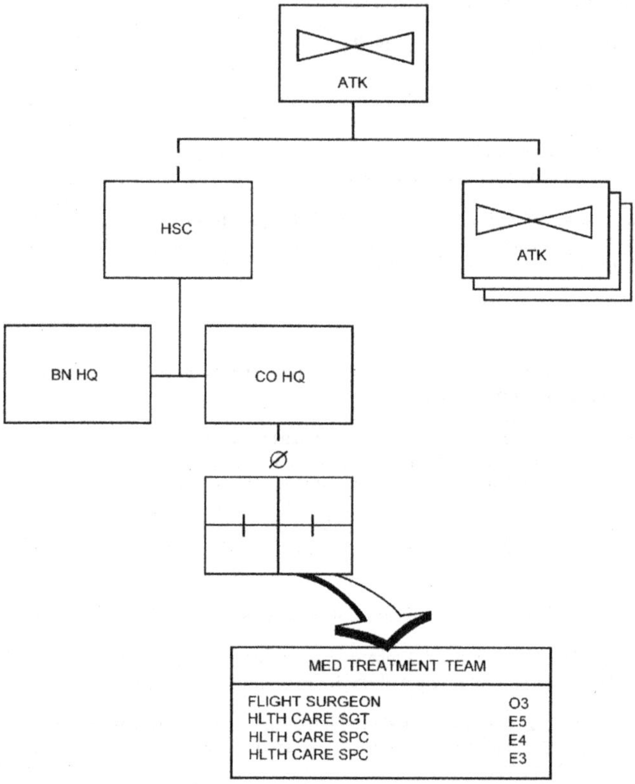

Figure 2-16. Medical section, headquarters and headquarters company, attack helicopter battalion.

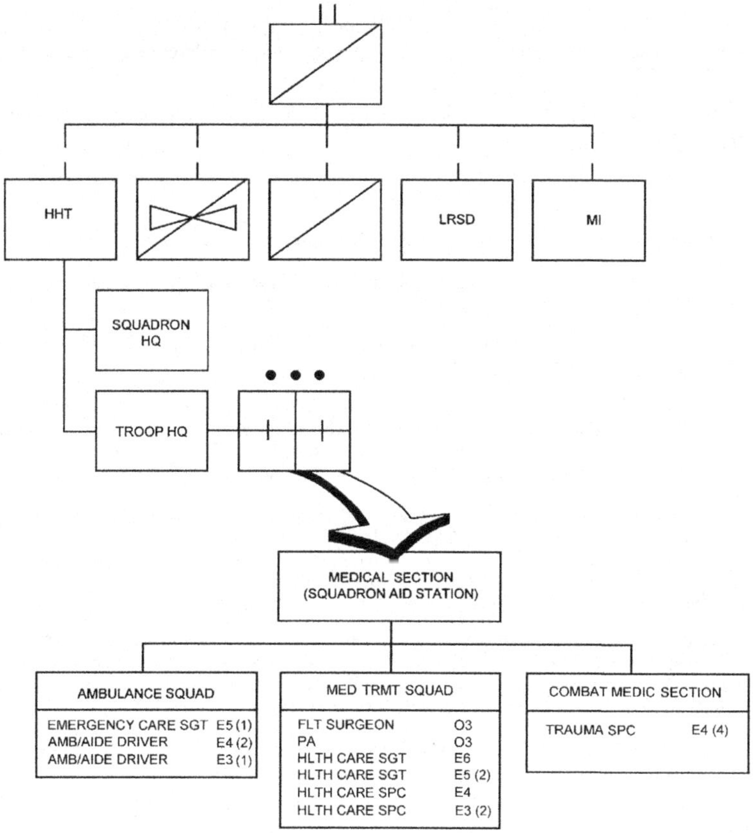

Figure 2-17. Medical platoon, headquarters and headquarters troop reconnaissance squadron.

(2) The ambulance squad has one emergency care SGT (MOS 91W20) and three ambulance/aide drivers (MOS 91W10).

(3) The combat medic element has one trauma specialist, MOS 91W10.

b. Section Sergeant. This NCO, also a health care SGT, assists the flight surgeon in accomplishing his duties. The specific duties of this NCO are the same as those described for the health care SGT in the DIVARTY HHB (refer to paragraph 2-34).

c. Health Care Specialists. The duties and functions of these specialists are the same as those discussed in paragraph 2-34.

d. Trauma Specialists. The duties and functions of trauma specialists are described in paragraph 2-20.

e. Ambulance Squad. Paragraph 2-21 describes duties of ambulance squad members.

f. Employment. The medical section establishes a BAS near the squadron headquarters and provides Echelon I CHS for members of the squadron.

(1) The section employs a treatment HMMWV, a cargo trailer, and two MESs (one trauma treatment set and one general sick call set).

(2) For communications, the section employs a telephone set (TA 312/PT) and is deployed in the headquarters and headquarters support company's wire communications net. It also employs an FM tactical radio and is deployed in the net designated by the squadron SOI. This section has access to the supporting medical company's tactical operations net for requesting Echelon II CHS.

g. Operations. Paragraphs 2-19 describes a BAS operation; these are equally applicable to the squadron BAS. Figures 2-8 and 2-9 show suggested layouts of a BAS.

h. Medical Evacuation. Evacuation of patients from the squadron aid station is provided by the supporting medical company.

i. Medical Supply. The medical section maintains a 2-day (48-hour) stockage level of medical supplies for the squadron. Routine requests for Class VIII resupply are submitted through command channels to the DMSO. Supplies may be picked up by the requesting unit or forwarded to the BAS during routine ambulance runs. For emergency resupply procedures, see paragraph 2-25.

j. Property Exchange. See paragraph 2-32.

2-39. Medical Section, Headquarters and Headquarters Company, Division Headquarters, L-Edition TOE

a. Organizations and Functions. The HHC division headquarters medical treatment team is organized as shown in Figure 2-18. Personnel staffing of this section includes a PA, a health care SGT, and two health care specialists.

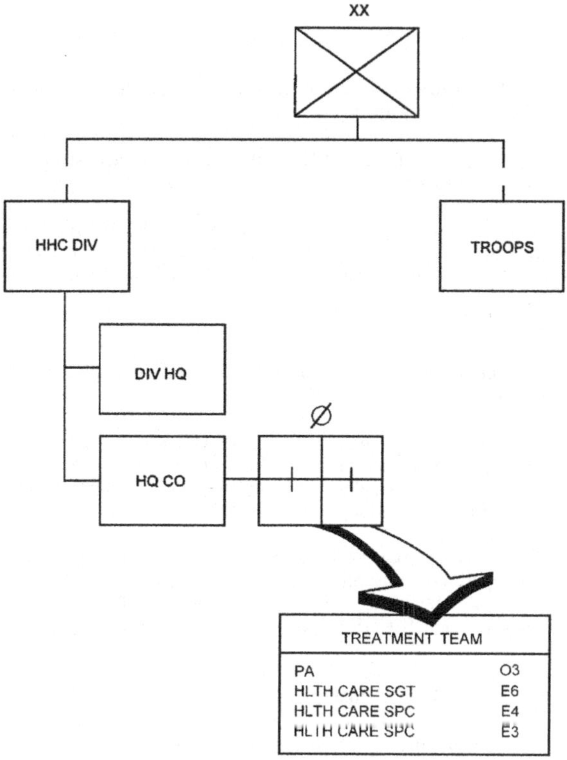

Figure 2-18. Medical section, headquarters and headquarters company, division headquarters.

(1) *Physician assistant*. The PA is responsible for the medical treatment provided by HHC medical personnel. He works under the clinical supervision of the division surgeon. In the division treatment team, the PA is the principal advisor to the division surgeon in the areas of PA affairs, executive medicine issues, and quality assurance/implementation. The specific duties of the PA are the same as those described in the DIVARTY HHB (refer to paragraph 2-34).

(2) *Health care sergeant*. Refer to paragraph 2-34.

(3) *Health care specialists*. Refer to paragraph 2-34.

b. Employment. The medical treatment team establishes an aid station near the division headquarters and provides Echelon I CHS for members of the division HHC.

(1) The section employs a treatment HMMWV, a cargo trailer, and two MESs (one trauma treatment set and one general sick call set).

(2) For communications, the section employs a telephone set (TA 312/PT) and is deployed in the HHC wire communications net. It also employs a FM tactical radio and is deployed in the net designated by the division SOI. This section has access to the MSMC's tactical operations net to request Echelon II CHS as required.

2-40. Combat Medic Section, Headquarters and Headquarters Company, Combat Engineer Battalion, L-Edition TOE

a. Organization and Functions. The combat medic section of the combat engineer battalion is organized as shown in Figure 2-19. Personnel staffing of this section includes a section SGT and ten trauma specialists. The combat medic section provides EMT and treatment of minor wounds and injuries. It coordinates for and/or requests MEDEVAC support as required.

b. Section Sergeant. The section SGT, MOS 91W30, prepares reports, requests general and medical supplies, maintains supply economy procedures, and maintains the ASL of expendable supplies. He supervises combat medic section personnel. He coordinates Echelon I CHS from supported maneuver battalion medical platoon and Echelon II CHS from the supporting medical companies. His duties also include—

- Assigning tasks to trauma specialists.

- Providing and/or coordinating for Class VIII resupply for trauma specialists when deployed with engineer platoon or squad.

- Conducting tactical and technical proficiency training for subordinate members of the section.

- Conducting sanitation inspections of troop living areas, food service areas, waste disposal areas, and potable water distribution points and equipment.

- Coordinating and conducting CLS training for the battalion.

- Providing medical planning input to the S1 on battalion operations.

c. Trauma Specialists. The duties and functions of trauma specialists are described in paragraph 2-20.

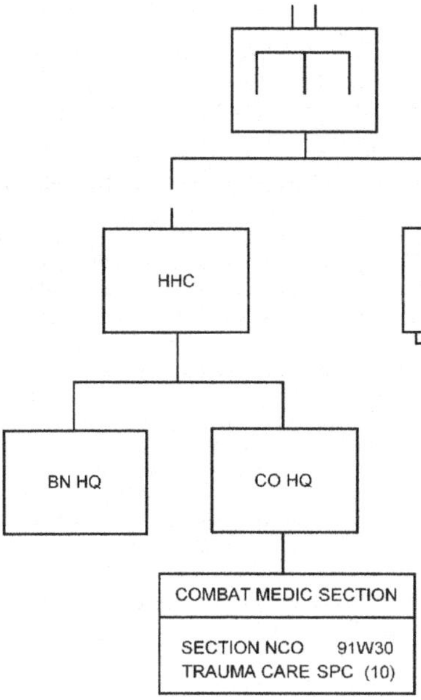

Figure 2-19. Medical platoon, headquarters and headquarters company, engineer battalion.

Section VII. ADDITIONAL MEDICAL ASSETS OPERATING IN THE BRIGADE AREA OF OPERATIONS (FORCE XXI/DIGITIZED DIVISION)

2-41. Treatment Squads/Teams from the Forward Support Medical Company, F-Edition TOE

The treatment squad provides emergency and routine sick call treatment to soldiers assigned to supported units. These teams can perform their functions while located in the FSMC area, or can operate independently of the FSMC for limited periods of time. The squad has the capability to split and operate as separate treatment teams (Teams Alpha and Bravo) for limited periods of time. While operating in these separate modes, they may operate two separate treatment stations. Normally, a squad or team deploys forward to augment or reinforce maneuver battalion medical platoons. Ambulance squads/teams may be deployed to AXP especially when there are extended evacuation routes. It can be assigned to reinforce or reconstitute battle losses of maneuver battalion medical platoons.

2-42. Forward Surgical Team, A-Edition TOE (Force XXI/Medical Reengineering Initiative)

Corps-level initial surgical support will be provided by the FST. The FST (Corps), TOE 08518LA00, and the FST (Airborne/Air Assault Division/ACR [Light]), TOE 08518LB00, are clinically standardized modules regardless of their assignment. These teams are comprised of 20 personnel and each has two operating room (OR) tables. The FST is organized into four functional areas—triage-trauma management, surgery, recovery, and administrative/operations. The mission of the FST is to provide a rapidly deployable immediate surgical capability enabling patients to withstand further evacuation. The requirement to project surgery forward increases as a result of the extended battlefield. This small, lightweight surgical team is designated to provide surgical augmentation to the FSMCs in support of the maneuver brigades, brigade TF, or the Interim Brigade Combat Team (IBCT). The FST is capable of continuous operations with divisional or nondivisional medical companies/troops for up to 72 hours; the ability to continue operations is limited by personnel fatigue/exhaustion and available supplies. The FST provides urgent, initial surgery for otherwise nontransportable patients. The FST's surgical capability is based on two OR tables with a surgical capacity of 24 OR table hours per day. Other capabilities include—

- Emergency medical treatment, to include assets to receive, triage, and prepare incoming patients for surgery.

- Surgery, including initial surgery and continued postoperative care for up to 30 critically wounded or injured patients over a period of 72 hours with the FST's organic MESs prior to resupply.

- Nursing care. Postoperative acute nursing care for up to eight patients, simultaneously, prior to further medical evacuation.

- Rapid strategic deployability. The team's personnel and equipment (less vehicles) are capable of deploying in one C-130 aircraft for initial entry missions, when required. The FST is capable of subsequent movement by helicopter sling-load operations.

- Tactical mobility. The team is 100 percent mobile with organic vehicles; it has a total of six HMMWVs.

For definitive information of the FST, see FM 8-10-25.

2-43. Forward Support Medical Evacuation Team, L-Edition TOE (Force XXI/Medical Reengineering Initiative)

The brigade may be augmented with a corps FSMT. When deployed forward to the BSA, the FSMT leader coordinates the air ambulance team's evacuation missions. The FSMT, assisted by the support operations section, provides real-time tactical information to the air ambulance crew about evacuation missions from the brigade combat team units/elements to supporting brigade MTFs. When air ambulances operate forward of the BSA, they will execute the A2C2 plan through the maneuver brigade S3. The FSB support operations section provides planning and coordination between aeromedical evacuation and the supported maneuver brigade. The brigade S3 provides the A2C2 plan that includes the air corridors, air control points, and communications checkpoints. The brigade S3 will provide updates as required. Air ambulances deployed to the BSA provide medical evacuation from forward areas (BAS) back to the BSA. Air ambulance evacuation from the point of injury will be METT-TC-dependent. Corps air ambulances providing GS evacuate from the BSA to supporting corps MTF. Aeromedical elements provide around the clock immediate response evacuation from either the BSA or their location based on METT-TC. To accomplish this, elements must maintain a close tie with the A2C2 system in the brigade. The brigade A2C2 element provides an airspace plan through the division OPORD/OPLAN A2C2 annex. The aircrew must also be familiar with the daily airspace control order and the airspace control plan. These documents contain all airspace control measures (ACM), to include free fire areas, no-fly fire areas, restricted operations zones, and established and standard Army aircraft flight routes. These routes and ACM change on a daily basis and cannot be integrated into the division OPORD. The BSS will ensure all A2C2 information is provided to corps aeromedical elements. The BSS does not generate A2C2 information, but does provide A2C2 planning information to division A2C2 elements. For definitive information on the corps air ambulance company and its FSMT that deploy forward into the brigades' AO, see FM 8-10-26.

2-44. Corps Ground Ambulance Company, Either the L-Edition TOE or the A-Edition TOE Supporting Force XXI/Medical Reengineering Initiative

The corps ground ambulance company is assigned to the corps MEDEVAC battalion. The basis of allocation within the combat zone is one per division supported. The current Army of Excellence ambulance company has four ambulance platoons with each platoon having 10 ambulances each. Under the MRI, the new MRI ground ambulance company will have a total of 24 ground ambulances. When deployed to the division, the ground ambulance company is attached to the MSB or division support battalion for Force XXI. The mission of the ground ambulance company in the division is to provide medical evacuation support to the division's maneuver brigades and to other divisional units and corps units operating in the division, as required. Normally, corps ground ambulances provide medical evacuation from the FSMC located in the BSA and from either the MSMC or the DSMC (Force XXI) to the supporting corps combat support hospital. The corps ground ambulance will reinforce the MEDEVAC assets in the medical companies of the division, as required.

2-45. Corps Combat Stress Control Augmentation, A-Edition TOE (Force XXI/Medical Reengineering Initiative)

The division may be augmented with additional CSC personnel, if requested. The base of allocation for the CSC medical detachment is one per division supported by the corps. The CSC medical detachment provides complete MH and combat stress preventive and treatment services in DS of division and corps personnel deployed forward. The new MRI detachment is a 43-person unit composed of a headquarters, a CSC preventive section and a CSC fitness section. The old Medical Force 2000 CSC medical detachment had 23 personnel and was designed to be a corps-level package to augment the organic MH sections of the divisions. Whereas, the new MRI CSC medical detachment retains the mission of providing DS to a division's maneuver brigades and general/reinforcing support to the DSA, including corps units in those areas. In addition, the detachment now augments area support in the corps immediately behind the division. The detachment must function with its elements widely dispersed, some working in and for the supported division and others working in the corps for the medical command/brigade. The CSC medical detachment personnel provide CSC planning, consultation, training, and staff advice to C2 headquarters and the units to which they are assigned/attached regarding—

- Combat and operational stressors affecting the troops.
- Mental readiness.
- Morale and cohesion.
- Potential for BF casualties.

The detachment and its personnel are dependent on units that they are attached for support, to include—

- Food service.
- Water distribution.
- Medical treatment.
- Logistical support including Class VIII items.
- Patient administration (detachment has one patient administration specialist, MOS 71G10, that works with the supporting unit).

For definitive information on the CSC medical detachment, Medical Force 2000 and MRI, see Change 1, FM 8-51.

CHAPTER 3

COMMAND AND CONTROL

Section I. PREPARATION FOR COMBAT HEALTH SUPPORT

3-1. Plans

a. Mission Analysis. Planning starts with mission analysis. The battalion begins mission analysis when the brigade provides a warning order (WARNO). The mission analysis is Step 1 of the military decision-making process (MDMP). See FM 101-5 for further discussion on the MDMP. For guidance on military decision making in abbreviated planning for a time-constrained situation, see FM 101-5. The battalion headquarters conducts concurrent planning with the brigade headquarters or after the brigade plan is developed. The battalion staff may receive additional information from the brigade staff elements to assist them with the planning process. This information is normally transmitted in a force text e-mail message via the tactical local area network (LAN). As part of the mission analysis and based on the battalion commander's intent and guidance, the medical platoon develops CHS estimates for supporting battalion operations. An understanding of the battalion's time lines or battle rhythm will assist the battalion medical platoon leader and field medical assistant in developing the CHS input, through the battalion S1, to the battalion's OPLAN/OPORD. The battalion surgeon and field medical assistant work with and through the battalion S1 for mission analysis input. See Chapter 5 and Appendix H of FM 101-5 for additional information on WARNOs. Mission analysis includes—

- Assessing CHS capabilities (organic and attached assets with current status and location).
- Assessing limitations (specify reason that CHS assets are not available).
- Identifying specified, implied, and essential CHS tasks in the brigade OPORD.

The following are examples of subject areas that should be addressed during mission analysis:

- Treatment (to include surgical requirements).
- Emergency dental treatment.
- Combat stress control.
- Preventive medicine.
- Medical evacuation support by air and ground ambulances (and nonmedical evacuation platforms, if necessary).
- Class VIII resupply.
- Medical maintenance.
- Nuclear, biological, and chemical operations.

- Threat to treatment and evacuation assets that is capable of causing CHS failure.
- Casualty estimates (number and types of casualties).
- Terrain effects on location, acquisition, and evacuation of casualties.
- Current medical status of battalion personnel.

b. Battalion Course of Actions. Battalion COAs development/analysis and wargaming are accomplished after mission analysis. Course of action development and wargaming result in the production of the OPORD and the CHS annex. During wargaming, the evacuation and treatment facets of the medical plan are synchronized with the overall battalion plan. The S1 will provide the overall casualty figures by battalion and, possibly, company. The questions of how many casualties, at what point in the fight (when), where they will occur, and how they are produced (direct fire, artillery, chemical, and so forth) can be forecasted based on input from the S1. During the wargaming, the S2 will portray enemy capabilities and likely actions. The S3 will focus on friendly actions. The medical platoon leader needs to pay careful attention to this exchange. This will be the best predictor of what, when, where, how, who, and other information that will be useful in adjusting the CHS plan. This information allows the medical platoon leader and field medical assistant options to select preplanned locations for positioning ambulances or treatment teams. Locating the treatment teams in the appropriate locations is of the utmost importance. The published movement planning factors for inside the division AO for wheeled and tracked vehicles in good terrain and weather are 8 kilometers and return in 1 hour (or 16 kilometers per hour). See FM 8-55 for additional planning guidance. The trauma specialist's goal is to get the casualty to ATM within 30 minutes. For an ambulance to leave the BAS and pick up a patient and return within 30 minutes, it must be within 4 kilometers of the soldier's point of injury. Keep in mind that this is under favorable conditions. Limited visibility, difficult or unfamiliar terrain, obstacles (friendly and enemy), and enemy actions will make the evacuation mission longer. If the BAS is farther than 4 kilometers away, it starts out as an impossibility. The METT-TC will govern specific solutions. Supporting the fight and maintaining a good support distance becomes a definite challenge because most of your evacuation routes can potentially be covered by enemy weapons and direct and indirect fire. For example, Russian manufactured AT-5 spandrel antitank-guided missiles have a 4-kilometer direct fire range. The medical platoon plan must take this threat into account. Information obtained from the S2 and S3 will also provide the triggers that will allow medical elements to occupy these positions at appropriate times and in a manner that reduces the risk from threat actions. This information is incorporated into the CHS plan and is published in the battalion order. The following are examples of key areas that will be analyzed during this process; they include—

- Casualty estimates broken down to the lowest level possible, by TF, by phase line, and so forth.
- Battalion- or brigade-directed actions as part of the squadron or brigade CHS plan.
- Current medical platoon equipment status (maintenance status on all the platoon's key items of equipment, both medical and nonmedical).

3-2. Operation Order

Once the battalion receives the brigade WARNO, it begins mission analysis and determines its tactical plan. This process continues until the full brigade OPORD is issued. Part of the mission analysis is to assess the brigade CHS plan for its adequacy of support to the battalion. If the medical platoon leader finds a problem, he briefs this to the battalion commander. The mission analysis brief will be after the staff has analyzed the full OPORD. Another part of the analysis is determining the employment and emplacement of medical treatment elements (BAS/treatment team) based on the brigade plan. This information is provided to the battalion S3 and he updates the battalion plan. This is normally accomplished prior to the brigade combined arms rehearsal. When the battalion commander approves the OPLAN, it becomes the OPORD. The OPLAN and OPORD are developed by the S3 section using input from each of the staff elements with the S1 being the staff coordinating element for CHS. The battalion OPORD is revised or updated based on mission changes. Table 3-1 is an example of an OPORD/OPLAN outline format.

a. Matrix Operation Order. A matrix OPORD may be used as an alternative to the standard five-paragraph OPORD. The purpose of the matrix OPORD is to cut orders production time and to provide subordinates more time for RECON, preparation, and rehearsal. There is no standard format for a matrix OPORD. Matrix orders expand on the execution matrix found on many operations overlays. The single-page format may include all signal information for the day of the operation and it can be placed in the corner of a map case for easy reference. Matrix orders are usually issued with standard operations, intelligence, and fire support overlays. Rather than a five-paragraph order outline format as seen in Table 3-1, the medical platoon leader is more likely to see and work with a matrix OPORD. Figure 3-1 is an example of a matrix OPORD.

b. Medical Support Matrix. The medical support matrix should be integrated with the tactical overlay. Figure 3-2 is a sample format for a medical support matrix. If deviation from the matrix occurs, the BAS location must be known at all times. The BAS should remain on location as long as practical. Extra first aid medical supplies can be issued to maneuver elements for resupply of CLS.

Table 3-1. Sample Outline Format for an Operation Order/Plan

(CLASSIFICATION)

TIME ZONE USED THROUGHOUT THE PLAN (ORDER)

REFERENCES:

TASK ORGANIZATION:

1. **SITUATION**
   A. ENEMY FORCE
   B. FRIENDLY FORCE
   C. ATTACHMENT AND DETACHMENTS
   D. ASSUMPTION (OPLAN ONLY)

2. **MISSION**

3. **EXECUTION**
   INTENT:
   A. CONCEPT OF OPERATION
      (1) MANEUVERS
      (2) FIRES
      (3) RECONNAISSANCE AND SURVEILLANCE
      (4) INTELLIGENCE
      (5) ENGINEER
      (6) AIR DEFENSE
      (7) INFORMATION OPERATIONS
   B. TASK TO MANEUVER UNITS
      (1) ENGINEER
      (2) AIR DEFENSE
      (3) FIELD ARTILLERY
   C. TASKS TO COMBAT SUPPORT UNITS
      (1) INTELLIGENCE
      (2) FIRE SUPPORT
      (3) SIGNAL
      (4) NBC
      (5) PROVOST MARSHAL
      (6) PSYOP
      (7) CIVIL MILITARY
      (8) AS REQUIRED
   D. COORDINATING INSTRUCTIONS
      (1) TIME OR CONDITION WHEN A PLAN OR ORDER BECOMES EFFECTIVE
      (2) COMMANDER'S CRITICAL INFORMATION
      (3) RISK REDUCTION CONTROL MEASURES
      (4) RULES OF ENGAGEMENT
      (5) ENVIRONMENTAL CONSIDERATIONS
      (6) FORCE PROTECTION AS REQUIRED
      (7) AS REQUIRED

4. **SERVICE SUPPORT**
   A. SUPPORT CONCEPT
   B. MATERIEL AND SERVICE
   C. MEDICAL EVACUATION AND HOSPITALIZATION
   D. PERSONNEL
   E. CIVIL MILITARY, AS REQUIRED
   F. AS REQUIRED

5. **COMMAND AND SIGNAL**
   A. COMMAND
   B. SIGNAL

ACKNOWLEDGE:
NAME (COMMANDER'S LAST NAME)
RANK (COMMANDER'S RANK)
OFFICIAL: NAME AND POSITION

ANNEXES

(CLASSIFICATION)

SEE FM 101-5 FOR DEFINITIVE INFORMATION ON OPERATION PLANS/ORDERS. ADDITIONAL INFORMATION IS PROVIDED IN APPENDIX 3 (PERSONNEL) OF ANNEX I, OR IN A SEPARATE APPENDIX FOR CHS.

TF 2-69 ARMOR DEFENSIVE MATRIX OPORD #1-88 DTG 011000A JAN 88 COPY 21 of 25 COPIES

IF CALL SIGN ____ (42.10) CDR 16 S3 91 FSO 07 MAIN 76 XO 34 S3A 23 FAC 37 PLDR 12 PSG 15 AJ1 (48.75) S2 39 S4 AA BHO 22 ADA V1T15 (42.00) AJ2 (45.00)			BN MORTAR/FS NET (52.40) BDE FS NET (38.55)		ATTACHMENTS/DETACHMENTS ~~A/1 68~~ ~~72 ENG~~		
UNIT/STATUS	TM MECH / G	TM TANK / G	Co C / G	Co D / G	SCOUTS / G	T2 ENGR / G	MORTARS / G
CALL SIGN/FREQ	IIM (40.25)	V6D (52.65)	NBY (50.10)	M9U (512.25)	D41 (142.89)	T5V (52.50)	ABE (52.40)
TASK ORG	1 2/8 2 2 1	1 3 3/4 2	1 2 3 3	III 1 2 4	EYES EARS 3/D 5	1 3	BLUE { FDC RED { FDC
IF MISSION: TF2-69 defends in sector 020600Jan 88 from coords to coords and coords to coords to destroy MRR. CDRS INTENT: Occupy deception positions along PL DOG; NLT 020400 move to prepared positions along PL CAT; destroy en in EA LION (Ave 1) or EA TIGER (Ave 2) allowing no penetration of PL MOUSE.							
⑤SP MOVE ORDER③	6	011400 / 2	4	7	011000 / 1	3	PRI TO TM TANK / 5
① OCCUPY PL DOG (DECEPTION) ORIENT ⑨	BP 2	BP 3	BP 1	BP 4	SCREEN IF FRONT ALONG PL BEAR	PREPARE OBSTACLES ALONG PL CAT	(COORDS) / PRI INIT TO SCOUTS
① PL CAT (BALANCED ATK) ⑥	PB6 / EA LION	BP 7 / EA TIGER	BP 5 / BA LION	BP 8 / EA TIGER		CONTINUE OBSTACLE & SURVIVABILITY POSITION WORK	NU213562 / PRI TM MECH when EN crosses PL DOG
④ MAIN ATK Ave 1 ②		o/o BP 11 / Ave 1		BP 9 / EA LION	Screen if right flank	PRI effort to BP 11	PRI TM MECH
⑧ MAIN ATK Ave 2 ⑩	o/o BP 12 / Ave 2	BP 7 / EA TIGER	BP 10 / EA TIGER	BP 8 / EA TIGER	Screen if left flank	PRI effort to BP 12	PRI TM TANK
FIRES 2 to FA PRI to 2-69 ·PRI Init sets ·FPFs: ARTY: TM Mech MORT: TM Tank Illum planned in AB 4004, 5, 6, 7 but o/o TF CDR FASCAM planned in AB 4004#5 o/o Bde Cdr	DECEPTION ·OCC BP 6 ·RECON BP 12 ·Conserve MIN ·Fire IGTS: AB 4004 AB 4006 [illegible] mine field AB 4006 after secure pass	SKYLINE VC ·OCC BP 7 ·RECON BP 11 ·OBSV MA 12 ·Fire TGTS AB 4005 AB 4007	OF TANKS ON ·OCC BP 5 ·Prep BP 10 ·OBSV NA 11 TO BP 10 at night; RPT Time	·OCC BP 8 ·Prep BP 9 ·OBSV NA 122 3/D ATN3 020500	·Clear TF Section NLT 011400 ·3/D OPCON until 020500 ·Fire Tgts AB 4002 AB 4003 ·Observe NAI 12		DISPLACE BY SECTION
OBSTACLES	Minefield AB 4006	Minefield AB 4007	Wire Obstl #1	Wire Obstl #2		Minefield 1-AB 4006 2-AB 4007	
FIGHTING POSN PRI (No.) LOC	#1(4) BP 6	#2(10) BP 7	#3(8) BP 5 #5(4) BP 10	#4(8) BP 8 #6(4) BP 9		#7(6) BP 5 #8(6) BP 8	
CDR BP 6	S3 BP 7	MAIN CP CP 2	CBT TNS CP 4	UMCP CP 6	FLD TNS BSA (CP 10)	LRP A CP 12	LRP B CP 8
ADA Yellow Tight	MOPP 2	OEG 50 cGy	SOI KTV 1062T	PYRO Main Atk AV 01 Red Star AV 02	AJ CODEWORD Sprint	PIR (1) EN (2) ALOC	STAND TO 0600 Daily

Figure 3-1. Matrix operation order.

UNITS	CROSS LD	CROSS PL RED	CROSS PL BLUE	ON OBJ DALLAS
CO A AND CO B (-)	TT 1 (NB 583492) (TOM)	TT 1 (NB 585501) (JIM)	TT 1 (NB 587507 (BILL)	TT 1 (NB 591510) (GREG)
CO C AND RESERVE	TT 2 (NB 581489) (VICKY)	TT 2 (NB 583499) (BETH)	TT 2 (NB 585505) (JANET)	TT 2 (NB 589508) (JOY)
RM 1: USE ROUTE STEVEN				
RM 2: USE ROUTE ANNA				

Figure 3-2. Medical support matrix.

c. Implementation of Combat Health Support Annex. For successful implementation of the CHS annex of the battalion plan, the CHS plan must be coordinated and synchronized with the battalion plan so that CHS requirements are met. The medical platoon leader may receive additional coordinating instructions from the BSS as the CHS annex is developed. This additional information permits the medical platoon leader to make informed decisions as he develops the CHS annex and affords some additional time for better development of good initial plans. For successful implementation of the CHS annex of the battalion's/ squadron's plan, the CHS plan must be coordinated and synchronized with the overall plan. To achieve optimal synchronization, the battalion CHS plan is rehearsed as an integral part of the combined arms plan at the combined arms rehearsal. The CHS rehearsal by itself is a technique that will increase understanding and synchronization, but is not as effective as when it is integrated into the combined arms rehearsal. The rehearsal of the CHS plan will allow battalion medical elements and leaders to analyze the tactical CHS plan to ascertain its feasibility, its common sense, and the adequacy of its C2 measures prior to execution.

3-3. Rehearsal

To achieve optimal synchronization, the battalion CHS plan is rehearsed as an integral part of the combined arms plan at the combined arms rehearsal. Medical platoon leaders and field medical assistants provide input to the battalion plan and develop the concept for the battalion CHS plan. During the decision-making/ orders process, they identify critical events and synchronize the CHS plans. In addition to medical locations on the CSS overlay, these plans indicate the triggers for CHS events. At the battalion rehearsal, battalion leaders practice their synchronized plans that include CHS. The sequence of events for the CHS portion of the battalion rehearsal includes—

- Ambulance teams practicing execution of triggers for area medical support responsibilities and triggers for movement of supporting FSMC ambulances.

- The medical platoon SGT detailing the concept and procedures for MEDEVAC (both ground and air ambulances) in the battalion.

- The battalion XO or S4 explaining triggers for BAS/treatment team movement and ensuring that the battalion HHC and maneuver companies understand when and where the BAS/treatment teams are located.

- The medical elements providing projected triggers and times they will be at projected locations.

The CSS/CHS annex of the battalion OPORD that includes map overlays is the conclusion of the medical planning efforts; the rehearsal is the culmination of the preparation phase for an operation. The medical platoon leader has the responsibility for rehearsing CHS operations. Rehearsals are done to achieve a common understanding and a picture of how the plan will be implemented.

- All plans must be completed prior to the battalion rehearsal.

- The CHS portion of the battalion rehearsals should focus on the events that are critical to mission accomplishment. A successful rehearsal ensures explicit understanding by subordinate medical personnel of their individual missions; how their missions relate to each other; and how each mission relates to the commander's plan. It is important for all medical echelons to see the total CHS concept.

- Rehearsing key CHS actions allows participants to become familiar with the operation and to visualize the "triggers" which identify the circumstances and timing for friendly actions. This visual impression helps them understand both their environment and their relationship to other units during the operation. The repetition of critical medical tasks during the rehearsal helps leaders remember the sequence of key actions within the operation and when they are executed.

- The battalion OPORD is then issued through effective troop leading procedures.

Section II. TROOP-LEADING PROCEDURES

3-4. Eight Steps of Troop Leading

The commander makes most tactical decisions. He then announces them in the form of orders that include his intent and concept of the operation. Based on these orders, the medical platoon leader uses troop-leading procedures to organize his time during planning and preparation and to translate the operation into instructions his soldiers can understand. He can then lead the platoon more effectively in the execution of the mission. Troop leading is a dynamic process that begins when the unit receives a new mission or is notified by a WARNO that a new mission is imminent. Whenever possible, troop-leading procedures are integrated and accomplished concurrently rather than sequentially. Time management is the key. The medical platoon leader normally uses one-third of the available time to plan, prepare, and issue the order; his field medical assistant, platoon SGT, ambulance squad leaders, and treatment team leaders then have the remaining two-thirds of the time available to prepare their ambulances and MES to support the operation. The following discussion focuses on the eight steps of troop leading procedures:

- Receive and analyze the mission.

- Issue a WARNO.
- Make a tentative plan.
- Initiate movement.
- Conduct RECON and coordination.
- Complete the plan.
- Issue the order.
- Supervise and refine.

a. Receive and Analyze the Mission. The medical platoon leader normally receives his orders as an oral OPORD or as a fragmentary order (FRAGO) updating a previously issued OPORD. Graphics may be copied from the S3's overlay or sent by digital transmission. Initial coordination within the battalion S1, medical platoon personnel, and with the supporting FSMC should be accomplished upon receipt of the mission.

NOTE

Before the OPORD or FRAGO arrives, the medical platoon leader may receive a series of WARNOs from the battalion S3 providing advance notice of an impending operation. The medical platoon leader should disseminate all pertinent information contained in the WARNO as quickly as possible after they are received. The battalion S4 may be the one who issues orders to the medical platoon since he has tactical control of the combat trains.

(1) Upon receipt of the WARNO, the medical platoon leader's first task is to extract the CHS mission based on the battalion commander's guidance and intent. The key to understanding the medical platoon CHS mission, as part of the battalion team, lies in two elements of the plan—the commander's intent and the purpose he envisions for the battalion and each company.

(2) The medical platoon leader's knowledge of the intent and purpose allows him to use his initiative and to be proactive and exploit battlefield opportunities to accomplish the CHS mission. If the medical platoon leader does not understand the intent or purpose, he should ask the commander for clarification.

(3) The medical platoon leader analyzes the mission using the factors of METT-TC. These factors allow the platoon leader to identify the platoon's purpose; the specified, implied, and essential tasks it must perform; and the time line by which the platoon will accomplish those tasks. The following outline

of METT-TC factors will assist the medical platoon leader in analyzing the mission and creating a time line. The medical platoon leaders need the answers to the questions pertaining to METT-TC.

(*a*) *Mission.*

- What is the battalion commander's intent?

- What are the current capabilities (organic and attached assets with current status and locations)?

- What are the specified, implied, and essential CHS tasks in the battalion and brigade OPORD?

- What are the limitations (CHS assets that are not available, specify reason)?

- What other tasks must be accomplished to ensure mission success (implied tasks)? Implied tasks are those that are not specified in the OPORD but that must be done to complete the mission. They do not include tasks that are covered in the unit SOP. The medical platoon leader identifies implied tasks by analyzing the enemy, the terrain, friendly troops available, and the operational graphics.

- What is the current patient status (for example, awaiting evacuation)?

- Are patient evacuation vehicles required to use contaminated routes?

- Are patient decontamination stations required?

- Where are the locations of the treatment team and of BAS or other MTF providing area medical support? (Current/projected?)

- What are the area medical support responsibilities?

(*b*) *Enemy.*

- What have been the enemy's recent activities?

- What is the composition of the enemy's forces?

- What are the capabilities of his weapons?

- What is the location of current and probable enemy positions?

- What is the enemy's most probable COA?

Enemy information is included in paragraph 1 of the OPORD. It is important that the medical platoon leader analyze this information in terms of how the medical platoon supports the operation. For example, the threat the enemy imposes on the battalion will influence how CHS operations are conducted based on the projected number of casualties the enemy will inflict on the battalion.

(*c*) *Terrain (and weather).* The medical platoon leader analyzes the terrain using the factors of OAKOC (obstacles; avenues of approach; key terrain; observation and fields of fire; and cover and concealment). Elements of this analysis include the following:

- *Obstacles.*
  - Where are natural and existing obstacles located, and how can they affect medical treatment and evacuation?
  - Where are likely areas for enemy-emplaced obstacles, and how can they affect maneuver?
  - Are there bypasses, or must obstacles be breached?
- *Avenues of approach.* Where are the best avenues of approach (mounted and dismounted) for enemy and friendly forces? These are considerations for determining evacuation routes and in planning for future locations for the BAS.
- *Key terrain.*
  - Where is the key terrain? Will FM communications be affected?
  - How can key terrain be used to support the mission?
- *Observation and fields of fire.*
  - Are these influenced by the key terrain that dominates avenues of approach?
  - Where can the enemy observe and engage battalion personnel (danger areas)?
  - Where are the natural firing positions that medical platoon personnel can use to defend against enemy attack?
- *Cover and concealment.*
  - What routes within the AO offer cover and concealment for placement of the BAS or a treatment team?
  - Do the natural firing positions in the AO offer cover and concealment for the platoon or enemy?

- *Weather.* The medical platoon leader can use these questions as he analyzes the impact of weather on the mission:

  - What are the light conditions (including percentage of night illumination) and visibility?

  - What are the times for beginning of morning nautical twilight (BMNT), sunrise, sunset, end of evening nautical twilight (EENT), moonrise, and moonset?

  - How has recent weather affected the AO?

  - How will fog, rain, dust, heat, snow, wind, or blowing sand affect the men and equipment during the mission?

NOTE

The effects of weather on smoke or NBC weapons/operations should also be considered.

(*d*) *Troop.*

- What is the supply status of ammunition, fuel, and other necessary items including Class VIII?

- What is the present physical condition of the soldiers, as well as of vehicles and equipment?

- What is the training status of the platoon?

- What is the state of morale?

- How much sleep have the men had?

- How much sleep will they be able to get before the operation begins?

- Does the platoon need any additional assets to support or accomplish its mission?

- What attachments are available to help the platoon accomplish its mission? What is the task organization in the WARNO or OPORD?

(*e*) *Time available.*

- What times were specified by the commander in the OPORD for such activities as movement, RECON, rehearsals, and LOGPAC operations?

- What priorities of work can the platoon accomplish (examples include security, maintenance, resupply, coordination, rehearsals, inspections, and sleep)?

NOTE

The medical platoon leader conducts reverse planning to ensure that all specified, implied, and essential tasks can be accomplished in the time available. He develops a reverse planning schedule (time line) beginning with actions on the objective and working backward through each step of the operation and preparation to the present time. This process also helps the platoon in making efficient use of planning and preparation time. Once the METT-TC analysis is complete, the medical platoon leader can then write the platoon mission statement based on the battalion CHS plan and answer the questions of WHO, WHAT, WHEN, WHERE, and WHY. This is a clear, concise statement of the purpose of the operation and the essential task(s) that will be crucial to its success. The essential tasks (the WHAT) should be stated in terms that relate to enemy forces, friendly forces, and/or the terrain (for example, "ESTABLISH BAS (-) ONE OR TWO TERRAIN FEATURES FROM THE MAIN BATTLE AREA"; "ESTABLISH THE BAS 1 TO 3 KILOMETERS FROM THE MAIN BATTLE AREA"; or "BE PREPARED FOR RAPID FORWARD DEPLOYMENT OF A TREATMENT TEAM IN SUPPORT OF COMPANY A"). The purpose (the WHY) explains how the platoon mission supports the commander's intent. The elements of WHO, WHERE, and WHEN add clarity to the mission statement.

NOTE

Simultaneous planning and preparation are key factors in effective time management during the troop-leading procedures. The next five steps (issue a WARNO; make a tentative plan; initiate movement; conduct RECON and coordination; and complete the plan) may occur simultaneously and/or in a different order. There may be multiple WARNOs.

b. Issue a Warning Order. The medical platoon leader alerts his platoon to the upcoming operation by issuing a WARNO that follows the five-paragraph OPORD format (see Table 3-1).

NOTE

The medical platoon leader will often do this from the battalion TOC during mission analysis.

Warning orders maximize subordinates' planning and preparation time by providing essential details of the impending operation and detailing major time line events that will support mission execution. The amount of detail included in a WARNO depends on the available time, the platoon's communications capability, and the information subordinates need to initiate proper planning and preparation. The WARNO may include the following information:

- Changes to task organization.
- Updated graphics (platoons equipped with intervehicular information systems or appliqué digital systems send new overlays).
- Enemy situation.
- Battalion mission.
- Commander's intent (if available).
- Combat health support mission.
- A tentative time line, to include the following:
  - Earliest time of movement.
  - Readiness condition and vehicle preparation schedule.
  - Reconnaissance.
  - Training/rehearsal schedule.

NOTE

Some individual and collective training may be initiated by the medical platoon leader before he issues the OPORD; this technique maximizes preparation time and allows the platoon to focus on tasks that will support the anticipated operations. For example, a medical platoon may train on treatment of different types of wounds or injuries that may be seen during the operations.

  - Time and location at the battalion OPORD will be issued, plus the platoon OPORD will also be briefed.
  - Service support instructions (if not included in the time line).

As critical information is received or updated, the medical platoon leader should issue subsequent or updated WARNOs to keep the platoon informed.

c. Make a Tentative Plan. The medical platoon leader begins developing his CHS plan when the battalion receives its first WARNO from the brigade. Based on the commander's intent, guidance, and the results of his mission analysis, the medical platoon leader develops a tentative plan that addresses all specified, implied, and essential tasks using the OPORD format. The tentative plan also covers RECON and coordination requirements between the platoon and adjacent and supporting units. The field medical assistant and the medical platoon SGT are excellent sources of ideas concerning the battalion CHS plan. The medical platoon leader can develop his COA and OPORD almost simultaneously with the battalion OPORD. Since the medical platoon is a battalion asset, most of their specified tasks will be developed during the battalion MDMP. The medical platoon leader can issue a WARNO from the battalion TOC during his participation in the MDMP. By the time the OPORD is given, he should have most of his platoon order prepared. The development of the platoon time line from backward planning should be his priority as soon as the battalion OPORD is prepared.

d. Initiate Movement. Many of the battalion-level operations require movement to forward assembly areas and to battle positions during the planning phase of an operation. This means that elements of the medical platoon will also move to these locations in support of the battalion units. Medical platoon elements move with the supported units according to the battalion plan. As often as possible, within the restraints of OPSEC, medical task organization should be done during hours of daylight, or as early as possible. In spite of multiple "own the night technologies," executing movements required by task organization is still more efficiently accomplished in daylight. Activities may include ensuring CHS for the company quartering party or beginning priorities of work.

e. Conduct Reconnaissance and Coordination. Effective RECON takes into account the factors of METT-TC and OAKOC from both friendly and enemy perspectives. As a minimum, the field medical assistant or the platoon SGT conducts a detailed map RECON to identify primary and alternate routes of MEDEVAC and preplanned sites for locating the BAS or a BAS (-). If time and security considerations permit and authorization is obtained from higher headquarters, an on-site ground RECON is the best way to survey the AO. In addition, the medical platoon leader should check with the S2 for an intelligence update. The medical platoon leader should take as many ambulance squad leaders as possible on his RECON. For offensive operations, the medical platoon leader should attempt to find a vantage point that will permit rapid accessibility to supported units while making use of natural terrain features to afford as much protection as possible for the BAS. Ground RECON for offensive operations usually is limited to checking routes to the start point (SP), the line of departure (LD), and the axis just beyond the LD. For defensive operations, the medical platoon leader should conduct a RECON of the unit position and the MEDEVAC routes to be used. Whenever tactically feasible, the platoon should make provisions to mark routes and locations for day and night operations during the RECON. See Chapter 4 for TTP on marking routes and positions. During the RECON (or during battalion-level rehearsals), the medical platoon leader or his representative should coordinate evacuation routes with supported companies.

f. Complete the Plan. The medical platoon leader refines the plan based on the results of the wargame, RECON, and coordination with the BSS' FSMC and supported battalion units. He then completes the plan using these results and any new information from his commander. He should keep the plan as simple as possible, at the same time ensuring that the CHS plan supports the commander's intent.

g. Issue the Order. Prior to the order, the platoon NCO should ensure that all subordinate elements have copies of the overlays correctly posted to the maps and a copy of any matrices from the battalion OPORD that covers key CHS actions. All platoon members receiving the OPORD verbally should be prepared to copy these instructions in the five-paragraph OPORD format. They should ask for explanations of any terminology or actions that they do not understand. If possible, the medical platoon leader issues the order from a vantage point overlooking the terrain on which the platoon will support the maneuver units. If not, he uses a terrain model, sand table, sketches, or his map to orient the platoon. He can also build a model of the AO using a briefing kit that contains such items as engineer tape, colored yarn, 3- by-5-inch index cards, and "micro" vehicle models. When time and security permit, the medical platoon leader issues the order to as many members of the platoon as possible. As a minimum, he assembles the treatment teams and ambulance squads, and combat medic section. He briefs the platoon using the five-paragraph OPORD format. He should then send the ambulance emergency care SGT to brief the trauma specialist assigned to each of the maneuver companies. To ensure complete understanding of the operation, the medical platoon leader and medical platoon personnel conduct confirmation briefings immediately after the OPORD is issued. The treatment team and ambulance team leaders brief the medical platoon leader to confirm their understanding of his intent, the specific tasks their team must perform, and the relationship between their tasks and those of other medical units/elements in the operation. If time permits, the medical platoon leader should lead the medical platoon in a walk-through using a sand table.

h. Supervise and Refine. Flexibility is the key to effective operations. The medical platoon leader must be able to refine his plan whenever new information becomes available. If he adjusts the plan, he must inform the platoon and supervise implementation of the changes. Once the operation has begun, the medical platoon leader must be able to direct his platoon in response to new situations and new orders. Platoon orders, back-briefs, rehearsals, and inspections are essential elements of the supervision process as the platoon prepares for the mission. The following paragraphs discuss these procedures in detail:

(1) *Team orders.* The medical platoon leader, the field medical assistant, and the platoon SGT make sure all ambulance team members have been briefed by squad leaders (heavy battalion and mechanized infantry medical platoons have track commanders [TC] for their M113 and M557 armored ambulance and treatment vehicles) and understand the platoon mission and concept of the operation.

(2) *Back-briefs.* The back-brief is, in effect, a reverse briefing process; those who receive an OPORD confirm their understanding of the order by repeating and explaining details of the operation for their leader or commander. In the medical platoon, the medical platoon leader should conduct back-briefs after the TC/team leaders have had a chance to review the OPORD but before the platoon rehearsal begins. The TC/team leaders brief the medical platoon leader on **how** their teams will accomplish the specific tasks assigned to them in the order.

NOTE

Although the back-brief is an effective means of clarifying the specifics of the plan, it does not require medical platoon personnel to practice or perform their assigned tasks. By itself, therefore, it is not an ideal rehearsal technique.

(3) *Rehearsals*. A rehearsal is a practice session conducted to prepare units for an upcoming operation or event. The medical platoon leader should never underestimate the value of rehearsals. Many units, in fact, consider rehearsals as a separate step (ninth) of troop-leading procedures. The medical platoon leader uses well-planned, efficiently run rehearsals to accomplish the following:

- Reinforce training and increase proficiency in critical tasks.
- Reveal weaknesses or problems in the plan.
- Synchronize the actions of subordinate elements.
- Confirm coordination requirements between the platoon and supporting medical units/elements.
- Improve each soldier's understanding of the concept of the operation, the direct fire plan, anticipated contingencies, and possible actions and reactions for various situations that may arise during the operation.

Effective rehearsals require personnel to perform required tasks, ideally under conditions that are as close as possible to those expected for the actual operation. Participants maneuver their actual vehicles or use vehicle models or simulations while interactively verbalizing their elements' actions. In a platoon-level rehearsal, the medical platoon leader will select the tasks to be practiced and will control execution of the rehearsal. The platoon can prepare for operations using reduced-force rehearsals and/or full-force rehearsals. The medical platoon leader conducts reduced-force rehearsals when time is limited or the tactical situation does not permit everyone to attend. Platoon members, who can take part, practice their actions on mock-ups, sand tables, or actual terrain (usually over a smaller area than in the actual operation). The full-force rehearsal is the most effective, but consumes the most time and resources. It involves every soldier who will participate in the operation. If possible, it should be conducted under the same conditions (such as weather, time of day, and terrain) that the platoon expects to encounter during actual operations. The medical platoon leader can choose among several techniques in conducting rehearsals, which should follow the crawl-walk-run training methodology to prepare the platoon for increasingly difficult conditions. Rehearsal techniques include the following:

- *Special rehearsal.* Individual and/or team tasks that will be critical to the success of the operation are rehearsed as necessary. The medical platoon leader may initiate special rehearsals when he issues the WARNO.
- *Map rehearsal.* This is usually conducted as part of a back-brief involving the TC or a complete ambulance team. The leader uses the map and overlay to guide participants as they back-brief their role in the operation. If necessary, he can use a sketch map.
- *Communications rehearsal.* This reduced-force or full-force rehearsal is conducted when the situation does not allow the platoon to gather at one location. Crewmen check their vehicles' communications systems and rehearse key elements of the platoon fire plan.

- *Key leader rehearsal.* Usually conducted as part of a larger force, this rehearsal involves leaders moving over the key terrain in wheeled vehicles while discussing the mission.

- *Sand table or terrain model.* The reduced-force or full-force technique employs a small-scale table or model that depicts graphic control measures and important terrain features for reference and orientation. Participants walk or move "micro" vehicles around the table or model to practice the actions of their own vehicles in relation to other members of the platoon.

- *Force on force.* This is used during a full-force rehearsal. The platoon may rehearse with sections or individual tanks going "force on force" against each other. Platoons should first rehearse with good visibility over open terrain. Rehearsals become increasingly realistic until they approximate those expected in the AO.

(4) *Inspections.* Inspections allow the medical platoon leader to check the platoon's operational readiness. The key goal is to ensure that soldiers and vehicles are fully prepared to execute the upcoming mission. Inspections also contribute to improved morale. It is essential that all leaders in the medical platoon know how to conduct precombat checks according to the platoon SOP. Procedures for a comprehensive inspection include the following:

- Perform before-operation maintenance checks and report or repair deficiencies.

- Upload vehicles according to platoon SOP. The standardization of load plans allows the medical platoon leader, the field medical assistant, and the platoon SGT to quickly check accountability of equipment. It also ensures standard locations of equipment in each vehicle; this can be an important advantage if the soldiers are forced to switch to a different vehicle during an operation.

- Review the supply status of rations, water, fuel, oil, ammunition, MES, first-aid kits, and batteries (for such items as flashlights, night vision devices, and NBC alarms). Direct resupply operations as necessary.

- Ensure vehicles are correctly camouflaged so they match the AO.

The medical platoon leader, the field medical assistant, and/or the platoon SGT should observe treatment and ambulance teams during preparation for CHS operations. They should conduct the inspection once the TC/squad or team leader report that their team(s) and vehicles are prepared. The precombat inspection must be a "hands on, show me that it works" event. This is the only way to ensure the platoon is properly prepared. If trauma specialists and/or ambulance teams cannot return to the platoon, then someone goes to inspect them.

3-5. Abbreviated Troop-Leading Procedures

When there is not enough time to conduct all eight troop-leading steps in detail, such as when a change of mission occurs after an operation is in progress, the medical platoon leader must understand how to trim the procedures to save time. Most steps of these abbreviated troop-leading procedures are done mentally, but the medical platoon leader skips none of the steps. Once the order is received, the medical platoon leader conducts a quick map RECON, analyzes the mission using the factors of METT-TC, and sends for the TC and team leaders. He makes sure they post the minimum required control measures on their maps and issues a FRAGO, covering the key elements of the enemy and friendly situations, the platoon mission, and the concept of the operation. The medical platoon leader and TC may also conduct a quick walk-through rehearsal of critical elements of the CHS plan using a hastily prepared terrain model or sand table. In some cases, there may not be enough time even for these shortened procedures. The platoon may have to move out and receive FRAGO by radio or at the next scheduled halt. It then becomes critical for the medical platoon leader to send FRAGO of his own to the TC and team leaders explaining the platoon's requirements and objectives in supporting the battalion maneuver plan. Digital and Global Positioning Systems (GPS) are valuable tools when the platoon is forced to use abbreviated troop-leading procedures and FRAGO. They allow the medical platoon leader to designate way points to assist in navigation and identifying evacuation routes. Other keys to success when abbreviated procedures are in effect include a well-trained platoon; clearly developed, thoroughly understood SOP; and an understanding by all members of the platoon of the current tactical situation (situational understanding). The medical platoon leader, the field medical assistant, and the platoon SGT must keep the platoon informed of the ever-changing enemy and friendly situations. They accomplish this by monitoring the battalion net and issuing frequent updates to the other platoon members using the radio and digital information systems. Whenever time is available, however, there is no substitute for effective, thorough troop-leading procedures. The odds of success increase still further when detailed planning and rehearsals are conducted prior to an operation, even if time is limited. Successful medical platoon leaders make the most of every available minute. Specified delegation of tasks will also speed this process up.

Section III. COMPUTERS, INFORMATION, DIGITIZATION, AND COMMUNICATIONS

3-6. Medical Standard Army Management Information System

The MC4 system when fielded will be a theater, automated CHS system, which links commanders, health care providers, and supporting elements, at all echelons, with integrated medical information. The MC4 system when developed and fielded will receive, store, process, transmit, and report medical C2, medical surveillance, casualty movement/tracking, medical treatment, medical situational understanding, and MEDLOG data across all levels of care. This will be achieved through the integration of a suite of medical information systems linked through the Army data telecommunications architecture. The MC4 system begins with the individual soldier and continues throughout the health care continuum. The best way to visualize the future MC4 system capability is as a piece of the Army digital computer network where all ten CHS functional areas have been digitized and CHS information is available to specified commands, supported units, and their personnel. See FM 8-10 for information on AMEDD functional areas. When fully developed, not only will the MC4 system provide Army commanders with CHS information, but will provide them with a seamless transition to the joint CHS environment. The MC4 system will consist of three basic components: software, hardware, and telecommunications systems.

a. Software capability.

(1) The joint TMIP will provide government off-the-shelf (GOTS)/commercial off-the-shelf (COTS) software and interoperability standards to support joint theater operations. The software provides an integrated medical information capability that will support all echelons of care in a theater of operations with links to the sustaining base. Medical capabilities provided by the software to support commanders in the theater will address medical C2 (including medical capability assessment, sustainability analysis, and MI); MEDLOG (including blood product management and medical maintenance management); casualty evacuation; and health care delivery.

(2) The MC4 system will support Army-unique requirements and any software needed to interface with Army information systems such as CSSCS, Global Combat Support System-Army (GCSS-A), FBCB2, Warrior Programs, and the Movement Tracking System. These systems will also be used in the medical platoons of the new IBCT. For additional IBCT medical platoon information, see Appendix D.

b. Hardware Systems. The hardware will consist of COTS automation equipment supporting the above software capabilities. Examples include, but are not limited to, computers, printers, networking devices, a digital patient record, and personal information carriers (PIC) that contain medical information.

c. Telecommunications Systems. The MC4 system will rely on current and proposed Army solutions for tactical, operational, and strategic telecommunications systems to transmit and receive digitized medical information throughout the theater and back to the sustaining base. There will be no separate AMEDD communication system. Telecommunications at brigade and below will be accomplished through the tactical internet; above brigade level, telecommunications will be accomplished through the Warfighter Information Network (WIN) architecture. The MC4 system will include hardware or software required to interface with current and emerging technologies supporting manual, wired, and wireless data transmission. At end-state, the MC4 system users will exchange data electronically via the WIN architecture. In the interim, commercial satellite and/or high frequency radios will be fielded to selected medical units (for example, Medical Detachment-Telemedicine [MDT], and so forth) receiving the MC4 system to support high bandwidth requirements until the WIN architecture is fully fielded. Personnel operating satellite assets are resourced in the MDT TOE and will be located with the MDT.

d. Patient Treatment Recording System. In the future under the MC4 system, medical information about each soldier of the maneuver battalion will be entered into a local database maintained at the supporting BAS or troop medical clinic. This information will include the soldier's immunization status, medical deployability status, and dental deployability status. Until a digital patient record and the PIC are fully functional and fielded, and in accordance with AR 40-66, a field medical record jacket (DD Form 2766) and its accompanying records will be maintained by the soldier's primary care provider. See Appendix B for definitive information on management of the individual health record in the field.

3-7. Information and Communications

a. The Medical Platoon Communications and Information Systems. Information and communications assets available to the BSS include those identified in Table 3-2. In the digitized medical platoon, each vehicle will have a FBCB2/position/navigation system (1 each). The FBCB2 system aids the medical platoon with maintaining real-time situational understanding of what is happening on the battlefield. The medical platoon exchanges information with forward deployed medical platoon elements, the FSMC, and the BSS.

b. Combat Health Support Functions on Force XXI Battle Command Brigade and Below System. The FBCB2 system is a hardware/software suite that digitizes C2 at brigade and below level. This system provides a seamless battle command capability for performance of missions throughout the operational continuum at the tactical level. The FBCB2 system is the implementation of information technology to provide increased battlefield operational capabilities. The system is positioned on specified platforms and will perform combat, CS and CSS functions for the planning and execution of operations. This system gives the medical platoon a CRP of the current CHS situation at BASs, AXPs, and the FSMC. For the first time, the medical organizations and elements are digitally linked to the platforms and organizations they support. The current CSS functionality on FBCB2 system gives the combatant a common relevant picture of the current CSS situation at his level of command and at subordinate levels. It also provides the personnel and logistics leaders situational understanding of CSS operations throughout their battle space. It provides an enhanced capability to synchronize support to customer units. Combat service support functionally on FBCB2 includes the following:

- Logistical Situational Report (LOGSITREP).
- Personnel Situational Report.
- Medical Situational Report (MEDSITREP).
- Situational understanding.
- Logistics call for support.
- Logistics task order.

Currently, the FBCB2 system also permits information to be entered using free text such as comments and other pertinent CSS information. This common battle space picture will enable CHS providers to maintain the operational tempo set by the maneuver commander. There are three medical screens incorporated into the CSS FBCB2 function. They are the medical functionality in the LOGSITREP, the MEDSITREP, and the MEDEVAC request. It is important that units use standard message and report formats to eliminate confusion. As the system is further developed and additional CHS screens are added, there will be less space for using free text. Figure 3-3 is the medical screen as seen on the CSS function of FBCB2. Descriptions of each screen are provided below.

(1) *Medical functionality in the logistical situational report.* This message provides visibility of selected Class VIII items at the BAS and FSMC stock levels, date and time group of the most recent report, and location of medical units. Recipients of the report are the FSC, the FSB support operations (HSSO), the BSS, and the DSS. This report is entered into the CSSCS by the BSS.

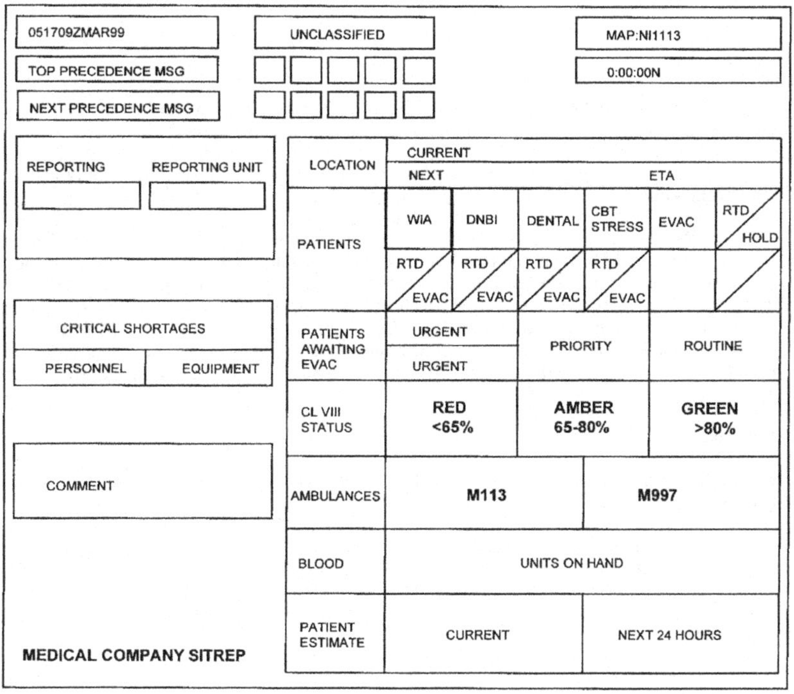

Figure 3-3. Sample of medical screen incorporated into the combat service support Force XXI Battle Command Brigade and Below function.

(2) *Medical Situational Report.* The FSMC and BAS prepare and submit this report. The recipients of the report are maneuver commanders and their S1, the FSC, the FSB support operations (HSSO), the BSS, and the DSS. The BSS and FSMC receive roll up from the BASs. The DSS receives a roll up of the FSMC reports. Adjacent units can receive information copies of the MEDSITREP. This message reports the following information:

(*a*) Current location and proposed next location with estimated time of arrival.

(*b*) Number of patients seen and classified as wounded in action, DNBI, dental, and combat stress. The field will also show the number of patients evacuated and the number RTD.

(*c*) Patient(s) awaiting MEDEVAC.

(*d*) The Class VIII status of the element/unit, the number of ambulances that are mission capable, and the number of units of blood and type on hand, if required.

(*e*) There will be a free text field for critical Class VIII or other supply shortages and commander's comments.

(3) *Medical evacuation request.* This request is currently embedded into FBCB2 and is a digitized standard 9-line MEDEVAC request. The current messaging is from the requestor to the medical platoon leader (with an information copy to the maneuver battalion commander). The medical platoon leader either responds or forwards the request to the FSMC commander who dispatches the appropriate MEDEVAC asset. Information copies of all MEDEVAC requests are sent to the BSS so they can maintain real-time situational understanding on the volume of requests. The FSMC commander sends an information copy to the BSS with after-action information that includes destination of evacuated patient(s).

3-8. Radio Nets

a. Battalion Communications. Battalion communications are sent over a variety of radio nets. Primary battalion communications nets are—

(1) *Command net.* A secure command net is used for C2 of the TF. All organic and attached units, including the FSO, forward air controller, and leaders of supporting elements, enter the battalion command net. Primarily, during the execution of the mission, only commanders transmit; all others monitor and transmit only essential information. The command operations net (see Figure 3-4) is controlled by the battalion main CP.

(2) *Operation and intelligence net.* The operations and intelligence (O&I) net is a secure net established to provide a mechanism for the battalion TF to accept routine items of information concerning O&I reporting without cluttering or interfering with the battalion command net.

(3) *Administrative/logistics net.* The administrative/logistics net is a tactical net, controlled by the combat train command post (CTCP), used to communicate the administrative and logistical requirements of the TF. All organic and attached units normally operate in this net.

(4) *Special radio nets.*

(*a*) The scout platoon net or a designated frequency may function as a surveillance net when required. The S2 and elements assigned surveillance missions operate on this net. Other elements enter or leave the net to pass information as required.

(*b*) The FSE and company fire support teams operate in the supporting FA command fire direction net and a designated fire direction net to coordinate FA fires for the battalion. The TACP operates in USAF tactical air-request and air-ground nets to control air strikes.

(*c*) Supporting air defense units monitor the early warning net. In the absence of collocated air defense support, the main CP will also monitor the division early warning net.

(*d*) Attached or OPCON support assets may operate in their parent unit nets, but they must also monitor the command net at all times.

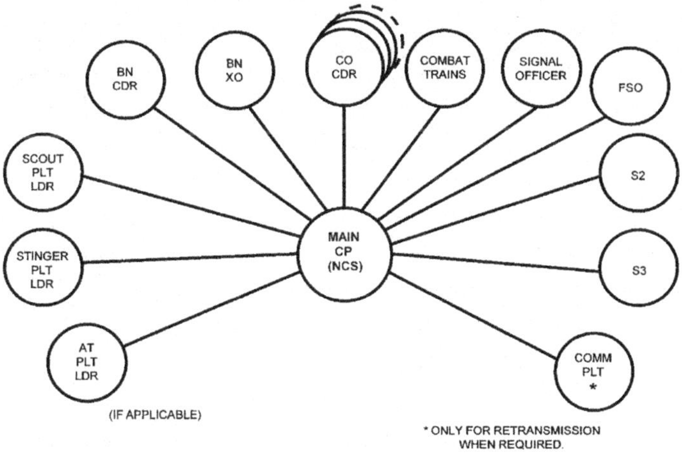

Figure 3-4. Example of a battalion command operations net.

b. Administrative/Logistics Radio Net. The administrative/logistics radio net (see Figure 3-5) is used for most CSS traffic. However, at battalion level, CSS communications can be via any combination of FM radio, mobile subscriber equipment (MSE), courier, computer, or wire. Lengthy reports should be sent by messenger, wire, computer, or tactical facsimile (FAX).

(1) The CTCP is the NCS for the administrative/logistics net. The S4, S1, HHC commander, battalion maintenance team (less light), support platoon leader, medical platoon leader, company XO or 1SG, and others (as required) operate in the battalion administrative/logistics net. The CTCP also operates in the brigade administrative/logistics net and in the battalion command net.

(2) The main CP and CTCP should be positioned, when wire is available and circumstances permit, so wire can be used as the main means of communication between them. Wire allows a constant flow of information between the CP. It also enhances the ability of the CTCP to stay abreast of the tactical situation and thus to provide better support. Wire communications produce no electronic signature and, therefore, are more secure than radio. When MSE is fielded, wire is needed only as a backup means of communications.

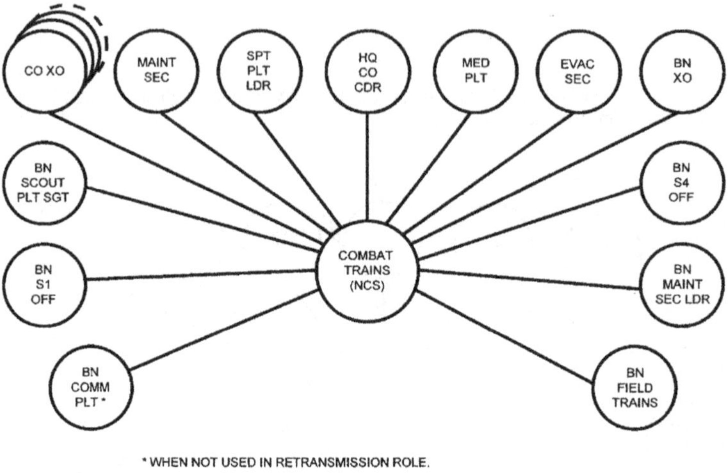

Figure 3-5. Example of an administrative/logistics net.

c. Medical Platoon Internal Operations Net. The platoon has access to the maneuver battalion HHC wire communications network for communications with all major elements of the battalion. Wireless communications for this section consists of a tactical FM radio mounted in the platoon headquarters vehicle. The medical platoon employs an FM radio network for CHS operations. The platoon headquarters section serves as the NCS for the platoon (see Figure 3-6). Table 3-2 lists the information and communications assets available to the Force XXI medical platoon.

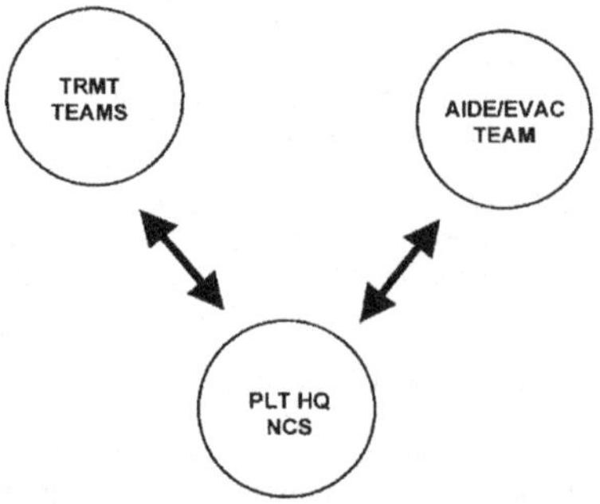

Figure 3-6. Example of the medical platoon internal operations net.

Table 3-2. Information and Communications Assets Available to the Force XXI Battalion Medical Platoon

RADIO SETS	
AN/VRC-89F	PLATOON HEADQUARTERS VEHICLE AND TREATMENT TEAM ALPHA (SURGEON) VEHICLE (1 EACH)
AN/VRC-88F	TREATMENT TEAM BRAVO
AN/VRC-90F	PLATOON AMBULANCE (1 EACH)
ROUTERS	**COMPUTER SYSTEMS**
TACTICAL LAN LAN ROUTER	MC4 LAPTOPS FOR EACH TREATMENT AND AMBULANCE VEHICLE
	FBCB2, 1 IN EACH VEHICLE ASSIGNED TO THE MEDICAL PLATOON
	OTHER SYSTEMS
FBCB2	MEDICAL PLATOON VEHICLE (1 EACH)
GPS	MEDICAL PLATOON VEHICLE (1 EACH)
EPLRS	MEDICAL PLATOON VEHICLE (1 EACH)
BCIS	MEDICAL PLATOON VEHICLE (1 EACH)
DVE	MEDICAL PLATOON VEHICLE (1 EACH)

d. Brigade Medical Operations Net. The brigade medical operations net under Army of Excellence units is the FSMC's command net. For Force XXI brigades and brigades with surgeon's sections assigned to the brigade, the brigade headquarters will maintain communications with medical elements supporting the brigade through its FM medical net. Single-channel ground and airborne radio system (SINCGARS) components provide the BSS with an AN/VRC-89 series radio set (FM) which has a receiver/transmitter (RT) capable of using two FM nets for reception and transmission. This permits the BSS to communicate with CHS elements via the administrative/logistic net (FM). The AN/VRC-89 series has two RTs (and one power amplifier). Two RTs allow the BSS to participate in two FM nets. These nets include the brigade administrative/logistics net and one each of the three medical platoons operations nets. The BSS also communicates using amplitude modulated (AM)-improved high frequency radios (IHFR) with its AN/GRC-213 or AN/GRC-193A radio. Another technique is to use the FSMC command net for brigadewide medical communications while using the administrative/logistics net for other CSS integration (see Figure 3-7). Situational understanding is monitored using the FBCB2 system and by face-to-face contact with other brigade staff members in the brigade TOC.

e. Mobile Subscriber Equipment. Mobile subscriber equipment will allow the BSS to communicate throughout the battlefield in either a mobile or static situation. As the Army continues to digitize the battlefield and modernize the force, the use of automation continues to develop. The MSE packet switching network gives units the ability to connect to division and corps LAN or wide area networks (WAN). A WAN is similar to the LAN but covers a larger distance. This allows units/CP to connect computer systems such as the CSSCS, maneuver control system, and FBCB2 system to an ethernet cable (coaxial) and send and receive information in an extremely efficient manner. Because of the limitations of a network constructed with coaxial cable, a WAN uses a combination of the MSE packet switch network and radio networks to distribute the data where necessary through the system. Packet switching does not use or take up existing telephone lines. Instead, telephone lines are freed up even more because information is being sent over a network on computers and related equipment. Using the common hardware/software facilitates the interface and exchange of information between the BSS and the medical platoons operating BAS, the FSMC, the DSS, the corps, and the division medical elements. See FM 63-2-2 for information concerning automated data processing continuity of the operations plan.

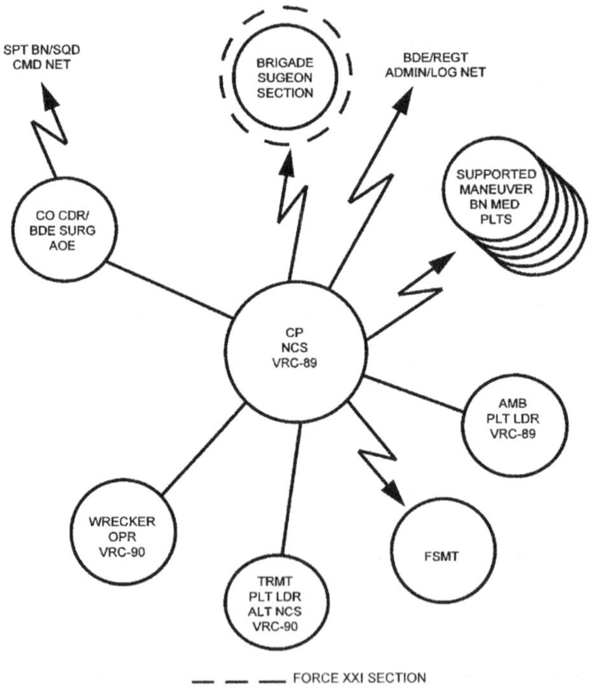

Figure 3-7. Medical company/troop command FM net/brigade medical operations net.

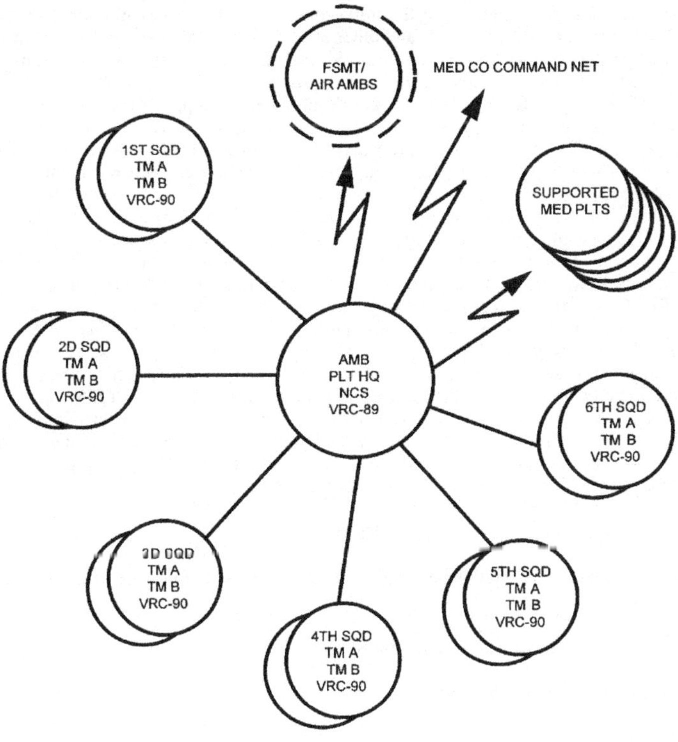

Figure 3-8. Dedicated medical evacuation FM net.

CHAPTER 4

COMBAT HEALTH SUPPORT OPERATIONS

4-1. Combat Health Support for Reconnaissance Operations

Reconnaissance precedes all military operations and verifies or refutes analyzed information in IPB products. The RECON information may require that the initial plans or COA be modified or discarded. The R&S systems are used to see the enemy, terrain, and other aspects of the battle space that will affect operations. As advanced warfighting systems are fielded, the R&S capabilities at all levels will be significantly enhanced but will never completely replace human collection requirements. The scout platoon is the primary R&S asset for the battalion/TF. This unit is specifically designed and equipped to conduct R&S operations and can do some limited intelligence analysis. It is the only element in the battalion/TF specifically trained in reconnaissance and can work up to 30 kilometers across the forward line of own troops (FLOT) forward of the battalion/TF. The key to synchronizing this asset is predicated on each member of the platoon understanding the battalion/TF commander's operational intent and concept. Members should understand their roles in obtaining the major decision point criteria. Detailed plans and orders are key to effective employment of the scout platoon. The continuous nature of the R&S process requires the commander to manage the employment of the scout platoon for R&S requirements tied to critical decision points. Reconnaissance missions require a detail list of exactly what is being tasked and a clear priority for collection. If RECON assets could find and report only one piece of information, what does the commander want it to be? For example, the answer to this questions is reflected in the instructions to the scouts. "Execute zone RECON forward of the battalion/TF, beginning at 0345 hours. Collect and report the following in priority. Phase I: Number 1, confirm or deny enemy antitank minefield at grid FT 456689; Number 2, determine enemy unit and type and disposition of hill 413, grid FT535731." This allows the scout platoon leader to collect vital information first and focus his efforts, enhancing the battalion/TF's planning process. Reconnaissance assets should not be launched without determining the initial RECON plan. The medical platoon leader will be involved with developing the CHS portion of the RECON plan. The RECON plan can be issued as a FRAGO and should include—

- Composition/task organization for RECON.
- Key facts (priority intelligence requirements) to be gathered by the RECON party and expected results.
- Movement routes/formations to the RECON location.
- Actions on reaching the location.
- Special instructions to members of the RECON party. Collection task should be specified.
- Any special equipment required (chemical detection kits, expedient or directional antennas, and so on).
- Contingency plans.
- Requirements for continued surveillance after the RECON.

- Indirect fire support for RECON.
- Security arrangements.
- Communications arrangements (nets, retransmission sites, and so on).
- Withdrawal, linkup, and dissemination plans.

Combat service support considerations for RECON include—

- Developing SOPs for load plans of outposts.
- Establishing CSS procedures for both mounted and dismounted RECON missions.
- Establishing CSS procedures for air versus ground insertion of RECON assets.
- Developing procedures for both aerial and ground sustainment.
- Developing resupply techniques to include—
    - Using multiple/false landing zones away from outposts as cache drop-off points.
    - Predetermining the locations and times for resupply of Classes I, III, IV, V, VII, and XI.
    - Establishing locations for caches on successive missions/insertions.

Combat health support considerations include—

- Determining the CHS requirements.
- Deploying trauma specialists and medical assets in DS of RECON operations, as required and appropriate.
- Planning for casualty evacuation/extraction operations.
- Developing a CHS SOP for supporting RECON elements deployed deep into enemy territory.
- Selecting evacuation sites (remembering that all cache sites are potential casualty evacuation sites).
- Developing TTPs for cross-FLOT casualty extraction.
- Establishing the time for pick up and the pickup point for aerial extraction of casualties. (The last known/reported location is normally the aerial pickup point and the best time is 30 minutes prior to BMNT or 30 minutes after EENT.)

- Developing a detailed plan for ground extraction, to include link up to quick reactionary force (QRF) and escort to casualty exchange point.

- Conducting rehearsals for day and night extractions.

- Requesting escort if a QRF is not established.

During RECON missions if a member or members of the platoon become casualties, initial care will be self-aid, CLS advanced first aid, or EMT from a trauma specialist. Trauma specialists may be deployed as a rider in one of the scout vehicle or an ambulance team may be in support. If casualties/patients require extraction/medical evacuation, it becomes a combat mission. Since this is a combat mission that may require offensive action, medical personnel are placed under the control of a QRF leader. Provided in Figure 4-1 is a technique for conducting cross-FLOT ground extraction and lists the overall considerations. Figure 4-2 is a technique for conducting cross-FLOT air extractions and lists the overall considerations. Since the RECON element is usually small, enemy contact could produce a significant number of casualties requiring extraction of the entire team. Either ground or air could be used to execute the extraction of the RECON team. As stated above, this is a combined arms operation. Medical personnel will participate in the planning, preparation, and execution of these missions. If medical evacuation vehicles are not sent with the extraction force, then one vehicle should be designated for casualties and augmented with a trauma specialist. For additional information on reconnaissance operations, see FMs 7-20, 7-30, 7-92, 17-98, and 71-3.

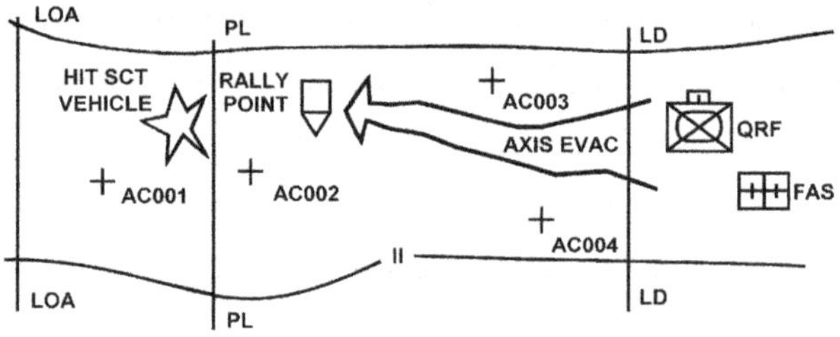

OVERALL CONSIDERATIONS

- EXTRACTION IS A COMBAT MISSION
- PLAN/REHEARSE QRF
- PLAN COMBAT MULTIPLIERS
- PLAN ARTILLERY SUPPORT THROUGHOUT ZONE/SECTOR
- PLAN FALSE/DISTRACTING ARTILLERY
- CLOSE AIR SUPPORT
- GRAPHICS ISSUED TO SUPPORT MISSION
- ALERT ALL RECON ASSETS
- USE RECON ASSETS IN SECTOR/ZONE TO VECTOR QRF TO RALLY POINT
- POSITION AID STATION NEAR QRF STAGING AREA
- PLAN MUST COVER MOVEMENT IN AND OUT OF SECTOR/ZONE
- LIMITED ATTACKS TO DISTRACT ENEMY/CLEAR ROUTES, ETC.

Figure 4-1. A technique for cross-FLOT ground extraction.

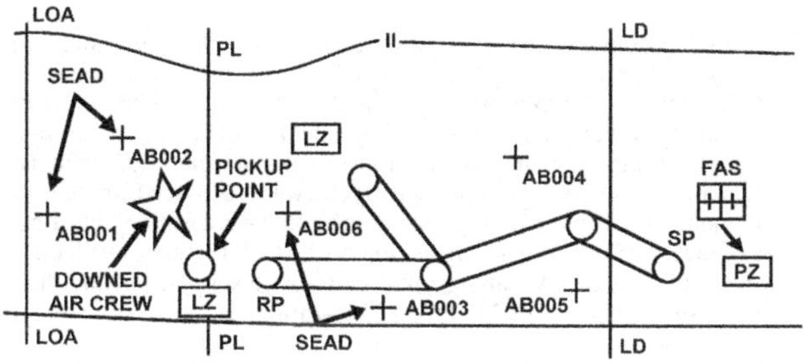

OVERALL CONSIDERATIONS

- EXTRACTION IS A COMBAT MISSION
- PLAN COMBAT MULTIPLIERS
- LETHAL SEAD—INGRESS/EGRESS
- NONLETHAL SEAD—TLQ-17/COMPASS CALL
- FALSE LZ/INSERTIONS
- FALSE ARTILLERY/SEAD
- CLOSE AIR SUPPORT
- ALERT ALL RECON ASSETS
- ISSUE INTEL UPDATE AT PZ
- USE RALLY POINTS/DOWNED CREW PICKUP POINTS
- POSITION AID STATION NEAR LZ

Figure 4-2. A technique for cross-FLOT air extraction.

4-2. Combat Health Support for the Offense and the Defense

a. Combat Health Support for the Offense.

(1) The offense is the decisive form of war, the commander's only means of attaining a positive goal, or of completely destroying an enemy force (FM 100-5). Rapid movement, deep penetrations, aggressive action, and the ability to sustain momentum regardless of counterfires and countermeasures characterize the offense.

(2) When developing the CHS plan to support the offense, the CHS planner must consider many factors (FM 8-55). The forms of maneuver, as well as the threat's capabilities, influence the character of the patient workload and its time and space distribution. The analysis of this workload determines the allocation of CHS resources and the location or relocation of MTF.

(3) Combat health support for offensive operations must be responsive to several essential characteristics. As operations achieve success, the areas of casualty density move away from the supporting MTF. This causes the routes of MEDEVAC to lengthen. Heaviest patient workloads occur during disruption of the threat's main defenses, at terrain or tactical barriers, during the assault on final objectives, and during threat counterattacks. The accurate prediction of these workload points by the CHS planner is essential if MEDEVAC operations are to be successful.

(4) As advancing combat formations extend control of the battle area, supporting medical elements have the opportunity to clear the battlefield. This facilitates the acquisition of the battle wounded and reduces the vital time elapsed between wounding and treatment. There are two basic problems confronting the supporting medical units and MEDEVAC elements. First, contact with the supported units must be maintained. Responsibility for the contact follows the normal CHS pattern—higher echelon evacuates from lower echelon. Contact is maintained by forward deployed air and ground evacuation resources. Secondly, the mobility of the MTF supporting the combat formations must be maintained. The requirement for prompt MEDEVAC of patients from forward MTF requires available ambulances to be echeloned well forward from the outset. Air and ground ambulance support beyond the capabilities of the FSMC is requested from the supporting corps MEDEVAC battalion.

(5) In traditional combat operations, the major casualty AO is normally the zone of the main attack. As the main attack accomplishes the primary task of the combat force, it receives first priority in the allocation of combat power. The allocation of combat forces dictates roughly the areas that are likely to have the greatest casualty density.

(*a*) In the division, CHS (Echelon I) for the brigade is provided by maneuver battalion medical platoons. The medical platoon operates the BAS or squadron aid station, places trauma specialists in DS of the maneuver companies and RECON troop, provides patient evacuation from forward areas, and deploys treatment teams in DS of battalion or squadron elements for up to 48 hours without resupply. The FSMC (Echelon II) located in the BSA provides MEDEVAC support from the BAS/squadron aid station to the BSA and reinforces treatment capabilities at BAS/squadron aid stations for limited periods of time. When combat operations commence, the medical platoon normally locates its BAS as far forward as combat operations permit. The BAS treatment squad can split into two treatment teams and operate as two separate aid stations (BAS [-]), normally not to exceed 24 hours. Treatment Team A operates the main aid station (MAS) and Treatment Team B operates the forward aid station (FAS). In continuous operations, when operating for longer periods, personnel efficiency and unit capability will tend to deteriorate. Each team employs treatment vehicle(s) with two MES—one trauma set and one general sick call set. The medical platoon will depend on CLS, trauma specialists, company health care SGT, and unit 1SG to assist with clearing the battlefield and getting injured or wounded soldiers to the CCP. From the forward areas and the CCP, armored ambulances will provide medical evacuation to the FAS and MAS. Based on casualty estimates, additional ambulances may be forward positioned at CCP. The CCP, Treatment Team A, and Treatment Team B must select sites that provide cover and concealment and afford some protection to their patients. For additional information on force protection, see Appendix E. Treatment teams must maintain as much mobility as possible while providing stabilization care and rapid medical evacuation. Treatment and evacuation elements must ensure that adequate stocks of Class VIII items are on-hand and, if necessary, request Class VIII resupply from the FSMC.

(*b*) The Force XXI medical platoon may deploy a treatment team forward in support of brigade RECON troops based on mission requirements. As the battle moves from the original area of contact, coordinated movements of treatment teams allow for continuous CHS. Once patients are received, a treatment team from the BAS or squadron aid station will care for and treat these patients until their MEDEVAC or appropriate disposition. The remaining treatment teams of the maneuver BAS and squadron aid stations move with the battle and provide CHS to the maneuver and RECON elements according to the brigade order.

(*c*) Careful coordination must be used when deciding when and which treatment team will receive patients. The plan must take into account maintaining the mobility of the medical platoon so that at least one treatment team will be able to follow the force as it maneuvers. This will keep evacuation routes and times to a minimum. Maneuver BAS/treatment teams provide area medical support to units without organic medical support operating in the maneuver battalion's AO. Brigade RECON will plan for area medical support for cross-FLOT operations. After MEDEVAC or appropriate disposition of their patients, the treatment team prepares for its next move. This echeloned displacement (leapfrog) technique provides for maximum utilization of medical platoon treatment teams and permits continuous uninterrupted CHS to maneuver battalions on the move. Each of the above actions must be coordinated with the CTCP and the FSMC.

(*d*) The BAS is under the tactical control of the battalion S4 and is normally deployed in the vicinity of combat trains. To reduce ambulance turnaround time in providing ATM to patients within 30 minutes of wounding, the BAS may split and place its treatment teams as close to maneuvering companies as tactically feasible. The battalion S4 closely coordinates locations for forward positioning CSS elements (including medical treatment elements) with the battalion S3. This is to ensure that the location of these elements is known by commanders of maneuvering and CS forces. Coordination ensures that CSS elements are not placed in the way of friendly maneuvering forces; in line of incoming or outgoing fires; or in areas subject to be overrun by rapidly advancing enemy forces. Treatment teams situated close to (within 1,000 meters of) maneuvering companies in contact must be prepared to withdraw to preplanned, alternate positions on short notice. They must also be aware of enemy capabilities. Some adversaries will possess weapons capable of direct fire at 4,000 meters.

(*e*) Trauma specialists are allocated to infantry and mechanized infantry companies on the basis of one trauma specialist per platoon and, under Force XXI, a health care SGT is allocated for each company. In armored units, the allocation is one health care SGT and one ambulance team per company. The location of the trauma specialist is of extreme importance for early acquisition and medical treatment of casualties. The mechanized infantry platoon trauma specialist normally locates with, or near, the element leader. When the platoon is moving on foot in the platoon column formation, he positions himself near the element leader trailing the base squad forward of the second team. This formation is the platoon's primary movement formation. When the platoon is mounted, the trauma specialist will usually ride in the same vehicle as the platoon SGT. The company senior trauma specialist collocates with the 1SG. When the company is engaged, he remains with the 1SG and provides medical advice as necessary. When a casualty occurs, first aid will be rendered by self-aid/buddy aid or by the CLS. The platoon/company trauma specialist will then go to the casualty's location or the casualty will be brought to the trauma specialist. The trauma specialist makes his assessment; administers initial medical care; initiates a DD Form 1380 (FMC); then requests evacuation or returns the individual to duty. The patient is evacuated from the point of injury or is transported on a nonmedical vehicle (coordinated by the 1SG) to a collection point. A vehicle from the medical platoon evacuation section (usually pre-positioned forward) picks up the patient and transports him to the BAS. As the tactical situation allows, the trauma specialist will provide medical treatment and prepare patients for evacuation. The ambulance team supporting the company works in coordination with the trauma specialists supporting the platoons. When a casualty occurs in a tank or an armored fighting vehicle, the evacuation team will move as close to the vehicle as possible, making full use of cover, concealment, and defilade. Assisted, if possible, by the vehicle's crew, they will extract the casualty from the vehicle and administer EMT. They move the patient to the treatment team or to a CCP to await further

MEDEVAC. The company health care SGT normally remains with the company CP, but may be used anywhere in the company, assisting the ambulance teams in some situations. The 1SG oversees the evacuation of casualties back to a CCP and may employ nonstandard medical platforms to accomplish his mission. The company health care SGT will position himself at the CCP and provide medical treatment, to include prioritizing patients for MEDEVAC to the BAS.

(6) When maneuvering companies anticipate large numbers of casualties, augmentation of the medical platoon with one or more treatment teams from the FSMC should be made. Augmenting treatment teams are under the tactical control of the battalion S4, but are under the technical control of the battalion surgeon. Medical treatment facilities should not be placed near targets of opportunity such as ammunition, POL distribution points, or other targets that may be considered lucrative by the opposing force. Considerations for the location of the BAS should include—

- Tactical situation/commander's plan.
- Expected areas of high casualty density.
- Security.
- Protection afforded by defilade.
- Convergence of lines of drift.
- Evacuation time and distance.
- Accessible evacuation routes.
- Avoidance of likely target areas such as bridges, fording locations, road junctions, and firing positions.
- Good hard stand on solid ground with good drainage.
- Near an open area suitable for helicopter landing.
- Available communication means.

(7) Types of operations in the offense include—

(*a*) *Movement to contact*. Medical evacuation support in movement to contact is keyed to the tactical plan. Prior deployment of FSMC ground ambulances with the maneuver battalion's organic medical platoons permits uninterrupted and effective MEDEVAC support from the BAS to the FSMC located in the BSA. Movement to contact operations is executed when there is little or no threat information. The FSMC and treatment teams from maneuver BAS in support of these operations must maintain their flexibility and be prepared to adjust CHS once contact is established.

(*b*) *Exploitation and pursuit.* Medical evacuation support of exploitation and pursuit operations resemble those discussed for the envelopment (paragraph 4-3*a*[2]). Since exploitation and pursuit operations can rarely be planned in detail, evacuation operations must adhere to TSOP and innovative C2. These actions are often characterized by—

- Fewer casualties
- Decentralized operations.
- Unsecured ground evacuation routes.
- Exceptionally long distances for evacuation.
- Increased reliance on convoys and air ambulances.
- More difficult communications.

(*c*) *Deliberate attack.* The deliberate attack is based on a more detailed knowledge of the threat disposition and likely actions. The brigade's actions in contact will be more predictable than the fluid situation found in the movement to contact, or exploitation or pursuit. Specific terrain and routes/ avenues of approach can be selected. Units can conduct at least a map RECON of their planned locations. While there may be CHS requirements during the approach, the assault on the objective will produce the greatest number of casualties. Some of the CHS considerations for the deliberate attack include—

- Higher percentage of casualties.
- Casualties will be more concentrated in time and space.
- Once the objective is secured, treatment teams can move to the objective instead of evacuating patients from the objective to the treatment teams.
- Use of air ambulance to overcome some obstacles may be required.
- Higher likelihood of wounded EPW.

(8) At the BAS, patients requiring further evacuation to the rear are stabilized for movement. Patient holding and food service are not available at the BAS; therefore, only procedures necessary to preserve life or limb, or enable a patient to be moved safely are performed at the BAS. Evacuation from the BAS or point of injury is performed by the FSMC's ambulance platoon and by the FSMT. Constant effort should be made to prevent unnecessary evacuation; patients with minor wounds or illnesses are treated and RTD as soon as possible. Patients with dental emergencies are evacuated to the FSMC for treatment. Ammunition, hand grenades, and light antitank weapons are not evacuated from the forward areas with a patient. Ambulatory patients may retain their rifle or pistol. Other weapons, such as machine guns and those items identified above, are collected and given to the soldier's 1SG or turned in to the battalion S4 for appropriate disposition according to command SOP/policy. All excess equipment collected

at the BAS is turned in to the battalion S4 or as directed by command SOP. Patients will always retain their protective mask. Other functions of the BAS include—

- Receiving and recording patients in the daily disposition log.
- Preparing FMC as required (see FM 8-10-6 for definitive information on initiating and completing a DD Form 1380).
- Verifying information contained on the FMC of all patients evacuated to the BAS.
- Requesting and monitoring MEDEVAC of patients.
- Monitoring personnel, when necessary, for NBC contamination prior to medical treatment.
- Notifying the S1 of all patients processed through the BAS, giving identification and disposition of patients.
- Supervising patient decontamination and treating NBC patients (see FMs 8-9, 4-02.283, 8-284, and 8-285).

NOTE

Patient decontamination is performed by a pretrained decontamination team. This team is composed of eight nonmedical personnel from supported units that work under the supervision of medical personnel. Patient decontamination teams perform best when they are trained and are permitted to exercise their skills with the supporting BAS. For definitive information on patient decontamination procedures, see FMs 8-10-1, 8-10-7, 4-02.283, 8-284, and 8-285.

b. Combat Health Support for the Defense. There are three forms of the defense: area defense, mobile defense, and retrograde. The area defense concentrates on denying threat access to designated terrain for a specific period of time, rather than on the outright destruction of the threat. The mobile defense focuses on denying the threat force by allowing him to advance to a point where he is exposed to a decisive counterattack by the striking force. The primary defeat mechanism, the counterattack, is supplemented by the fires of the fixing force. The third form of defense is the retrograde. The retrograde is an organized movement to the rear and away from the threat. The threat could force these operations or a commander can execute them voluntarily. Delay, withdrawal, and retirement are the three forms of retrograde operation.

(1) Combat health support provided for the defense may reflects lower casualty rates, but threat actions and the maneuver of combat forces complicate forward area patient acquisition.

Medical personnel are permitted much less time to reach the patient, complete vital EMT, and remove him from the battle site. Increased casualties among exposed medical personnel further reduce the medical treatment and evacuation capabilities. Heaviest patient workloads, including those produced by threat artillery and NBC weapons, can be expected during the preparation or initial phase of the threat attack and in the counterattack phase. The threat attack can disrupt ground and air routes and delay evacuation of patients to and from treatment elements. The depth and dispersion of the defense create significant time and distance problems for evacuation assets. Combat elements could be forced to withdraw while carrying their remaining patients to the rear. The threat exercises the initiative early in the operation, which could preclude accurate prediction of initial areas of casualty density. This makes the effective integration of air assets into the MEDEVAC plan essential. The use of air ambulances must be coordinated with the FSMT, normally positioned in the BSA under Force XXI and/or with either the aviation brigade or the BSA in AOE divisions. See FM 8-10-26 for additional information on the air ambulance company and FSMT.

(2) The CHS requirements for retrogrades can vary widely depending upon the tactical plan, the threat reaction, and the METT-TC factors. Firm rules that apply equally to all types of retrograde operations are not feasible, but considerations include—

- Requirement for maximum security and secrecy in movement.

- Influence of refugee movement conducted in friendly territory, which could impede MEDEVAC missions.

- Integration of evacuation routes and obstacle plans.

- Difficulties in controlling and coordinating movements of the force that could produce lucrative targets for the threat.

- Movements at night or during periods of limited visibility.

- Time and means available to remove patients from the battlefield.

In stable situations and in the advance, time is important only as it affects the physical well-being of the wounded. In retrograde operations, time is more important. As available time decreases, CHS managers at all echelons closely evaluate the capability to collect, treat, and evacuate all patients.

- Medical evacuation routes required for the movement of troops and materiel. This causes patient evacuation in retrograde movements to be more difficult than in any other type of operation. The threat could disrupt C3. Successful MEDEVAC requires including ambulances on the priority list for movement; providing for the transportation of the slightly wounded in cargo vehicles; and providing guidance to subordinate commanders defining their responsibilities in collecting and evacuating patients. Special emphasis must be placed on the triage of patients and consideration given to the type of transportation assets available for evacuation.

- Decisions concerning patients left behind. When the patient load exceeds the means to move them, the tactical commander must make the decision as to whether patients are to be left behind. The medical staff officer keeps the tactical commander informed so that he can make timely decisions. Medical personnel and supplies must be left with patients who cannot be evacuated. (Refer to FM 8-10 for additional information.) Every effort will be made to avoid this "solution."

4-3. Combat Health Support for Maneuver and Enabling Operations

a. Choices of Maneuver.

(1) *Penetration*. In this tactic, the attack passes through the threat's principal defensive position, ruptures it, and neutralizes or destroys the threat forces. Of all forms of offensive maneuver, the penetration of main threat defenses normally produces the heaviest patient workload. The commander's plan for the penetration normally has three phases.

- Breaching the enemy's main defensive position.
- Widening the gap created to secure the flanks by enveloping one or both of the newly exposed flanks.
- Seizing the objective with its associated subsequent exploitation.

During breach operations, plans must be in place for clearing casualties off the battlefield since increased number of casualties should be anticipated. Each company team requires an armored ambulance for casualty evacuation. If the engineer company is the breaching force, it must have one armored ambulance in DS. The BAS splints into two treatment teams, Alpha and Bravo, while dividing the two supporting FSMC ambulances. The treatment teams pre-position with the TF prior to LD. Treatment Team Bravo moves forward behind one of the company teams and is designated as the FAS. Treatment Team Alpha (MAS) follows the TF formation. The AXP with a treatment team from the FSMC must be integrated into the TF scheme of maneuver. Without the AXP, the TF medical elements will lose their ability to move as patients collect at the FAS and MAS. The AXP moves forward with the TF combat trains. The TF combat trains should move within 4 kilometers, but no more that 10 kilometers behind the lead elements of the TF. Once the breach is completed, the FAS moves through the breach to the other side, while the MAS moves to the position previously occupied by the FAS. The AXP moves forward to the position previously occupied by the MAS. Patient acquisition starts slowly, but becomes more rapid as the attack progresses. Evacuation routes lengthen as the operation progresses. Heavy preparatory fires which can evoke heavy return fire often precede the penetration maneuver. These threat fires could modify the decision to place evacuation assets as far forward as possible. The FSMC can reinforce the penetration force medical elements. Patient evacuation could be slow and difficult due to a bottleneck at the penetration. Medical evacuation support problems multiply when some combat units remain near the point of original penetration. This is done to hold or widen the gap in threat defenses while the bulk of tactical combat forces exploit or pursue the threat. Treatment elements are placed near each shoulder of the penetration; ground evacuation cannot take place across an avenue of heavy combat traffic. Because of the heavy traffic, the area of the penetration is normally a target for both conventional and NBC weapons. The trigger to push treatment team/BASs through the penetration and where they will go must be identified in the OPORD.

(2) *Envelopment*. In the envelopment, the main or enveloping attack passes around or over the threat's principal defensive positions. The purpose is to seize objectives which cut the threat's escape routes and subject him to destruction in place from flank to rear. Since the envelopment maneuver involves no direct breach of the threat's principal defensive positions, the MEDEVAC system is not confronted with a heavy workload in the opening phase. However, ambulances are positioned well forward in all echelons of CHS to quickly evacuate the patients generated by suddenly occurring contact. Medical treatment facilities moving with their respective formations assist with clearing the battlefield to reduce delays in treatment. After triage and treatment, the patients are evacuated to MTF in the rear by supporting ground ambulances from the FSMC. When the isolated nature of the envelopment maneuver precludes prompt evacuation, the patients are carried forward with the treatment element. This must be planned for in detail and is an extreme measure when no other option is feasible. Expect an increase in mortality from wounds. Again, nonmedical vehicles could be pressed into emergency use for this purpose. When patients must be carried forward with the enveloping forces, CHS commanders use halts at assembly areas and phase lines to arrange combat protection for ground ambulance convoys through unsecured areas. Further, the commander should take advantage of friendly fires and suppression of threat air defenses to call for prearranged air ambulance support missions, or emergency use of medium-lift helicopter backhaul capabilities.

(3) *Infiltration*.

(*a*) Infiltration is a choice of maneuver used during offensive operations. The division can attack after infiltration or use it as a means of obtaining intelligence and harassing the threat. Though it is not restricted to small units or dismounted actions, the division employs these techniques with a portion of its units, in conjunction with offensive operations conducted by the remainder of its units.

(*b*) Combat health support of infiltration is restricted by the amount of medical equipment, supplies, and transportation assets that can be introduced into the attack area. No deployment of BAS treatment teams without their organic transportation should be attempted. Elements of unit-level CHS should be accompanied by their organic vehicles, and ambulances should receive priority for deployment. It may be necessary to man-carry enough BAS equipment into the attack area to provide EMT and ATM; this, however, results in degrading mobility. When the element is committed without its ambulances, patients are evacuated to the BAS by litter bearer teams. These litter teams must be designated and equipped by the commanders in their orders. Noise, light, and litter discipline during evacuation in an infiltration depends on how the casualty was wounded. Disease and nonbattle injury soldiers may not have been noticed by the enemy. If the casualty is a battle injury, the enemy has already detected that element. Once the enemy has detected and engaged the force, causing casualties, maximum allowable use of standard and nonstandard evacuation platforms should be used. This will increase lift capabilities and save time and soldier's lives. Patient evacuation from the BAS and medical resupply of the force may be provided by litter bearers, depending upon distances and degree of secrecy required.

(*c*) When airborne and air assault forces are used, infiltrating elements can land at various points within the threat's rear area and proceed on foot to designated attack positions. As in surface movement, the amount of medical equipment taken could be limited. In airborne operations, the evacuation of patients will be by litter bearers or ground ambulances to collecting points or the BAS and then by FSMC ambulances to the clearing station operated by the FSMC treatment platoon. In air assault operations, the evacuation is by litter bearers to collecting points or the BAS and then by air ambulances to a clearing station. Once the combat element begins the assault on the objective, secrecy is no longer important and its isolated location requires CHS characteristic to airborne and air assault operations until ground linkup.

(4) *Turning movement*. A turning movement is a variation of the envelopment in which the attacking force passes around or over the threat's principal defense positions to secure objectives deep in the threat's rear; thus, forcing the threat to abandon his position or divert major forces to meet the threat. As stated above, the turning movement is a variant to the envelopment in which the attacker attempts to avoid the defense entirely; rather, the attacker seeks to secure key terrain deep in the threat's rear and along his LOC. Faced with a major threat to his rear, the threat is thus "turned" out of his defensive positions and forced to attack rearward at a disadvantage. Medical evacuation support to the turning movement is provided basically in the same manner as to the envelopment. As the operation is conducted in the threat's rear area, LOC and evacuation routes could be unsecured, resulting in delays in resupply and evacuation.

b. Enabling Operations.

(1) *Passage of lines*. This situation presents a challenge for the CHS planner. There will be a number of MEDEVAC units using the same air and road networks. Coordination and synchronization are essential if confusion and overevacuation are to be avoided. The medical units of the force manning the line should provide area support to the force passing through. This allows continued mobility for the moving force. The below information facilitates this coordination.

- Radio frequencies and call signs.
- Operation plans and TSOPs.
- Location of MTFs.
- Location of CCP and AXP.
- Main supply route, forward arming and refueling points, and A2C2 data.

(2) *Security operations*. Security operations obtain information about the enemy and provide reaction time, maneuver space, and protection to the main body. Security operations are characterized by aggressive RECON to reduce terrain and enemy unknowns, to gain and maintain contact with the enemy to ensure continuous information, and to provide early and accurate reporting of information to the protected force. See FM 17-95 for definitive information on security operations. The discussion below focuses on how CHS is provided for security operations. Security operations include the following missions: cover, screen, guard, and area security.

(*a*) *Cover.* The covering forces are dependent upon organic resources found in the maneuver battalion medical platoon for initial support. The level of command for the covering force determines the responsibility for the subsequent evacuation plan. In a corps covering force, for example, the corps CHS structure has the responsibility for establishing and operating the MEDEVAC system to support the forward deployed corps forces. This is done to prevent the tactical combat force following the covering forces from becoming overloaded with patients prior to the hand off and passage of lines. The use of CCPs, AXPs, and nonmedical transportation assets (casualty evacuation) to move the wounded is essential. The covering force battle could be extremely violent. Patient loads will be high and the distance to MTFs can be much longer than usual. The effectiveness of the MEDEVAC system depends upon the forward positioning of a number of ground ambulances and the effective integration of corps air ambulances into the evacuation plan.

(*b*) *Screen*. The primary purpose of a screen is to provide early warning to the main body. Screen missions are defensive in nature and largely accomplished by establishing a series of observation posts and conducting patrols to ensure adequate surveillance of the assigned sector. The screen provides the protected force with the least protection of any security mission. Combat health support will be provided by organic medical elements and ambulances teams deployed from the supporting medical company.

(*c*) *Advance, flank, and rear guards*. A guard force accomplishes all the task of a screening force. Additionally, a guard force prevents enemy ground observation of and direct fire against the main body. A guard force reconnoiters, attacks, defends, and delays as necessary to accomplish its mission. A guard force normally operates within the range of main body indirect-fire weapons. A guard force is deployed over a narrower front than a screen to permit concentration of combat power. These forces normally receive MEDEVAC support through the attachment of evacuation teams. The teams evacuate patients to predesignated CCPs along a main axis of advance or to the nearest treatment element providing area support. Employment of air ambulances provides a measure of agility and flexibility.

(*d*) *Area security*. Area security is a form of security that includes RECON and security by designated personnel, airfields, unit convoys, facilities, MSR, LOC, equipment, and critical points. Area security operations are conducted to deny the enemy the ability to influence actions in a specific area or to deny the enemy use of an area for his own purpose. This may entail occupying and establishing a 360-degree perimeter around the area being secured, or taking actions to destroy enemy forces already present. The area to be secured may range from specific points (bridges, defiles) to areas such as terrain features (ridgelines, hills) to large population centers and adjacent areas. Combat health support will be provided by organic and attached medical elements. In area security, Echelon I CHS is provided by organic or attached treatment teams. Echelon II CHS is provided by the supporting medical company via DS and on an area support basis. Depending on the type of area security operations being conducted, both air and ground ambulances may be employed.

(3) *River crossing operations*. The river barrier itself exerts decisive influence on the use of medical units. An attack across a river line creates a CHS delivery problem comparable to that of the amphibious assault. Combat health support elements cross as soon as combat operations permit. Early crossing of treatment elements reduces turnaround time for all crossing equipment that is used to load patients on the far shore. Maximum use of air ambulance assets is made to prevent excessive patient buildup in far shore treatment facilities. Near shore MTFs are placed as far forward as assault operations and protective considerations permit to reduce ambulance shuttle distances from off-loading points. For detailed information on river crossing operations, refer to FM 90-13. Rescuing casualties in the water must be considered by the TF medical planner.

(4) *Reconnaissance operations*. The RECON in force is an *attack* to discover and test the threat's position and strength or to develop other intelligence. The tactical combat force usually probes with multiple combat units of limited size, retaining sufficient reserves to quickly exploit known threat weaknesses. Combat health support techniques follow those discussed above for a movement to contact. Ambulances are positioned well forward and moved at night to enhance secrecy. The echeloning of ambulances is an indication to the threat that an attack is imminent due to the forward placement of CHS. Clearing stations are not established until a significant patient workload develops. Patients received at BASs of reconnoitering units are evacuated to clearing stations as early as practical, or are carried forward with the force until a suitable opportunity for evacuation presents itself. The maximum possible use of air ambulance assets is made to cover extended distances and to overcome potentially unsecured ground evacuation routes.

(5) *Unified action*. The majority of the operations occurring at the present time are joint, interagency, or multinational operations. During the initial planning stages of the operations, the CHS planner must determine who is responsible for providing MEDEVAC support to the force. The CHS planner must also ensure that duplications in support do not exist; that guidelines are established as to eligible beneficiaries; when individuals are to be returned to their own nation's health care delivery system; and what mechanisms exist for reimbursement of services. For additional information, refer to FM 8-42 and Joint Publication 4-02.

(6) *Integrated warfare operations*. Medical evacuation in an NBC environment is discussed in FMs 8-10-6 and 8-10-7.

4-4. Combat Health Support During Night Operations

The battalion surgeon and medical platoon members must anticipate that brigades do a substantial amount of work at night or in limited visibility. They must ensure that the platoon TSOPs (see Appendix F for an example of a TSOP) are available and used throughout the squadron for providing MEDEVAC and treatment at night. Real-life trauma care at night will be enhanced by the ability to use white light (visible light) at the earliest opportunity. Therefore, medical units/elements must establish standard procedures to use white light without compromising the tactical environment. This means training to erect shelters as soon as possible and routinely during hours of darkness. Personnel must understand that some shelters block visible light, but that the same shelters glow when viewed through night vision goggles (NVG). In some extremely mobile situations, ambulance/vehicles could be used to enclose patients and care providers, thus allowing treatment to proceed under white light conditions.

a. *Conducting Night Operations*. Medical leaders must understand night operations technology and their capabilities for conducting night operations. The brigade, battalion, and squadron surgeons should know how to use both far infrared (IR) devices (and how their capabilities can enhance CHS operations at night) such as the combat identification panel (CIP) and near IR devices, such as the BUDD light and the Phoenix light. See the discussion below on IR and night vision devices. The surgeons need to know the status and amount of equipment on-hand and to identify equipment needed. They must plan the SOPs and METT-TC-specific techniques necessary to perform the CHS mission. For these types of operations, the commander should be advised to consider—

- Appropriating civilian buildings to reduce light and thermal signatures.
- Lightproofing shelters.
- Using nonvisible spectrum light in conjunction with night vision devices.
- Reducing noise signature to a minimum.

In addition, units are susceptible to a night attack. This further slows logistics and CHS activities. Use of chemical lights may be applicable. However, overuse of chemical lights degrades light discipline and security. Chemical lights are visible from a distance of a kilometer or more. Possible techniques for medical units/elements include—

- Chemical lights to light CP areas, thus eliminating generator noise and thermal signature.
- Magnetic holders to allow placement of color chemical lights on vehicles.
- Chemical lights to illuminate areas of vehicle engine compartment for night repairs.
- Chemical light holders to regulate the amount and direction of light.

b. *Combat Health Support Considerations.*

(1) Light discipline requirements affect CHS operations much as they do supply and maintenance operations. Medical units/elements will use additional fuel to run a vehicle-mounted night site. Treatment operations require lightproof shelters. Patient acquisition is more difficult. Units should employ some sort of casualty-marking system such as luminous tape.

(2) Limited visibility slows MEDEVAC. This requires additional ground ambulances to compensate. In the offense, ambulances move forward with BASs. However, personnel have to accomplish this movement carefully to avoid signaling the threat. Personnel use predesignated AXPs. Medical evacuation by air ambulance is difficult and requires precise grid coordinates as well as prearranged signals and frequencies. As in daylight, CHS operations conducted at night require active participation of all involved units. Operational procedures must include near and far recognition, signaling, predetermined marking of CCPs, routes, and MTFs. Maximum use of modern navigation tools, such as the GPS, IR, and night vision devices, will enhance the ability of medical units/personnel to carry out CHS in support of night missions. Night operating procedures must be routine and practiced as a part of routine operating procedures. This is especially true for medical units/personnel since they have a 24-hour responsibility under all conditions, not just combat operations.

c. *Infrared and Night Vision Devices.*

(1) A far IR device, such as the CIP, is a "quick fix" device for friendly identification. The thermal taped-covered CIP provides an aid in distinguishing friendly from threat vehicles when thermal sights are used. Combat identification panels do not replace current acquisition, identification, or

engagement procedures. They provide a device visible through thermal sights to increase situational understanding and provide a safety net at the normal engagement range. These devices can be used to further identify medical vehicle and units.

(2) Near IR devices that aid in C2 may be used for signaling and marking devices. The IR beam is an effective means to increase situational understanding and combat effectiveness and improve identification. These devices reduce the fratricide risk when used for marking obstacles, seized terrain, and breached sites. Additionally, these lights are super signaling devices (that is, configuration of certain patterns to indicate unit identification, turn on/off to signal accomplishment of a task, cross a phase line, and signal from one ground position to another specific position, or from ground to air). They are also useful in specialized units such as pathfinders for marking pickup, drop, or landing zones. These are excellent devices for near recognition signaling to guide incoming evacuation vehicles.

(a) BUDD light. The BUDD light operates using active near IR light viewed through image-intensifying devices. These image-intensifying devices are only effective during nighttime conditions. Near IR devices can be directional or omni-directional and emit a steady pulse or codable pulse. The BUDD light is a compact near IR source, using a standard 9-volt (BA-3090) battery as its power source. Both the BUDD light and its power source will fit in the palm of your hand. The average life span of the battery power for a BUDD light is 8 hours of continuous use. The near IR pulse emitted by the BUDD light is similar to a strobe light and pulses every 2 seconds. It is invisible to the naked eye and thermal imagers. The pulse is clearly visible out to 4 kilometers under optimal conditions when pointing the beam directly at the viewer. The directional characteristic of the beam makes it possible to limit observation by an enemy. If used to mark vehicles, care should be taken to minimize the light illuminating the vehicle's surface. The enemy has to have image-intensifying devices to see the lights directly; however, they may see the light being reflected off of vehicles when the lights are employed in a directional mode. This device is most effective for C2 purposes. The BUDD light is also very useful for dismounted operations at night.

(b) Phoenix light. The Phoenix light operates using active near IR light viewed through image-intensifying devices. The Phoenix light can be used as a codable IR beacon. The light is powered by a standard 9-volt (BA-3090) battery. The Phoenix light is ideal for use when positive identification at night must be made out to 4 kilometers under optimal conditions. The IR beacon has a range equal to the BUDD light. One advantage is the ability to code many beacons with different codes (sequence of flashes—including Morse code—up to 4 seconds), enabling anyone to be distinguished in a group. A programmed sequence will repeat until canceled or when the battery expires (same as a BUDD light). Operating instructions include connecting the battery to the Phoenix light. Using a metal object, a coin is best, make connection across the two pins on top of the light. A microminiature red indicator flashes the sequence as the code is entered. At the end of the 4-second memory, a green microminiature indicator will flash, indicating the end of the input sequence. The Phoenix light is now emitting the desired code. To check the code, make a connection across the pins. The green microminiature indicator will flash the code. To change the code, disconnect the battery and repeat the instructions. The Phoenix light also can be used during dismounted operations. The programming of a code can assist in distinguishing one unit from another. An active Phoenix light or BUDD light can be covered or uncovered as necessary to ensure the light is visible only when necessary

(c) Night vision devices. There are numerous types of night vision devices in the Army inventory but this subparagraph will focus on what the squadron medical platoon has on its TOE. Each vehicle in the medical platoon will have two night vision devices. The wheeled vehicle driver will use either the AN/PVS-7B (discussed below) or the driver's vision enhancer (DVE). The DVE is a thermal imaging system capable of operating in degraded visibility conditions such as fog, dust, smoke, and darkness. In conditions of reduced visibility, the DVE allows a vehicle to maintain speeds up to 55 to 60 percent of those attained during normal daylight operations. Unlike traditional night vision devices that magnify ambient light, the DVE generates a picture based on very minute variances in temperature in the surrounding environment. It gives the operator visibility to the horizon in total darkness and the ability to recognize a 22-inch object at a distance of 360 feet. It can elevate 35 degrees, depress 5 degrees, and rotate 170 degrees in either direction. The DVE consists of a sensor module, display control module, positioning module, wiring harness, and mounting equipment. A combat DVE and a tactical wheeled vehicle DVE will be available. The ambulance and treatment vehicle drivers will use a DVE if available or continue to wear NVG. The NVG (AN/PVS-7B, a hand-held, head-mounted, or helmet-mounted night vision system) enable walking, driving, weapons firing, short-range surveillance, map reading, treatment of patients, and vehicle maintenance in both moonlight and starlight. It has an IR projector that provides illumination at close ranges and that can be used for signaling. There is a high light-level shutoff if the device is exposed to damaging levels of bright light. There is a compass that attaches to the device and allows for reading an azimuth through the goggles. This device has a weight of 1.5 pounds and operates on two AA batteries. The ambulance and treatment vehicle when fielded will have IR headlights. These IR headlights can be used for assisting drivers who wear NVG and for signaling. As with all lights, extreme caution must be taken in tactical situations. The IR headlights are typically very bright to personnel wearing NVG.

d. Example Techniques for Using Chemical Lights for Marking and Signaling.

NOTE

Techniques are only limited to available equipment and imagination.
The METT-TC should always take precedence.

(1) For marking, chemical lights can be placed inside standard military short or long pickets to mark routes and positions. The concave side of the picket contains the chemical light and the convex side faces the most likely direction of enemy observation. This technique controls the direction of the light while assisting with such things as MEDEVAC routes, supported unit collection points, AXP, or link-up point identification.

(2) For signaling, tying a chemical light to a length of cord or string and twirling it overhead in a circle is an unmistakable signal. This only needs to be used until recognition (radio) is established. This technique makes use of widely available common supplies. It is especially useful for a unit guiding an incoming ground or air ambulance.

(3) For marking casualty locations, for example, MOUT operations.

(4) For marking triage areas at the BAS to identify patient triage categories.

APPENDIX A

THE GENEVA CONVENTIONS

A-1. Effects of Geneva Conventions on Combat Health Support

The conduct of armed hostilities on land is regulated by both written and unwritten law. This land warfare law is derived from two principal sources—custom and lawmaking treaties such as The Hague and Geneva Conventions. The rights and duties set forth in these conventions are part of the supreme law of the land; a violation of any one of them is a serious offense. The Geneva Conventions are four separate international treaties, signed in 1949, and are respectively entitled:

a. Geneva Convention for the Amelioration of the Conditions of the Wounded and Sick in Armed Forces in the Field (GWS).

b. Geneva Convention for the Amelioration of the Condition of Wounded, Sick, and Shipwrecked Members of Armed Forces at Sea (GWS Sea).

c. Geneva Convention Relative to the Treatment of Prisoners of War (GPW).

d. Geneva Convention Relative to the Protection of Civilian Persons in Time of War (GC).

See FMs 8-10 and 27-10 for definitive information on the Geneva Conventions and the laws associated with land warfare.

A-2. Geneva Wounded and Sick

a. Custodial and medical responsibilities must be carried out for persons (military or civilian) who are wounded as a result of military operations regardless of their nationality or legal status.

NOTE

Persons whose legal status is in doubt are accorded protection and treatment as prisoners of war until their legal status is determined.

b. Collection and treatment of the sick and wounded are responsibilities of medical personnel. The custodial and accounting functions are responsibilities of military police.

A-3. Identification and Protection of Medical Personnel under Geneva Wounded and Sick

Medical personnel who become captured are not considered prisoners of war but retained personnel.

a. Protected personnel include—

(1) Army Medical Department personnel exclusively engaged in the—

- Search for or collection, transport, or treatment of the wounded or sick.
- Prevention of disease.
- Administration of medical units and establishments (for example, this includes personnel such as the office staff, ambulance drivers, cooks, and cleaners that form an integral part of the unit or establishment).
- Veterinary staff functions relating to the administration of medical units and establishments.

(2) Non-AMEDD personnel who have received special medical training, if carrying out their auxiliary medical duties when captured by the enemy are protected. Once in enemy hands, they become prisoners of war when not doing medical work.

(3) Chaplains.

b. Each protected individual must—

(1) Carry a special water-resistant, pocket-sized identity card (DD Form 1934) that—

- Bears the red cross on a white background (the distinctive emblem of the Geneva Conventions).
- Is worded in the national language of the issuing force.
- Contains the surname and first name (at least), date of birth, rank, social security number, protected capacity serving, photograph, signature, and/or fingerprints of carrier.
- Is embossed with the stamp of the appropriate military authority (AR 640-3).

(2) Wear on the left arm a water-resistant armlet bearing the red cross emblem of the Geneva Convention (DA Pamphlet [Pam] 27-1 and FM 27-10).

This paragraph implements STANAG 2454.

A-4. Identification of Medical Units, Facilities, and Vehicles under Geneva Wounded and Sick

a. Identify—

(1) All medical units and facilities except veterinary units. Medical facilities also include the nonpatient care areas, such as those for dining, maintenance, and administration.

(2) Air and surface (ground and water) medical vehicles.

b. How:

(1) Display the distinctive flag of the Geneva Conventions (red cross on a white background) over the unit/facility and in other places on the unit/facility as necessary to adequately identify it. (The other emblem recognized by terms of the Geneva Conventions is the red crescent. Emblems not recognized by the Geneva Conventions but used by other countries, such as the red shield of David by Israel, should also be respected.)

(2) Mark with the distinctive Geneva emblem (red cross on a white background).

(3) The GWS protects from attack any medical vehicle appropriately marked and exclusively employed for the evacuation of the sick and wounded or for the transport of medical personnel and equipment. The GWS prohibits the use of medical vehicles marked with the distinctive emblems for transporting nonmedical troops and equipment.

This paragraph implements STANAG 2931.

A-5. Camouflage of the Geneva Emblem

The NATO STANAG 2931 provides for camouflage of the Geneva emblem on medical facilities where the lack of camouflage might compromise tactical operations. Medical facilities on land, supporting forces of other nations, will display or camouflage the Geneva emblem in accordance with national regulations and procedures. When failure to camouflage would endanger or compromise tactical operations, the camouflage of medical facilities may be ordered by a NATO commander of at least brigade level or equivalent. Such an order is to be temporary and local in nature and countermanded as soon as the circumstances permit. It is not envisaged that large, fixed medical facilities would be camouflaged. The STANAG defines "medical facilities" as "medical units, medical vehicles, and medical aircraft on the ground."

NOTE

Under tactical conditions, the need for concealment may outweigh the needs for recognition (AR 750-1).

A-6. Defense of Self and Patients under Care

a. Protected personnel are—

(1) Authorized to be armed with only individual small arms. (Army Regulation 71-32 provides the doctrine that governs what types of small arms medical personnel are authorized [limited to

pistols or rifles, or authorized substitutes].) These small arms may only be used for defensive purposes. The presence of machine guns, grenade launchers, booby traps, hand grenades, light antitank weapons, or mines in or around a medical unit would seriously jeopardize its entitlement to protected status under the GWS. The deliberate arming of a medical unit with such items could constitute an act harmful to the enemy and cause the medical unit to lose its protected status under the Conventions. This conclusion is not altered in the case of mines regardless of the method by which they are detonated, nor is it altered by the location of the medical unit. If the local non-AMEDD commander situates a medical unit where enemy attacks may imperil its safety, then that commander should provide adequate protection for the medical unit and its personnel.

(2) Permitted to fire only when they or their patients are under direct attack in violation of the GWS. Use of arms by AMEDD personnel for other than protection of themselves or their patients violates the GWS provisions governing the protected status of AMEDD personnel and results in the loss of protected status. Army Regulation 350-41 states the AMEDD personnel and non-AMEDD personnel in medical units will not be required to train or qualify with weapons other than individual or small arms weapons. However, AMEDD personnel attending training at NCO education system courses will receive weapons instruction that is part of the curriculum. This will ensure that successful completion of the course is not jeopardized by failure to attend the weapons training portion of the curriculum.

(3) Responsible for their own defense when operating at locations which preclude their being incorporated within defensive perimeters of nonmedical units. In addition to relying on their special status, medical units can provide for their defense by employing passive defense measures. Passive measures are those taken to reduce the probability of and to minimize the effects of damage caused by hostile action. Examples of these measures are the preparation of individual fighting positions within the immediate unit area; noise and light discipline; posting perimeter sentries; and channeling traffic in the unit area.

b. Protected personnel (under overall security defense plans) will NOT be required—

- To man or help man the perimeter defense of nonmedical units such as unit trains, logistical areas, or base clusters.

- To take offensive action against enemy troops.

- To perform actions that will cause loss of protected status and result in inadequate care of our sick and wounded prisoners of war. The platoon leader must clearly articulate this to all levels of command. The misuse of CHS vehicles/equipment will void all protection granted under the Geneva Conventions.

A-7. Geneva Prisoners of War

a. United States military forces are responsible for EPW from the moment of capture.

b. The echelon commander and medical unit commanders jointly exercise responsibilities for the custody and treatment of the sick, injured, or wounded enemy personnel and detained civilian personnel.

c. The sick, injured, or wounded prisoners are treated and evacuated through normal medical channels but are physically segregated from US and allied patients. Persons other than medical personnel are provided by the echelon commander to guard them. Evacuation of these EPW patients from the combat zone is initiated as soon as their medical conditions permit.

d. When intelligence indicates that large number of EPW may result from an operation, medical units may require reinforcement to support the anticipated additional EPW patient workload. Procedures for estimating the medical workload involved in the treatment and care of EPW patients are described in FM 8-55.

e. Enemy medical personnel are considered retained personnel rather than prisoners of war. They are to be employed to the maximum extent possible in such CHS duties as caring for detained or EPW patients, preferably those of their own armed forces. Captured medical supplies should be used in the care of these patients.

A-8. Geneva Civilian Persons

a. When the US is the occupying power, US forces have the responsibility to ensure that all civilian and refugee subsistence and health service needs are provided.

b. Sick or injured civilian persons resulting from military operations are provided initial medical treatment, as required, in conformance with established theater policies; then, they are transferred to appropriate civil control authorities as soon as possible. When such persons are evacuated, proper accommodations must be provided, including satisfactory conditions of hygiene, health, safety, and nutrition (Articles 49 and 55). In conditions of armed conflict and to the extent practicable, the Army must seek to fulfill the above commitments, as well as to protect and assist civilians and refugees under its control.

A-9. Compliance with the Geneva Conventions

a. As the US is a signatory to the Geneva Conventions, all medical personnel should thoroughly understand the provisions that apply to CHS activities. Violation of these Conventions can result in the loss of the protection afforded by them or prosecution. Medical personnel should inform the tactical commander of the consequences of violating the provisions of these Conventions.

b. The following acts are inconsistent with an individual or facility claiming protected status under the Geneva Conventions:

- Medical personnel are used to man or help man the perimeter of nonmedical facilities, such as unit trains, logistics areas, or base clusters.

- Medical personnel are used to man any offensive-type weapons or weapons systems.

- Medical personnel are ordered to engage enemy forces other than in self-defense or in the defense of patients and MTFs.

- Crew-served weapons are mounted on a medical vehicle.
- Mines or booby traps are placed in and around medical units and facilities.
- Hand grenades, light antitank weapons, grenade launchers, or any weapons other than rifles and pistols are issued to a medical unit or its personnel.
- The site of a medical unit is used as an observation post, a fuel dump, or an ammunition storage site.

c. Possible consequences of violations described in *b* above are—

- Loss of protected status for the medical unit and personnel.
- Medical facilities attacked and destroyed by the enemy.
- Medical personnel being considered prisoners of war rather than retained persons when captured.
- Combat health support capabilities decremented.
- Prosecution for violations of the law of war.

d. Other examples of violations of the Geneva Conventions include—

- Making medical treatment decisions for the wounded and sick on any basis other than medical priority, urgency, or severity of wounds.
- Allowing the interrogation of enemy wounded or sick even though medically not recommended.
- Allowing anyone to mistreat, torture, kill, or in any way harm a wounded or sick enemy soldier.
- Marking nonmedical unit facilities and vehicles with the distinctive emblem, or making any other unlawful use of this emblem.
- Using medical vehicles marked with distinctive Geneva Conventions emblem for transporting nonmedical troops, equipment, and supplies.
- Using a medical vehicle as a TOC.

e. Possible consequences of violations described in *d* above are—

- Criminal prosecution for war crimes.

- Medical personnel being considered prisoners of war rather than retained persons when captured.

NOTE

The use of smoke and obscurants by medical personnel is not a violation of the Geneva Conventions (see FMs 3-50 and 8-10-6 for information on the use of smoke).

APPENDIX B

MANAGEMENT OF INDIVIDUAL HEALTH RECORDS IN THE FIELD

B-1. General

a. This appendix provides guidance on the maintenance of the soldier's individual health record (HREC) and civilian employee medical records (CEMR) in the field. The governing regulation is AR 40-66.

b. Health records are maintained by the MTF that provides primary care for the soldier.

c. Unit commanders will ensure that HRECs are always available to AMEDD personnel who require such records in the performance of their duties. Unit commanders will also ensure that the information in the HRECs is kept private and confidential in accordance with law and regulations governing patient records administration.

d. Health records located at an Echelon I MTFs are maintained by unit medical personnel. The AMEDD officer-in-charge serves as the custodian of the HRECs and CEMRs. Army Medical Department officers are in charge of the HRECs and CEMRs for the members of the units and civilian employees for whom they supply primary medical care. They are also in charge of the HRECs, CEMRs, and the records of other individuals that are receiving treatment from the MTF. Health records are important for the conservation and improvement of the patient's health. Therefore, AMEDD officers will ensure that all pertinent information is promptly entered in the HREC/CEMR in their custody. If any such pertinent information has been omitted, the AMEDD officer will take immediate action to obtain such information from the proper authority and include it in the HREC/CEMR.

B-2. Health Records of Deployed Soldiers

a. Health Records. The HREC (DA Form 3444 or DA Form 8005 Series [Medical and Dental Treatment Records]) of deployed soldiers and the CEMR of deployed civilians will not accompany them to the combat area.

(1) The supporting MTF will initiate a DD Form 2766 (Adult Preventive and Chronic Care Flowsheet), DD Form 2766C (Adult Prevention and Chronic Care Flowsheet [Continuation Sheet]), DD Form 2795 (Pre-Deployment Health Assessment Questionnaire), and DD Form 2796 (Post-Deployment Health Assessment Questionnaire). If an individual deploys, the DD Form 2766 and DD Form 2766C will be photocopied prior to deployment and the copy will be kept in the medical record. The original DD Form 2766 and any DD Forms 2766C will accompany the individual to the field. The DD Form 2766 serves as the treatment folder for the individual that is deployed; other forms, such as DD Form 2766C, DD Form 2795, DD Form 2796, and Standard Form (SF) Form 600 (Health Record—Chronological Record of Medical Care) will be filed on the fastener inside DD Form 2766. The photocopies of the DD Form 2766 and DD Form 2766C will be removed and shredded when the originals are placed back into the HREC or CEMR. Forms that had been filed inside the DD Form 2766 folder will be removed and place in the HREC or CEMR (in the regular treatment folder).

(2) When processing individuals for deployment, the MTF and dental treatment facility (DTF) will audit each individual's HREC or CEMR and record essential health and dental care information on DD Form 2766. If a HREC or CEMR is not available, DD Form 2766 will be completed based on individual interviews and any other locally available data. A HREC may not be available for most Individual Ready Reserve, Individual Mobilization Augmentees, and retired personnel because these HREC may remain on file at the Army Reserve-Personnel Command (AR-PERSCOM) or the Department of Veterans Affairs.

(3) Upon notification of deployment, all military personnel will complete DD Form 2795.

(*a*) The individual being screened will fill out the section entitled *Demographics* on page 1, and the section entitled *Health Assessment* on page 2. These sections are self-explanatory.

(*b*) The health assessment administrator will fill out the boxed area on page 1 entitled *Administrator Use Only*, and will answer the user's questions on filling out the form. The administrator will document the deployment location as well as the completion date of the pre-deployment evaluation on DD Form 2766, Block 11—*Pre-/Post-Deployment History*. This does not apply to classified operations.

(*c*) The health care provider will fill out the section entitled *Pre-Deployment Health Provider Review* on page 2.

(*d*) A copy of the form will be filed on the fastener inside the DD Form 2766 folder; one copy will remain in the HREC, and the original form will be sent to the Army Medical Surveillance Activity, ATTN: MCHB-TS-EDM/Deployment Surveillance, Building T-20, Room 213, 6825 16th Street NW, Washington, DC 20307-5000.

(4) Department of Defense Directive (DODD) 6490.2 and Department of Defense Instructions (DODI) 6490.3 state that to the extent applicable, medical surveillance activities will include essential DOD civilian and contractor personnel directly supporting deployed forces, consistent with plans established under DODI 1400.32 and DODI 3020.37. If DD Form 2795 is used for civilians, a copy of the form will be filed on the fastener inside the DD Form 2766 folder; one copy will remain in the CEMR, and the original form will be sent to the Army Medical Surveillance Activity.

(5) If the deployed individual is taking part in a classified operation, the pre-deployment evaluation (DD Form 2795) is still required, but the form will be maintained only in the personnel folder.

(6) The completed DD Form 2766 and a copy of any printout from an automated immunization tracking system will be provided to the individual's command, or to the individual if he or she is an individual replacement, and then handed off to the MTF in the AO responsible for providing primary medical care to that individual. That MTF will maintain the DD Form 2766 as an outpatient field file for reference as needed. The MTF will ensure that the ABO/Rh blood type from a verified blood bank typing is recorded in Block 10. The field file will consist of, in part, DD Form 2766, DD Form 2795, and possibly DD Form 2766C, DD Form 2796, SF 600, SF 558 (Medical Record—Emergency Care and Treatment), SF 603 (Health Record—Dental), or DD Form 1380. These forms will be filed on the fastener inside the DD Form 2766. For detailed information on how to complete the DD Form 1380, see Appendix C, FM 8-10-6.

(7) If DD Form 2766 is not available, the individual's field file may be managed as a "drop" file (forms not attached) and integrated into the DD Form 2766 when it is available.

b. Forwarded Deployed Force. If time permits, follow guidance in *a*(1), (2), and (3) above. If not, consolidate HREC in-country and process when time permits.

c. Limited Contingency Operations. Retain the HREC at the MTF and DTF providing primary care. If the servicing primary care facility closes, forward the HREC to the MTF or DTF indicated by the servicing medical department activity (MEDDAC) and dental activity. If full mobilization occurs, follow guidance in *a*(1), (2), and (3) above.

d. Units That Do Not Process Through a Mobilization Station. Units that do not process through a mobilization station before deployment or otherwise do not have access to an MTF or DTF will follow the procedures in *b* above.

B-3. Use of Field Files/DD Form 2766

a. If a member's primary MTF changes, the field file/DD Form 2766 should be moved to the gaining MTF.

b. If a member requires admission to the hospital, every attempt will be made to forward the field file/DD Form 2766. The file will be returned to the member's primary MTF if disposition is RTD.

B-4. Storage of Health Records and Civilian Employee Medical Records

Forward deployed (Echelon I and Echelon II) MTFs will secure field chest or field file containers in quantities sufficient for the troop and civilian employee population supported. They will maintain the DD Form 2766 for each individual receiving primary medical care from their MTF.

B-5. Establishment and Management of the Field File in the Operational Area

a. A DD Form 2766 and the medical records identified above will be maintained by medical companies operating an Echelon II MTF or the medical platoon/section that operates an Echelon I MTF, or will be handed off to the MTF providing their primary care.

b. Supported units will be required to provide the primary care MTF a battle roster of personnel assigned. This roster should be provided when personnel assignment changes are made or upon request.

c. The MTF, when possible, will attempt to ensure that the HREC or CEMR accompanies the medically evacuated individual.

d. If an individual's primary MTF changes, the HREC or CEMR will be transferred to the gaining MTF.

e. If an individual requires hospital admission, every attempt will be made to forward the HREC or CEMR to the admitting hospital.

f. When the MTF determines that an individual was evacuated without the DD Form 2766 and other medical records in the file, then the individual's DD Form 2766 and other medical records are forwarded to the medical C2 headquarters responsible for regulating patients out of the AO. The medical C2 headquarters forwards the outpatient field file to the hospital where the patient was evacuated. The hospital patient administration section will attach the file to the inpatient chart and the file is evacuated with the patient out of the AO or theater.

B-6. Health Assessments after Deployment

a. All military personnel will complete DD Form 2796 prior to leaving the AO.

(1) The individual being screened will fill out the section entitled *Demographics* on page 1 and the section entitled *Health Assessment* on page 2. These sections are self-explanatory.

(2) The health assessment administrator will fill out the boxed area on page 1 entitled *Administrator Use Only* and will answer the user's questions on filling out the form. The administrator will document the deployment location (if this information is missing) and the completion date of the post-deployment evaluation on DD Form 2766, Block 11—*Pre-/Post-Deployment History*. This does not apply to classified operations.

(3) The health care provider will fill out the section entitled *Post-Deployment Health Provider Review* on page 2.

b. If a situation does not allow this health screening prior to departure, the individual's commander will ensure that the health assessment is completed and submitted to the local MTF commander within 30 days of the individual's return. The local MTF commander will ensure that a procedure is in place for submitting the original DD Form 2796 to the Army Medical Surveillance Activity and for filing a copy in the HREC.

c. If the DD Form 2796 is completed prior to leaving the AO, a copy of the form will be filed in the DD Form 2766 folder until it can be integrated into the HREC. The original DD Form 2796 will be submitted to the Army Medical Surveillance Activity, ATTN: MCHB-TS-EDM/Deployment Surveillance, Building T-20, Room 213, 6825 16th Street NW, Washington, DC 20307-5000.

d. The post-deployment assessment of Reserve Component personnel must be completed prior to release from active duty if not completed before redeployment. Reserve Component personnel who have been deployed will also complete DD Form 2697 (Report of Medical Assessment) according to AR 40-501. Reserve Component personnel who are called to active duty but never actually deployed will only complete DD Form 2697.

e. If DD Form 2796 is used for civilians, the form will be completed prior to leaving the AO. If a situation does not allow this health screening prior to departure, the individual's commander will ensure that the health assessment is completed within 30 days of the individual's return. If the DD Form 2796 is completed prior to leaving the AO, a copy of the form will be filed in the DD Form 2766 folder until it can be integrated into the CEMR. The local commander will ensure that a procedure is in place for submitting the original DD Form 2796 to the Army Medical Surveillance Activity and for filing the copy in the CEMR.

f. If the deployed individual is taking part in a classified operation, the post-deployment evaluation (DD Form 2796) is still required, but the form will be maintained only in the personnel folder.

B-7. Field Record Administration after Hostilities Cease

a. Field files/DD Form 2766 will be integrated with the HREC or CEMR after demobilization at the home station or at mobilization stations.

(1) On return to the MTF (post-deployment), forms, such as SF 600, will be removed from the DD Form 2766 folder and placed with the other SF 600 in the medical record.

(2) DD Form 2795 and DD Form 2796 will be removed from the DD Form 2766 folder and placed as shown in Figures 5-1, 5-2, or 7-1 of AR 40-66. If a previously photocopied DD Form 2795 is contained in the record, only one of the DD Forms 2795 will be kept; the other will be removed and shredded.

(3) The photocopies of the DD Form 2766 and DD Form 2766C will also be removed and shredded when the originals are placed back into the record. Field files/DD Form 2766 will be forwarded to AR-PERSCOM for those members whose HREC is maintained at AR-PERSCOM.

b. Each continental United States (CONUS) MTF must request records from AR-PERSCOM for those members who remain on active duty and are assigned for support upon demobilization.

c. Field files will be integrated with the HREC maintained at home station or mobilization station. Field files will be forwarded to Army Reserve Personnel Center (ARPERCEN) for those members whose HREC is maintained at ARPERCEN.

d. Each CONUS MTF must request records from ARPERCEN for those soldiers who remain on active duty and are assigned for support upon demobilization.

APPENDIX C

COMBAT LIFESAVER

C-1. Role of the Combat Lifesaver

a. Immediate far-forward first aid is essential on a widely dispersed and fluid battlefield to prevent soldiers from dying of wounds. Medical personnel may not be able to reach and apply EMT to all wounded soldiers at all points on the battlefield in a timely manner. The CLS is a nonmedical soldier trained to provide advanced first aid/lifesaving procedures beyond the level of self-aid or buddy aid. The CLS is not intended to take the place of medical personnel, but to slow deterioration of a wounded soldier's condition until medical personnel arrive. Functioning as a CLS for the soldier is a secondary mission undertaken only when the tactical situation permits. Even though this is secondary to his primary mission, the CLS has proven to be very effective in saving wounded soldiers lives. The CLS program is implemented according to AR 350-41.

b. The AirLand Battle doctrine was developed for a widely dispersed, rapidly moving battlefield. As was determined for the AirLand Battle doctrine and applicable to the Force XXI doctrine, there are constraints on the rapid acquisition of casualties and medical treatment. These constraints in many cases limit the ability of medical personnel to provide immediate, far-forward medical treatment. The plan developed to provide care for soldiers under AirLand Battlefield doctrine included the CLS. Under Force XXI doctrine, the CLS is an intricate part of providing care for the wounded soldier.

c. The CLS is a bridge between the self-aid/buddy aid training provided all soldiers and the medical training given to the trauma specialist. The CLS is given additional first aid training and training in selected medical tasks (such as initiating an intravenous infusion and providing limited care to a soldier suffering from BF).

C-2. Training the Combat Lifesaver

A correspondence course has been developed for training both active duty and Reserve Component personnel. The course is offered only in a group study mode and with training taking place at the unit level. Classroom instruction is provided by qualified instructors selected by the battalion commander or battalion/ squadron surgeon. Testing is performed at the unit level using the written and performance tests furnished in the correspondence course. Training and testing will be conducted according to the tasks, conditions, and standards established by the Academy of Health Sciences, AMEDDC&S, and published in correspondence course training materials. Students who successfully complete the written and performance tests will receive promotion points and be certified as a CLS. The course consists of student subcourse texts, student examination, and an instructor's manual.

C-3. Administering the Combat Lifesaver Course

a. Equipment and Supplies. Arrange for equipment and supplies as early as possible. The purchase of some items, such as intravenous infusion trainers and rescue breathing mannequins, may be required. The local Training and Audiovisual Support Center may have these items available. Training items will not be provided by either the Army Institute for Professional Development (AIPD) or the Academy of Health Sciences, AMEDDC&S.

b. *Enrollment.* Enrollment request should be sent to AIPD at least 6 weeks prior to beginning of the course according to instructions found in DA Pam 350-59. To establish this group enrollment, AIPD must receive the following:

(1) A request for training should be signed by the battalion commander or an LTC or higher and identify the primary instructor (group leader). All instructors of the Combat Lifesaver Course (Course Number: 081 F11) must meet the following criteria: hold a primary of MOS 91W or 18D, or be a licensed paramedic (state or national), registered nurse, PA, or physician. These requirements reflect the level of expertise necessary to resolve medical emergencies associated with the tasks to be taught.

(2) One DA Form 145 (Army Correspondence Course Enrollment Application) enrolling the primary instructor in IS0826, along with a list of assistant instructors, if any, is included. The DA Form 145 should be signed by a responsible official who has the authority to requisition or acquire the necessary medical supplies to support the training.

(3) A roster of students enrolling in IS0824 and IS0825 is provided. The roster must include each student's full name, rank, social security number, and component code. Ideally, no more than 15 to 20 students should be assigned to each group.

c. *Facilities.* Facilities must be reserved well in advance. The facilities chosen should allow clear observation of demonstrations and provide room for student practice. Handwashing devices are required.

d. *Course Material.* The course consists of two subcourses (shipped in one box), 40 credit hours consisting of self-study materials and approximately 3 days of classroom instruction and testing materials. Testing includes both proctored multiple-choice and performance examinations. Only one examination response (IS0827) is returned to AIPD for grading upon completion of IS0824 and IS0825. This is a GO or NO/GO course.

e. *Recertification.* Combat lifesaver tasks are perishable skills. Combat lifesavers must be recertified every 12 months on the performance-tested tasks in the CLS course. The printed material in the correspondence course can be locally reproduced and used for sustainment training. Do not reenroll soldiers requiring recertification. Unit instructors can accomplish this. The instructor's guide also includes a test appropriate to recertification. It is the responsibility of the S1, not the medical platoon leader, to ensure that personnel matters concerning the CLS program are resolved.

f. *Program Managers.* Program managers are not authorized to augment correspondence course material, change the length of the course, or increase or delete items contained in the CLS aid bag.

g. *Sustainment Training.* To the extent needed to sustain skill proficiency, CLS will be exercised during home station training activities (to include field-training exercises) and during deployment for training (to include rotations through combat training centers).

h. *Additional Information.* For additional information, write to Student Service Division at the Army Institute for Professional Development, US Army Training Support Center, Newport News, VA 23628-0001, or call DSN 927-3322/2127 or commercial 757-878-3322/2127.

C-4. Combat Lifesaver Aid Bag

a. Each certified CLS will be issued a CLS aid bag. The aid bag will be packed in accordance with the prescribed packing list and will be secured as a sensitive item (for example, weapon or night vision devices) at unit level. The aid bags will be issued to the CLS only upon deployment (training and actual).

b. It is the responsibility of each CLS to ensure that—

(1) His aid bag is stocked according to the prescribed packing list.

(2) All stocked items are serviceable.

(3) Items have not exceeded their expiration dates.

C-5. Class VIII Resupply of Combat Lifesaver Aid Bag

a. During garrison operations, Class VIII resupply of the CLS aid bags is conducted in the following manner:

(1) Units with assigned medical personnel will request Class VIII resupply for CLS aid bags.

(2) Divisional, brigade, and regimental units without assigned medical personnel request Class VIII support through the DMSO, brigade medical supply section, or regimental medical supply section. Force XXI units will request Class VIII resupply from either the DSMC medical supply section or the FSMC medical supply section.

(3) Nondivisional units request Class VIII support from the installation medical supply activity, which is normally the medical supply office of the medical department activity or a US Army medical center.

b. During field operations, resupply is accomplished in the same manner as during garrison operations with two exceptions.

(1) Nondivisional units obtain resupply support from the nearest medical unit available.

(2) Emergency resupply is provided to all units by the nearest medical unit capable of doing so.

c. Field medical units maintain stocks of medical material with which to effect resupply of CLS aid bags during field operations.

APPENDIX D

INTERIM BRIGADE COMBAT TEAM

Section I. INFANTRY BATTALION MEDICAL PLATOON

D-1. Medical Platoon

The medical platoon of the infantry battalion is organized with a headquarters section, a treatment squad, an evacuation section, and a combat medic section (see Figure D-1). For more detailed information on the functions and operations of the medical platoon, see Chapter 2, Sections IV and VI.

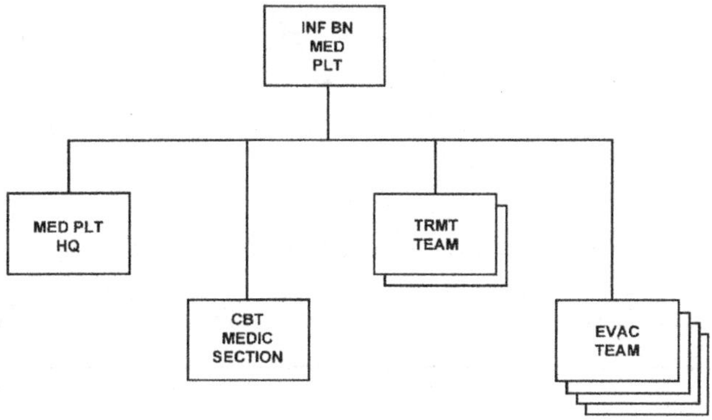

Figure D-1. Medical platoon infantry battalion.

a. Platoon Headquarters. The headquarters section operates under the direction of the medical platoon leader/battalion surgeon who is responsible for overseeing platoon operations. The platoon headquarters section is comprised of a field medical assistant and the platoon SGT. It is normally collocated with a treatment team or a treatment squad to form the BAS. The CP includes the plans and operations functions performed by the field medical assistant. The platoon has access to the infantry battalion HHC wire communication network for communications with all major elements of the battalion. Wireless communications for this section consists of a tactical FM radio mounted in the platoon headquarters vehicle. The medical platoon employs an FM radio network for CHS operations. The platoon headquarters section serves as the NCS for the platoon (see Figure D-2). Table D-1 lists the information and communications assets available to the platoon.

(1) The field medical assistant, an MS officer, is the operations/readiness officer for the platoon. He is the principal assistant to the platoon leader for operations, administration, and logistics. The field medical assistant coordinates CHS operations with the infantry battalion S1 and S4, and MEDEVAC with the brigade support medical company (BSMC).

(2) The platoon SGT assists in supervising the operations of the platoon. He also serves as the evacuation section SGT.

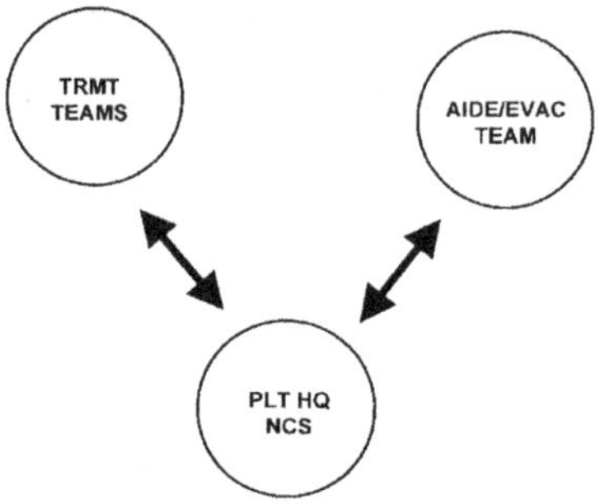

Figure D-2. Medical platoon internal communications net.

Table D-1. Information and Communications Assets Available to the Infantry Battalion Medical Platoon

RADIO SETS	
AN/VRC-89F	PLATOON HEADQUARTERS VEHICLE AND TREATMENT TEAM ALPHA (SURGEON) VEHICLE (1 EACH)
AN/VRC-88F	TREATMENT TEAM BRAVO
AN/VRC-90F	PLATOON AMBULANCE (1 EACH)
ROUTERS	**COMPUTER SYSTEMS**
TACTICAL LAN LAN ROUTER	MC4 LAPTOPS FOR EACH TREATMENT AND AMBULANCE VEHICLE
	MC4 (DISMOUNTED, HAND-HELD)
	PIC
	READER/WRITER (1 EACH, TRAUMA SPECIALIST)
	FBCB2, 1 IN EACH VEHICLE ASSIGNED TO THE MEDICAL PLATOON
	OTHER SYSTEMS
FBCB2	MEDICAL PLATOON VEHICLE (1 EACH)
GPS	MEDICAL PLATOON VEHICLE (1 EACH)
EPLR	MEDICAL PLATOON VEHICLE (1 EACH)
BCIS	MEDICAL PLATOON VEHICLE (1 EACH)
DVE	MEDICAL PLATOON VEHICLE (1 EACH)

b. Treatment Squad. The treatment squad consists of two treatment teams (Teams Alpha and Bravo). They operate the BAS and provide Echelon I medical care and treatment. This includes sick call, EMT, and ATM. Team Alpha is staffed with an operational medicine officer (primary care physician/ battalion surgeon), a health care SGT, and two health care specialists. Team Bravo is staffed with a PA, a health care SGT, and two health care specialists. The physician, PA, and health care SGT and specialists are trained to provide EMT and assist with ATM procedures, commensurate with their occupational specialties. The treatment teams can operate for limited times in split-based operations in DS of battalion units. The teams can also operate in split-based operations when the BAS must move to a new location. One team remains at current location and continues to treat patients while the other team moves to the new location and establishes patient care capabilities. Once the jump team has established a treatment capability at the new location, the other team evacuates or returns to duty all patients and moves to the new location.

(1) The medical platoon leader is a working physician on Treatment Team Alpha. He is also the battalion surgeon. In this role, he is a special staff officer and advisor to the battalion commander on employment of the medical platoon and on the health of the battalion. He is also the supervising physician (field surgeon) of the medical platoon's treatment squad. This officer is responsible for all CHS provided by the platoon. His responsibilities include—

- Planning and directing CHS for the infantry battalion. He does this in conjunction with the battalion S1, who is the coordinating staff officer responsible to the commander for health and welfare of the troops.
- Advising the infantry battalion commander and his staff on CHS operations and the medical threat.
- Supervising the administration, discipline, maintenance of equipment, supply functions, organizational training, and employment of medical platoon personnel.
- Examining, diagnosing, treating, and prescribing courses of treatment for patients, to include ATM.
- Training CLS.
- Supervising the battalion CSC program.
- Planning and conducting humanitarian assistance programs, when directed.
- Providing PVNTMED support for the battalion. Requesting PVNTMED support from the brigade for PVNTMED requirements beyond his capabilities.
- Planning and overseeing PVNTMED training for battalion personnel.
- Advising the commander on the health of the battalion.
- Supervising the training of unit field sanitation teams.

(2) The PA performs patient health care and administrative duties. The PA is ATM-qualified and works under the clinical supervision of the medical officer. The PA assumes the duties of the battalion surgeon/medical platoon leader in his absence. He performs the following duties:

- Establishes and conducts treatment team operations when deployed in split-based operations.

- Treats, within his ability, sick or injured patients. He refers those patients requiring treatment beyond his capability to the supervising physician.

- Provides EMT and ATM for wounded and DNBI patients.

- Conducts training for battalion personnel in first aid procedures (self-aid, buddy aid, and CLS), field sanitation, evacuation of the sick and wounded, and the medical aspects of injury prevention.

- Assists in the conduct of the battalion CSC program, to include individual and leader training on the prevention of BF and other stress-related conditions.

- Trains medical personnel in EMT procedures.

- Advising the command on PVNTMED concerns and conducting PVNTMED activities within the capabilities of medical platoon personnel. Assist in training unit field sanitation teams.

c. Combat Medic Section. Trauma specialists are allocated to the companies of the infantry battalion on the basis of one trauma specialist per platoon. The platoon trauma specialist normally locates with, or near, the platoon leader or platoon SGT. When the platoon is moving on foot in the platoon column formation, he positions himself near the element leader trailing the base squad forward of the second team. When the platoon is mounted, the trauma specialist will normally ride in the same vehicle as the platoon SGT. A health care SGT is allocated on the basis of one per infantry company. The company health care SGT normally collocates with the 1SG. When the company is engaged, he remains with the 1SG and provides medical advice as necessary. As the tactical situation allows, he will manage the company CCP, provide medical treatment, and prepare patients for MEDEVAC.

d. Evacuation Section. Medical platoon ambulances provide evacuation and en route care from the soldier's point of injury or the company's CCP to the BAS. The ambulance team supporting the company works in coordination with the trauma specialists supporting the platoons. When a casualty occurs in a fighting vehicle, the evacuation team will move as close to the vehicle as possible, making full use of cover, concealment, and defilade. Assisted, if possible by the vehicle's crew, they will extract the casualty from the vehicle and administer EMT. In mass casualty situations, nonmedical vehicles may be used to assist in casualty evacuation as directed by the supported commander. Plans for the use of nonmedical vehicles to perform casualty evacuation should be included in the infantry battalion's TSOP and OPORD. Patients are evacuated from the BAS to the BSMC by BSMC ground ambulances or FSMT aeromedical evacuation aircraft. During entry operations, based on the current concept, air ambulances will not be available for the first 96 hours.

D-2. Medical Communications for Combat Casualty Care

a. The MC4 system will be a theater, automated CHS system, which will link commanders, health care providers, and supporting elements, at all echelons, with integrated medical information. The battalion will have one MC4 (dismounted) with two laptops and one server. The system will provide digital enablers to connect, both vertically and horizontally, all ten CHS functional areas. When developed, the MC4 system will receive, store, process, transmit, and report medical C2, medical surveillance, casualty movement/tracking, medical treatment, medical understanding, and CHL data across all echelons of care. This will be achieved through the integration of a suite of medical information systems linked through the Army data telecommunications architecture. The MC4 system begins with the individual soldier and continues throughout the health care continuum. The best way to visualize the MC4 system capability is as a piece of the Army digital computer network where all ten CHS functional areas have been digitized and this CHS information is available to specified commands, supported units, and their personnel. When fully developed, not only will the MC4 system provide Army commanders with CHS information, but will provide them with a seamless transition to the joint health service support environment.

b. The MC4 system will consist of three basic components: software, hardware, and telecommunications systems.

(1) *Software system.*

(*a*) The joint TMIP will provide GOTS/COTS software and interoperability standards to support joint theater operations. The software provides an integrated medical information capability that will support all echelons of care in a theater of operations with links to the sustaining base. Medical capabilities provided by the software to support commanders in the theater will address—medical C2 (including medical capability assessment, sustainability analysis, and medical intelligence); CHL (including blood product management and medical maintenance management); patient evacuation; and health care delivery.

(*b*) The MC4 system will support Army-unique requirements and any software needed to interface with Army information systems such as CSSCS, GCSS-A, FBCB2, Warrior Programs, and the Movement Tracking System.

(2) *Hardware systems.* The hardware will consist of COTS automation equipment supporting the above software capabilities. Examples include, but are not limited to, computers, printers, networking devices, and the PIC.

(3) *Telecommunications systems.* The MC4 system will rely on current and proposed Army solutions for tactical, operational, and strategic telecommunications systems to transmit and receive digitized medical information throughout the theater and back to the sustaining base. There will be no separate AMEDD communication system. Telecommunications at brigade and below will be accomplished through the tactical internet; above brigade level, telecommunications will be accomplished through the WIN architecture. The MC4 system will include hardware or software required to interface with current and emerging technologies supporting manual, wired, and wireless data transmission. At end-state, the MC4 system users will exchange data electronically via the WIN architecture. In the interim, commercial

satellite and/or high frequency radio will be fielded to selected medical units (for example, MDT and so forth) receiving the MC4 system to support high bandwidth requirements until the WIN architecture is fully fielded. Personnel operating satellite assets are resourced in the MDT TOE and will be located with the MDT.

(4) *Patient treatment recording system.* In the future under MC4, medical information about each soldier will be entered into a local database maintained at the supporting BAS or troop medical clinic. This information will include the soldier's immunization status, medical deployability status, and dental deployability status. Until a digital patient record and the PIC are fully functional and fielded, and in accordance with AR 40-66, a field medical record jacket, DD Form 2766, and its accompanying records will be maintained by the soldier's primary care provider. See Appendix B for definitive information on management of the individual health records in the field.

c. When fielded, the MC4 system will be employed by the medical platoon in the conduct of its mission. The medical platoon will employ FBCB2 and other communications enablers for preparing and submitting daily patient feeder reports, transmitting medical surveillance information, requesting supplies, and conducting other administrative activities (see Table D-1 for types of medical platoon communication systems and enablers).

- The medical platoon headquarters will employ an FM radio as the platoon NCS.

- The treatment squad will employ FM radios and notebook computers to conduct treatment team/squad communications. The notebook computers will be used to read and enter patient data on the PIC; to conduct teleconsultation (TCON) and TMEN activities; provide patient flow/disposition information; and DNBI information to the battalion S1 and higher echelon medical leadership. The physician will provide TCON to the PA and receive TCON from the medical company or other medical activities out of theater through the MDT. The notebooks will be linked to others through FM radios. The platoon physician and PA will provide TMEN for ambulance personnel.

- The evacuation section will employ FM radios and notebook computers to conduct MEDEVAC communications. The notebook computers will be used to read and enter patient care information on the PIC. They will also use this device to receive TMEN from the physician and PA. The notebooks will be linked to the FM radios for communications operations.

- The trauma specialists will employ a hand-held device for reading and entering patient care information on the PIC.

D-3. Battalion Combat Health Support Planning

a. *Planning Considerations.* The battalion surgeon assisted by the field medical assistant and the platoon SGT is responsible for the CHS plan for the infantry battalion. As operational requirements or the mission changes, the CHS plan must be updated. The following factors should be considered:

- Commander's information requirements.

- Results of the mission analysis.
- Commander's intent.
- Planning guidance.
- Courses of actions.
- Tactical plan.
- Enemy.
- Terrain.
- Troops supported (unit and attached).
- Weather.
- Threat (including medical threat).
- Operational conditions and constraints.
- Civilian populace in the AO.
- Medical personnel status.
- Equipment status of the medical platoon.
- Supply status including Class VIII
- Communications capability.
- Nuclear, biological, and chemical defense (including radiation OEG).
- Patient decontamination.
- Medical platoon's training status.
- Unit field sanitation team training.
- Unit personnel first aid training status, to include CLS.
- Casualty estimates.
- Medical evacuation requirements and capabilities.

- Nonmedical support requirements from the battalion.
- Area support requirements.
- Special operations support requirement.
- Mass casualty operations in accordance with the TSOP.
- Medical records and reports requirements.
- Policy and procedure updates.

The foundation of the battalion CHS plan is the battalion commander's guidance and the brigade CHS plan.

b. *Plans.* See Chapter 3, paragraph 3-1.

c. *Operation Orders.* See Chapter 3, paragraph 3-2.

d. *Rehearsals.* See Chapter 3, paragraph 3-3.

D-4. Combat Health Support During Night Operations

See Chapter 4, paragraph 4-4.

D-5. Combat Health Support Tactical Standing Operating Procedures

The battalion surgeon/medical platoon leader is responsible for the development of the CHS annex for the battalion TSOP. The purpose of a TSOP is to establish routine protocols. The TSOP should not be dependent upon METT-TC factors. If a specific decision is required each time, it should not be included in the TSOP. The battalion TSOP is based on its higher headquarters TSOP and serves as the foundation for subordinate units to develop their TSOP. The battalion CHS annex to the TSOP should be clear and concise, yet provide sufficient detail of any procedural requirements. The CHS annex to the TSOP must reflect procedural guidance that supports current mission and doctrinal requirements. The CHS annex to the battalion TSOP should be maintained and reviewed at least every 6 months and revised as required. Most importantly, the TSOP must be used during training and understood at all levels prior to deployment or it has no real value.

Section II. RECONNAISSANCE, SURVEILLANCE, AND TARGET ACQUISITION SQUADRON

D-6. Medical Platoon

The medical platoon of the RSTA squadron provides Echelon I CHS for the squadron. The platoon is organized with a headquarters section, a treatment squad and an evacuation section (see Figure D-3). For more detailed information on the functions and operations of the medical platoon, see Chapter 4.

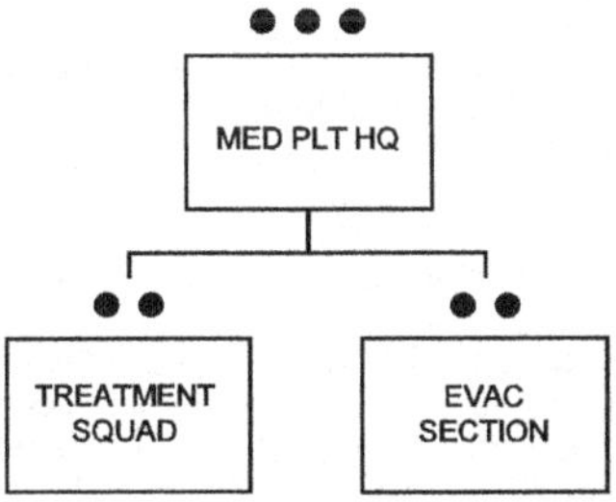

Figure D-3. Medical platoon, reconnaissance, surveillance, and target acquisition squadron.

D-7. Platoon Headquarters

a. The platoon headquarters section, under the direction of the platoon leader/surgeon that is assigned to the treatment squad, provides the C3 and logistics for the platoon. The platoon headquarters section is comprised of a field medical assistant, the platoon SGT and a medical specialist. It is normally collocated with a treatment team/squad to form the squadron aid station. The CP includes the plans and operations functions performed by the field medical assistant. The platoon has access to the HHT and the supported troop wire communication network for communications with all major elements of the squadron. Wireless communications for this section consists of a tactical FM radio mounted in the platoon headquarters vehicle. The medical platoon employs an FM radio network for CHS operations. The platoon headquarters section serves as the NCS for the platoon. Table D-1 lists the information and communications assets available to the platoon.

b. The medical platoon leader is a working physician on Treatment Team Alpha. He is the medical advisor to the squadron commander and his staff. He is also the supervising physician (field surgeon) of the medical platoon's treatment teams. This officer is responsible for all medical treatment provided by the platoon. His responsibilities include—

- Planning and directing CHS for the RSTA squadron.
- Advising the squadron commander and his staff on CHS operations and the medical threat.
- Supervising the administration, discipline, maintenance of equipment, supply functions, organizational training, and employment of assigned or attached personnel.
- Examining, diagnosing, treating, and prescribing courses of treatment for patients, to include ATM.
- Training the squadron's CLS.
- Supervising the squadron's CSC program, to include individual and leader training on the prevention of BF and other stress-related conditions.
- Planning and conducting humanitarian assistance programs when directed.
- Overseeing and coordinating the MEDEVAC of patients.

c. The field medical assistant, an MS officer, is the operations/readiness officer for the platoon. He is the principal assistant to the platoon leader for operations, administration, and logistics. The field medical assistant coordinates CHS operations within the squadron and with supporting brigade or corps medical elements. He coordinates MEDEVAC with the BSMC. The platoon SGT assists in supervising the operations of the platoon. He also serves as the evacuation section SGT.

D-8. Treatment Squad

The treatment squad has two treatment teams (Teams Alpha and Bravo) that are the basic medical treatment elements of the squadron aid station. They provide Echelon I medical care and treatment. This includes sick call, EMT, ATM, and triage for the management of mass casualty situations. Team Alpha is staffed with an operational medicine officer (primary care physician/squadron surgeon), a treatment squad leader (E-6), a health care SGT (E-5), and one health care specialist (E-3). Team Bravo is staffed with a PA, a health care SGT, and two health care specialists. The physician and PA are trained in ATM procedures, commensurate with their occupational positions or specialties. The PA assumes the duties of the squadron surgeon in his absence. The PA performs general technical health care and administrative duties. The PA works under the clinical supervision of the medical officer. He performs the following duties:

- Establishes and conducts treatment team operations when deployed to other locations away from the squadron aid station.
- Treats, within his ability, sick or injured patients. He refers those patients requiring treatment beyond his capability to the supervising physician.
- Provides EMT and ATM for wounded and DNBI patients.

- Conducts training for squadron personnel in first-aid procedures (self-aid, buddy aid, and CLS), field sanitation, evacuation of the sick and wounded, and the medical aspects of injury prevention.

- Assists in the conduct of the squadron CSC program, to include individual and leader training on the prevention of BF and other stress-related conditions.

- Trains medical personnel in EMT procedures.

D-9 Evacuation Section

The RSTA squadron evacuation section employs interim armored vehicle (IAV) ambulances and provide MEDEVAC through DS or on an area support basis. Each RECON troop will normally have one ambulance team in DS. There are three ambulance teams providing DS and one area support ambulance team that is positioned with the squadron aid station. Each ambulance team consists of an aide/evacuation NCO (E-5) and two ambulance aide/evacuation drivers (E-4 and E-3). Ambulance teams provide MEDEVAC and en route care from either the soldier's point of injury or a CCP to the squadron aid station/ treatment team or brigade medical element providing area support. In mass casualty situations, nonmedical vehicles may be used to assist in casualty evacuation as directed by the commander. Plans for the use of nonmedical vehicles to perform casualty evacuation should be included in the RSTA squadron TSOP and OPORD. The HHT and surveillance and target acquisition troop are provided Echelon I medical treatment and MEDEVAC support on an area support basis by the area support ambulance team. They will also operate dispersed throughout the RSTA squadron AO in support of RECON troops. These dispersed ambulances will evacuate to the nearest supporting Echelon I MTFs based on the OPORD and according to preplanned and coordinated area medical support responsibilities.

D-10. Squadron Combat Health Support Planning

a. The squadron surgeon, assisted by the field medical assistant and the platoon SGT, is responsible for the CHS plan for the RSTA squadron. Squadron CHS operations involve all of the factors that must be considered in the initial developmental stages of the squadron CHS plan. The CHS plan is updated to meet tactical or CHS operations requirements. The following factors should be considered:

- Information requirements (current task organization structure, medical troop strengths, projected weather and environmental factors, and maintenance status of medical equipment).

- Results of the mission analysis.

- Commander's intent.

- Planning guidance.

- Courses of actions.

- Tactical plan.

- Enemy.
- Terrain.
- Troops.
- Weather.
- Threat (including medical threat).
- Operational conditions and constraints.
- Civilian populace in the AO.
- Medical personnel status.
- Equipment status of the medical platoon.
- Supply status including Class VIII.
- Communications capability.
- Nuclear, biological, and chemical defense including OEG.
- Nuclear, biological, and chemical casualty considerations.
- Training status.
- Casualty estimates.
- Medical evacuation requirements and capabilities.
- Nonmedical support requirements from the squadron.
- Area support requirements.
- Special operations support requirement
- Medical records and reports requirements.
- Phases of operations.
- Policy and procedure updates.

b. The foundation of the squadron CHS plan is the squadron commander's guidance and the brigade CHS plan.

D-11. Squadron Operation Plans and Operation Orders

a. The brigade headquarters gives mission orders to the squadron headquarters. The medical platoon may receive additional coordinating instructions from the BSS. These coordinating instructions are normally transmitted in a force text e-mail message via the tactical LAN. As part of the mission analysis and based on the squadron commander's intent and guidance, the medical platoon develops CHS estimates for supporting squadron operations. An understanding of the squadron RECON troop time lines or battle rhythm will assist the squadron medical platoon leader and field medical assistant in developing the CHS input through the squadron S1 to the squadron OPLAN/OPORD.

b. Squadron COA development/analysis and wargaming are accomplished after mission analysis. Course of action development and wargaming result in the production of the OPORD and the CHS annex.

c. Once the RSTA squadron receives the brigade WARNO, it begins mission analysis. Based on its analysis and the full brigade order that follows, the squadron determines its tactical plan. Part of determining its plan is the placement of medical treatment elements (squadron aid station/treatment team). The brigade CHS plan will include the plan for Echelon II support as well as any tasking of support to the squadron medical platoon. The RSTA squadron medical platoon leader will assess the platoon's adequacy as part of the mission analysis. He will bring any shortfalls to the squadron commander's attention during the mission analysis briefing. The platoon leader and/or the field medical assistant then participate in COA development and wargaming to produce the squadron plan. When the commander approves the OPLAN, it becomes the OPORD. The OPLAN and OPORD are developed by the S3 section, using input from each of the staff elements, with the S1 being the staff coordinating element for CHS. The RSTA squadron CHS plan is revised or updated based on mission analysis or changes in CHS requirements. The medical platoon leader maintains current information on the following subject areas that include—

- Patient status board (for example, awaiting evacuation).
- Dirty routes/patient decontamination sites.
- Location of squadron aid station/treatment team and of BAS or other MTF providing area medical support to RSTA squadron elements (current/projected).
- Area medical support responsibilities.

D-12. Rehearsal

For successful implementation of the CHS annex of the RSTA squadron plan, the CHS plan must be coordinated and synchronized with the squadron plan so that CHS requirements are met. To achieve optimal synchronization, the CHS plan is rehearsed as an integral part of the combined arms plan at the combined arms rehearsal. See Chapter 3 for definitive information on rehearsal.

D-13. Combat Health Support for the Offense and the Defense

See Chapter 4 for definitive information on CHS for offensive and defensive operations.

D-14. Reconnaissance, Surveillance, and Target Acquisition Squadron Operations and Combat Health Support

a. The unique mission of the RSTA squadron will cause it to deploy its troops over a very large AO. These long distances will require careful planning and well-coordinated and aggressive CHS operations. This squadron will rely heavily on first aid and CLSs.

b. Many RECON and counter-RECON missions will occur at night. Combat health support for these missions must be planned in detail for limited visibility conditions. The acquisition and MEDEVAC of a RSTA squadron casualty will require a team effort on the part of the troops, the medical platoon, and the squadron staff. Often, trauma specialists will not be able to successfully execute this alone in the RSTA squadron AO. In most cases, it will require a combined arms effort.

D-15. Combat Health Support During Night Operations

The squadron surgeon and medical platoon members must anticipate that the brigade does a substantial amount of its work at night or in limited visibility. They must ensure that the platoon TSOP is available and used throughout the squadron for providing MEDEVAC and treatment at night. See Chapter 4, paragraph 4-4, for definitive information of night operations.

Section III. OVERVIEW OF MEDICAL FORCE STRUCTURE IN SUPPORT OF THE INTERIM BRIGADE

D-16. Combat Health Support for the Interim Brigade

a. Combat health support to the brigade is focused on the stabilization of wounds and injuries, and early evacuation of casualties out of the brigade's AO. Self-aid/buddy aid and the CLS is essential first aid, and early initial medical treatment provided by the trauma specialist is critical for reducing the morbidity and mortality rates. Additionally to reduce the morbidity and mortality rates, both EMT and ATM are performed at the BAS and the BSMC to ensure appropriate treatment and to stabilize the wounded or traumatized patient. After the initial first aid or medical treatment, an ambulance crew evacuates the patient to the supporting Echelon I treatment team/BAS.

b. The BSMC of the brigade support battalion is capable of providing Echelon I area medical support to BSA elements, back-up Echelon I support to forward maneuver battalions, and Echelon II CHS to all elements of the brigade. The BSMC is normally augmented with a surgical capability provided by an attached corps FST. Casualties that cannot be returned to duty by the brigade medical assets will be evacuated to a corps hospital or supporting MTF. After the first 96 hours of brigade operations, the brigade may be augmented with a corps FSMT. When deployed forward to the BSA, the BSMC commander coordinates the air ambulance team's evacuation missions.

c. The BSMC, assisted by the support operations section, provides real-time tactical information to the air ambulance crew about evacuation missions from the brigade combat team units/elements to supporting brigade MTFs. When air ambulances operate forward of the BSA, they will execute the A2C2 plan through the maneuver brigade S3. The BSB support operations section provides planning and coordination between aeromedical evacuation and the supported maneuver brigade. The brigade S3 provides the A2C2 plan that includes the air corridors, air control points, and communications checkpoints. The brigade S3 will provide updates as required. Air ambulances deployed to the BSA provide MEDEVAC from forward areas (BAS) back to the BSA. Air ambulance evacuation from the point of injury will be METT-TC dependent. Corps air ambulances may also evacuate from the BSA to supporting corps MTF. Corps aeromedical elements will operate from BSA providing around the clock immediate response evacuation aircraft. To accomplish this, elements must maintain a close tie with the A2C2 system in the brigade. The brigade A2C2 element provides an airspace plan through the brigade OPORD/OPLAN A2C2 annex. The aircrew must also be familiar with the daily airspace control order and the airspace control plan. These documents contain all ACM, to include free fire areas, no-fly fire areas, restricted operations zones, and established and standard Army aircraft flight routes. These route and ACM change on a daily basis and cannot be integrated into the brigade OPORD. The BSS will ensure all A2C2 information is provided to corps aeromedical elements. The BSS does not generate A2C2 information, but does provide A2C2 planning information to division A2C2 elements.

D-17. Brigade Surgeon's Section

The BSS is assigned to the HHC of the brigade and operates out of the brigade TOC. It is the brigade's primary planning cell for CHS and works closely with the S3 and his staff in the planning process. The BSS staff is responsible to the brigade commander for staff supervision of CHS within the brigade. The BSS is also responsible for coordinating GS and DS relationships of organic medical units and medical units/ elements whether under OPCON or attached to the brigade. The BSS also will—

- Advise the brigade commander on the health of the command.
- Monitor force health protection issues.
- Provide basic sick call support to headquarters personnel.
- Provide technical control/assistance to brigade medical personnel.
- Advise/make recommendations on risk reduction.
- Review/coordinate for either external support or augmentation.

D-18. Brigade Support Medical Company

The BSMC provides Echelon II CHS to all the brigade units operating within the brigade's AO. It also provides/ensures Echelon I CHS to all the brigade units without organic medical support. The BSMC will be organized into a company headquarters, a treatment platoon, an ambulance platoon, a PVNTMED section, and a MH section. The area support section in the treatment platoon contains an area support treatment squad (two treatment teams), which operates the brigade clearing station located in the BSA. These treatment teams will utilize HMMWVs with trailers, or an acceptable variant consistent with the interim brigade concept. The holding squad provides a holding capability of 20 cots. The primary use of the cots is to hold those patients awaiting MEDEVAC out of the brigade AO. An additional mission for the patient holding squad is to hold and provide nursing care for soldiers expected to RTD within 24 to 72 hours. The area support squad provides limited dental support, as well as limited laboratory and x-ray support to the BSMC. The treatment platoon will include two treatment teams to provide mobile Echelon I support within the brigade AO. These treatment teams will utilize IAVs (treatment variant) with trailers. The ambulance platoon will consist of six ambulance teams in IAV ambulances, and four ambulance teams in HMMWV ambulances. The teams in IAV ambulances will be deployed in support of the medical platoons in the maneuver battalions. The IAV ambulances will be used to evacuate patients from BAS to AXP (located between the BAS and the BSMC) where the patients are transferred to BSMC HMMWV ambulances and evacuated to the brigade clearing station operated by the BSMC.

D-19. Forward Surgical Team

Corps-level initial surgical support will be provided by the FST. This team is comprised of 20 personnel and has two operating tables.

D-20. Medical Force Protection

The BSMC medical force protection assets include a PVNTMED section and a MH section. Both sections have two-man teams. The PVNTMED section has an environmental science officer and a PVNTMED specialist. The PVNTMED team provides medical force protection through technical advice to the commander. They ensure the health of the brigade by performing sanitary inspections of food service, field sites, latrine, bathing, and other activities. The team is also responsible for coordination and oversight of medical surveillance. This includes early recognition of potential epidemics or biological warfare agent employment; and monitoring field water supplies, to include sample collection for potential NBC contamination. The MH team includes a behavioral science officer and a MH specialist. The mission of this team is combat operational stress control; they focus on the medical force protection aspects of treatment and prevention of combat stress. This includes the rapid reversal of dysfunctional stress reactions (BF). These medical force protection capabilities are essential to enhancing soldier survivability across the spectrum of battlefield contingencies through continual health hazard assessment, stress reduction, and minimizing disease and injury.

D-21. Medical Logistics (Class VIII/Blood)

The BSMC has limited Class VIII/ blood management capability. During deployment, lodgment, and early buildup phases, medical units operate from planned, prescribed loads and from existing pre-positioned war reserve materiel identified in applicable contingency plans. The BSMC will deploy with supplies to support a 72-hour, self-sustainment mission within the brigade (support for maneuver units is included). Resupply (Class VIII/blood requisitions) for the BSMC will be conducted via a functional module of MC4 electronic requisitions sent to the medical logistics management center, operating out of the AO. The medical logistics management center will send release orders for materiel to the appropriate MEDLOG activity. Class VIII resupply will flow via the battlefield distribution system under the direction of the brigade support battalion support operations section. Unique to the brigade is the range and complexity of MEDLOG support. The probability of corps-level CHS (surgery, hospitalization, intensive care, and extensive blood usage) within the brigade footprint will result in a significant increase in the variety and urgency of medical supplies and equipment. Priority of transportation of critical Class VIII materiel will have to be recognized and supported throughout the distribution pipeline. Blood support to the operation will follow the same procedures.

D-22. Infantry Battalion Medical Platoon

See paragraph D-1.

D-23. Reconnaissance, Surveillance, and Target Acquisition Squadron

See paragraphs D-4 through D-13.

D-24. Field Artillery Battalion

Combat health support for the FA battalion is provided by its medical section that is assigned to the battalion's headquarters service battery. The medical section has a combat medic section with a trauma specialist assigned to support each firing battery. There are three ambulance teams. Normally, one ambulance team is in DS of each firing battery and a treatment team. Echelon II care is provided by the BSMC.

D-25. Engineer Company

Combat health support for the engineer company is provided by its combat medic section. This section is under the supervision of a health care SGT and each platoon is supported by a trauma specialist. Echelon I medical treatment and MEDEVAC support will be provided on an area support basis from brigade medical assets. Echelon II care is provided by the BSMC.

D-26. Antitank Company

Combat health support for the antitank company is provided by its organic medical section that consists of a senior company medic and an individual trauma specialist for each platoon. The antitank platoons will be employed in the proximity of other brigade units that have organic medical treatment teams. Company personnel will receive additional Echelon I from brigade organic medical treatment teams. Echelon II CHS is provided by the BSMC of the brigade support battalion on an area support basis.

D-27. Units Without Organic Medical Personnel

Units without organic medical personnel will utilize self-aid/buddy aid and the CLS according to current directives. These units will normally operate in proximity to area medical support assets. Based on the concepts of employment for these units, personnel will receive Echelons I and II CHS on an area support basis from organic brigade medical assets.

APPENDIX E

FORCE PROTECTION

Section I. UNDERSTANDING FORCE PROTECTION

E-1. Protection

Protection has four components: force protection, field discipline, safety, and fratricide avoidance. Force protection, the primary component, minimizes the effects of enemy firepower (including weapons of mass destruction), maneuver, and information. Field discipline precludes losses from hostile environments. Safety reduces the inherent risk of nonbattle deaths and injuries. Fratricide avoidance minimizes the inadvertent killing or maiming of friendly troops by friendly fires.

E-2 Force Protection

Force protection consists of those actions necessary to prevent or mitigate hostile actions against DOD personnel (to include family members), resources, facilities, and critical information. It coordinates and synchronizes offensive and defensive measures to enable the joint force to perform while degrading opportunities for the enemy. It includes air, space, and missile defense; NBC defense; antiterrorism; defensive information operations; and security to operational forces and means. Force protection does not include actions to defeat the enemy or protect against accidents, weather, and disease. The increased emphasis on force protection at every echelon stems from the conventional dominance of Army forces. Often unable to challenge the Army in conventional combat, opponents seek to frustrate Army operations by resorting to unconventional means, weapons, or tactics. Force protection counters these threats. Force protection at all levels minimizes losses to hostile action. Skillful and aggressive counter-reconnaissance keeps adversaries from seeing the friendly forces while they are defining enemy actions and positions. Effective OPSEC keeps the adversary from exploiting friendly information. Proper dispersion helps reduce losses from enemy fires and terrorist action. Camouflage discipline, security operations, and field fortifications do the same. At the operational level, base protection and lodgment area security contribute to force protection. Theater missile defense forces protect installations and civilian populations from over-the-horizon strikes by conventional warheads and weapons of mass destruction. Army air and missile defense units complement the joint air component's control of the air.

E-3. Field Discipline

Field discipline, a second component of protection, guards soldiers from the physical and psychological effects of the environment. A hostile environment can sap soldier strength and morale far more quickly than enemy action. Soldiers adapt to the point that they out perform their fully acclimatized opponents, who may be native to the region. This mastery can only stem from thorough preparation and training in field craft skills. Commanders take every possible active measure and precaution to keep soldiers healthy and maintain their morale. Such actions include securing equipment and supplies from loss or damage. Commanders ensure systems are in place for adequate CHS, quick return of minor casualties, and PVNTMED. They provide effective systems for maintenance, evacuation, and rapid replacement or repair of equipment. Tactical commanders take care of their soldiers' basic health needs and prevent unnecessary exposure to debilitating conditions.

E-4. Safety

Safety is a third component of protection. Operational conditions often impose significant risks to soldiers' lives and health and make equipment operation difficult. Trained crews and operators must know the capabilities and limitations of their weapon systems. Commanders must know how to employ them. In designing operations, commanders consider the margins of human endurance. They balance the benefits and risks of sustained, high-tempo operations. In combat, fatigue extends reaction times and reduces alertness. Fatal accidents, loss of combat power, and missed tactical opportunities can follow. Command attention to safety and high levels of discipline lessen those risks, particularly as soldiers reach exhaustion. Safe operations come from enforcing standards during training. While taking calculated risks, commanders assume the obligation to embed safety in the conduct of all operations.

E-5. Fratricide Avoidance

A fourth component of protection is fratricide avoidance. *Fratricide* is the unintentional killing or wounding of friendly personnel by friendly firepower. The destructive power and range of modern weapons, coupled with the high intensity and rapid tempo of combat, increase the potential for fratricide. Tactical maneuvers, terrain, and weather conditions may increase the danger of fratricide as well. Commanders seek to lower the probability of fratricide without discouraging boldness and audacity. Situational understanding, positive weapons control, control of troop movements, and disciplined operational procedures, coupled with good leadership, can do this. Reducing fratricide increases soldiers' willingness to act boldly, confident that misdirected friendly fires will not kill them.

Section II. RISK MANAGEMENT, SECURITY MEASURES, AND TERRORISM CONSIDERATIONS

E-6. Risk Management

Risk is the chance of injury or death for individuals and damage to or loss of vehicles and equipment. Risks and/or the potential for risks are always present in every combat and training situation the medical platoon faces. Risk management must take place at all levels of the chain of command during each phase of every operation; it is an integral part of all tactical planning. The battalion commander, company commanders, staff, and all soldiers must know how to use risk management, coupled with fratricide reduction measures, to ensure that the mission is executed in the safest possible environment within mission constraints. For additional information on risk management, refer to FM 100-14.

E-7. Preventive and Protective Security Measures

Preventive and protective security measures should be taken by military units and individual soldiers to protect themselves and their ability to accomplish their mission while deployed. These measures include

OPSEC, communications security, and antiterrorism considerations. The commander develops an antiterrorism plan to institute passive defense measures. The commander must constantly evaluate security plans and measures against the terrorist threat in order to effectively identify security requirements. Terrorism can occur within the US or overseas. Commanders must remain vigilant regardless of where the unit is physically located. Unit training should include orientation to the terrorist threat and countermeasures to be taken at both the individual level and the unit level. For an in-depth discussion of CHS for combatting terrorism operations, refer to FM 8-42.

E-8. Terrorism Considerations

Terrorism considerations that commanders and staffs should consider when they address terrorism is the several relevant characteristics of terrorists and their activities. The first consideration is that anyone can be a victim. (Some terrorists still operate under cultural restraints, such as a desire to avoid harming women, but the planner cannot count on that.) Essentially, there are no innocents. Secondly, attacks which may appear to be senseless and random are not. To the perpetrators, their attacks make perfect sense. Acts, such as bombing public places of assembly and shooting into crowded restaurants, heighten public anxiety. This is the terrorists' immediate objective. Third, the terrorists need to publicize their attack. If no one knows about it, it will not produce fear. The need for publicity often drives the target selection; the greater the symbolic value of the target, the more publicity the attack brings to the terrorists and the more fear it generates. Finally, a leader planning for antiterrorism must understand that he cannot protect every possible target all of the time. He must also understand that terrorists will likely shift from more protected targets to less protected ones. This is the key to defensive measures.

a. Medical units/elements have specific protections afforded to them under the provisions of the Geneva Conventions. The CHS commander/leader must understand that these protections probably will not be recognized nor adhered to by terrorist elements. The CHS commander/leader in developing his force security plan should not consider the Geneva Conventions as a protection from attack by terrorist elements.

b. Terrorists rely on surprise and the victim's confusion at the time of the incident. Antiterrorism involves physical security, OPSEC, and the practice of personal protective measures by all personnel. Commanders and staffs must plan their response to terrorist threats and incidents. Combatting terrorism is an aspect of force protection and is the responsibility of commanders at all levels at all times. Properly planned and executed, the Army antiterrorism program will reduce the probability of surprise while discouraging attack by raising the risk to the attackers. For additional information on combatting terrorism, see Appendix O of FM 8-10-6.

APPENDIX F

TACTICAL STANDING OPERATING PROCEDURE

F-1. General

All medical platoon/sections assigned to combat arms, CS, and CSS units must establish a TSOP for conducting CHS operations supporting their units for all operational contingencies. These TSOP should be detailed and cover all aspects of unit CHS operations. This is an example TSOP that can be used by both the division and nondivisional medical platoons and sections to guide them in the development and refinement of their TSOP.

F-2. Sample Tactical Standing Operating Procedure

The sample shown is from a medical platoon assigned to an infantry battalion and it would be included in an annex of the battalion's TSOP. This TSOP supports the battalion for all operational contingencies and is written for wartime and other operations. There is not a standard format for all TSOP; however, it is recommended that the annex follow the format used by its higher headquarters.

DEPARTMENT OF THE ARMY
Headquarters and Headquarters Company
______Battalion, ______ Infantry
____________ Division
Fort ________, State/APO________, ZIP CODE_______

ANNEX O
MEDICAL PLATOON
TACTICAL STANDING OPERATING PROCEDURES

I. PURPOSE

The purpose of this TSOP is to provide guidelines, policies, and procedures. The implementation of this document will enhance the effectiveness of training and provide specific procedures for routine tasks during CHS operations. This TSOP has been prepared to standardize operations and CHS procedures for the battalion medical platoon in time of war and other operations.

A. Scope. The scope of this TSOP addresses the mission, organization, equipment, and medical platoon operations.

B. Applicability. This SOP applies to all personnel assigned to the ________ Infantry Medical Platoon.

C. Accountability. All personnel assigned to the medical platoon as a part of their initial orientation are required to become familiar with and have a working knowledge of this TSOP. Thereafter, all medical platoon personnel in leadership positions will review the TSOP every 90 days and update or recommend changes as required. Personnel not in leadership positions are required to review the TSOP a minimum of every 6 months or as necessary when conducting operations.

GEORGE I. DOE 2LT, MS Medical Platoon Leader	GREG I. ROCK SFC, USA Medical Platoon Sergeant

II. GENERAL

A. The medical platoon operates the BAS and provides Echelon I CHS for the battalion. Medical platoon personnel are under the command leadership of the battalion commander and the HHC commander. The battalion surgeon/medical platoon leader is a member of the battalion staff. He is responsible to the battalion commander for directing battalion CHS operations. The battalion surgeon/medical platoon leader is responsible to the brigade surgeon for technical oversight and supervision of all CHS activities within the battalion.

B. The medical platoon is dependent on the FSMC of the supporting FSB for Echelon II CHS. This includes medical evacuation from the BAS to the FSMC, patient holding, operational and CSC support, Class VIII resupply, PVNTMED support, medical maintenance, x-ray, laboratory, and operational dental care. The medical platoon requests augmentation/reinforcing support from the FSB's medical company.

III. ORGANIZATION AND MISSION

A. Organization. The medical platoon is organized as shown in Figures F-1.

1. Platoon headquarters.

a. The headquarters section, under the direction of the battalion surgeon, provides C3 and resupply for the platoon. The field medical assistant and the platoon SGT man the platoon headquarters. It is normally collocated with the treatment squad to form the BAS. The CP includes the plans and operations functions performed by the field medical assistant.

b. The field medical assistant, an MS Corps officer, is the operations/readiness officer for the platoon. He is the principal assistant to the battalion surgeon for operations, administration, and logistics. The field medical assistant coordinates CHS operations with the battalion S3 and S4 and coordinates patient evacuation with the FSMC. When a physician or PA are not assigned, he performs the duties of medical platoon leader.

c. The platoon SGT assists the platoon leader and supervises the operations of the platoon. He also serves as the ambulance section SGT. This NCO prepares reports; requests general supplies as well as medical supplies; advises on supply economy procedures; and maintains ASL of expendable supplies. He supervises the activities and functions of the ambulance section, to include operator maintenance of ambulances and equipment; OPSEC; and EMT.

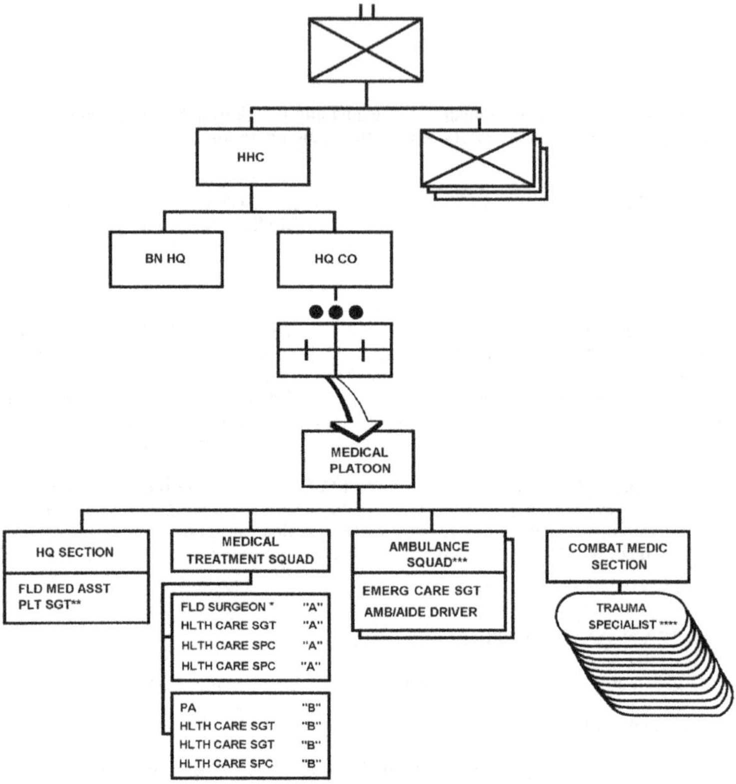

* ALSO BATTALION SURGEON
** ALSO SERVES AS AMBULANCE SECTION SGT
*** 4 AMBULANCE TEAMS
**** AIRBORNE BATTALIONS HAVE 12 TRAUMA SPECIALISTS ASSIGNED TO THE COMBAT MEDIC SECTION AND AIR ASSAULT BATTALIONS HAVE 10 TRAUMA SPECIALISTS ASSIGNED TO THE COMBAT MEDIC SECTION
"A" ALFA TREATMENT TEAM
"B" BRAVO TREATMENT TEAM

Figure F-1. Medical platoon, headquarters and headquarters company light infantry battalion.

2. Treatment squad. This squad is staffed with an operational medical officer (primary care physician/battalion surgeon), a PA, two health care SGT, and four health care specialists. The squad's physician, PA, and health care SGT are all trained in ATM procedures, commensurate with their occupational positions/specialties.

3. Combat medic section. A total of 12 trauma specialists are assigned to the combat medic section. To foster good interpersonal relations and morale of combat troops, every effort should be made to attach the same trauma specialists to the same unit they habitually support each time the unit deploys. However, during lulls in combat operations, they should return to the medical platoon for consultation and proficiency training.

4. Ambulance squads. Medical platoon ambulance squads have two emergency care SGTs, two emergency care specialists, and four ambulance drivers/aides assigned. Four ambulances are assigned to the medical platoon's ambulance squads.

B. Mission. The mission of the medical platoon is to provide Echelon I CHS for the battalion. This includes medical treatment, medical evacuation, and clearing the battlefield. It includes PVNTMED activities to counter either disease or combat and operational stress disorders. It includes EMT and ATM to save lives, limbs, or sight and to stabilize the wounded or injured patient for further evacuation. This also includes maintaining accurate field health records as well as the permanent health record in a garrison setting.

IV. MEDICAL EVACUATION OF SICK AND WOUNDED

A. Purpose. This policy and these procedures established for evacuation of sick and wounded are consistent with doctrine published in the following manuals.

- FM 8-10-1, The Medical Company—Tactics, Techniques, and Procedures.
- FM 8-10-6, Medical Evacuation in a Theater of Operations—Tactics, Techniques, and Procedures.
- FM 8-10-26, Employment of the Medical Company (Air Ambulance).
- FM 8-55, Planning for Health Service Support.

B. General.

1. Evacuation is based on the principle that rear higher echelon medical units are responsible for evacuating patients from supported units. Lower echelon supported and supporting units must ensure evacuation support plans are complete and current by close, direct coordination. See FM 8-10-6 for an in-depth discussion of medical evacuation; for additional information, refer to FMs 4-02.4, 8-10, 8-10-1, 8-10-26, 8-42, 8-55, 63-20, and 63-21.

2. Patients are evacuated no further to the rear than that necessary to obtain the medical care that will return them to duty. Patients are evacuated by the means of transportation that most clearly meets the treatment demands of their wounds, injury, or illness.

3. Allied military personnel, treated or held in a division MTF within reasonable proximity of their own national facility, are classified and processed as follows:

a. Allied military personnel requiring further treatment, but in stable condition for immediate transfer, are returned to their own national medical facility, as coordinated through liaison with the corps or division surgeon.

b. Allied military personnel requiring further stabilization are retained in US medical channels until they can be safely transferred to their own national MTFs. Arrangements for reception of the patient by the gaining MTF are completed prior to the evacuation.

c. The preferred method for evacuation of neuropsychiatric and BF casualties who can be managed without medications or physical restraints is a nonambulance ground vehicle. If physical restraints and/or medications are required during transportation, ground ambulance is preferred. An air ambulance should only be used if no other means of evacuation is available. Physical restraints are used only during transport and medications are given only if needed for reasons of safety. Those neuropsychiatric and BF patients with life- or limb-threatening conditions are evacuated by the most expedient means available. See FMs 8-10-6 and 8-51 for additional information on medical evacuation of psychiatric patient and BF soldiers.

C. Responsibilities for Medical Evacuation.

1. The medical platoon leader—

a. Develops an evacuation plan which will best support the operations being conducted.

b. Prepares/obtains the necessary maps of the AO and overlays from the S3.

c. Does reconnaissance of MEDEVAC routes, either map or on the ground.

d. Provides ambulance teams with strip maps; briefs the plan; and rehearses the MEDEVAC plan with the ambulance section when time permits.

e. Identifies and coordinates with the battalion TOC on the location of primary and alternate helicopter landing sites that are established.

f. Coordinates with the FSMC on medical evacuation and location of the AXP.

g. Oversees medical evacuation operations to ensure expedient evacuation from the battlefield.

2. The medical platoon SGT—

a. Ensures that evacuation wheeled assets are maintained and preventive maintenance checks and services (PMCS) are accomplished in accordance with 10/20 standards.

NOTE

The 10/20 indicate the technical manuals authorized for every item of equipment at unit level.

b. Ensures that ambulances are properly stocked with requisite Class VIII supplies and equipment.

c. Ensures ambulance communication and computer equipment is functional and that all personnel are familiar with SOI.

3. The evacuation squad leader—

a. Participates in route reconnaissance with the medical platoon leader or SGT.

b. Identifies locations of company aid post and CCP.

c. Ensures ambulances are maintained in a state of readiness to perform their evacuation mission.

d. Maintains vehicles in accordance with 10/20 standards and ensures PMCS are performed.

e. Ensures computers and communications equipment are functioning.

f. Keeps the squad updated on road conditions and the threat levels.

4. The senior (line) company trauma specialist—

a. Establishes a company aid post/CCP.

b. Requests medical evacuation of wounded or injured soldiers via the company's communications assets.

c. Coordinates with the BAS as required for additional medical evacuation assets.

d. Maintains aid bags in accordance with published packing lists.

e. Maintains prescribed Class VIII supplies on hand.

f. Request Class VIII resupplies for CLS, as required.

5. Companies (A, B, C, and HHC)—

a. Collect and secure casualty's sensitive items.

b. Remain custodian (1SG) of casualties' personal effects that do not accompany patient through the evacuation process.

c. Ensure the casualty's protective mask accompanies the soldier when he is evacuated from the battalion.

d. Ensure unit personnel are trained on manual carries and on the use of litter carries to transport casualties. Units must be prepared to assist medical personnel as required.

e. Complete the DA Form 1156, Casualty Feeder Report, and DA Form 1155, Witness Statement on Individual, when required. (Medical personnel do not complete these forms.)

6. The battalion S4—

a. Is involved in developing the mass casualty plan and the use of nonstandard vehicles to evacuated casualties.

b. Is responsible for coordinating with graves registration personnel for the transport of deceased personnel (BSA).

c. Provides pouches for human remains (body bags).

d. Provides transportation assets for deceased personnel.

D. Procedures for Medical Evacuation.

1. One ground ambulance, M997, is normally placed in direct support of each company.

2. One ground ambulance, M997, remains with the aid station for supporting the HHC and area support.

3. The two cargo trucks, M998s, used to haul BAS equipment and supplies, may be used for medical evacuation with the augmentation of medical personnel to ride with the patients and provide en route care.

4. Medical platoon ambulances are responsible for medical evacuation of patients from forward areas back to the BAS.

5. The FSMC positions one ambulance, M997, forward with the BAS that is responsible for evacuating patients back to an AXP or to the FSMC in the BSA.

6. One technique used to reduce evacuation times is to pre-position ambulances with the supported company.

7. Another technique is to have on call ambulance support. For this technique, units submit medical evacuation request via radio or landline. The BAS will dispatch an ambulance, M997, to the company aid post/CCP to retrieve the casualty. In some cases, the senior company trauma specialist and the 1SG, using the 1SG's vehicle, may evacuate the casualty from forward areas back to the CCP.

a. Procedures for information collection and MEDEVAC request preparation are performed in accordance with the 9-line MEDEVAC request identified in FM 8-10-6. Air ambulances may not fly forward of the BAS because of threat antiaircraft capabilities. This mission will be performed by a ground ambulance and the following information is collected, to include—

(1) What is the location of the pickup site?

(2) Radio frequency and call sign.

(3) Number of patients by precedence (URGENT, URGENT-SURGICAL, PRIORITY, ROUTINE, and CONVENIENCE).

(4) Special equipment required.

(5) Number of patients by type (litter or ambulatory).

(6) Number and types of wounds (peacetime).

(7) Security of pickup point (wartime).

(8) Method of marking pick-up site if required.

(9) Patient nationality and status.

(10) Nuclear, biological and/or chemical contamination (wartime) or terrain description (peacetime).

b. Upon receipt of the mission, the ambulance team on call will—

- Obtain road clearance from the S3 for the MSR/evacuation route to the requesting unit.

- Perform a radio check with the BAS and/or field trains upon departure from BAS, upon arrival at pickup site, and upon departure from pickup site. Convey problems encountered en route and difficulties regarding en route patient care will also be forwarded.

c. Actions at the unit CCP include—

- Securing (if not already done) site with driver and other available personnel.

- Parking vehicle facing the perimeter (doors to the rear).

- Surveying patients and performing triage (this is accomplished by the senior company trauma specialist on location). Assigning evacuation categories (for example, URGENT, URGENT SURGICAL, PRIORITY, ROUTINE, and TACTICAL CONVENIENCE). Separating NBC-contaminated casualties. Practicing appropriate protective measures in an NBC environment.

- Organizing litter teams (senior company trauma specialist).

- Checking each DD Form 1380 (FMC) to ensure it is completed according to AR 40-66 (senior company trauma specialist).

- Personnel at the CCP assisting with loading patients as directed by the ambulance driver in accordance with Chapter 10, FM 8-10-6.

d. Actions by the ambulance team upon departure of the CCP include—

- Notifying the BAS and/or the field trains as to the time of departure and ETA at the BAS. Stating number and type of patients, to include ambulatory and litter. Discussing potential en route medical complications with physician or PA.

- Moving along prescribed movement routes, taking advantage of cover and concealment. Verifying evacuation routes as tactical situation dictates.

e. Actions of the BAS upon arrival of patients.

- Patients are off loaded and provided EMT/ATM as required; the physician or PA will determine if the patient requires further evacuation. Patient information is entered into the treatment and disposition log and also provided to the battalion S1. If the patient is an URGENT or URGENT-SURGICAL precedence, an air ambulance is requested.

- If a ground ambulance is to be used, the ambulance team from the FSMC is notified of their mission. Normally to minimize the time that the supporting FSMC ambulance is away from the BAS, the FSMC will establish an AXP somewhere between the supported unit and the BSA. This AXP is where the FSMC ambulance team that is supporting the BAS evacuates the patient(s). At the AXP, the patient is transferred to another ground ambulance then evacuated the remaining distance to the FSMC. The medical platoon leader will inform the battalion staff and the patient's company commander on the disposition of his unit member.

- From the BSA, corps ground or air ambulances are responsible for evacuating to the corps combat support hospital.

- Aeromedical evacuations are conducted in accordance with FMs 8-10-6 and 8-10-26.

E. Control of Property and Equipment.

1. Soldiers evacuated from their unit to the BAS, as a minimum, have their protective mask and clothing.

2. Any property and equipment arriving with casualties other than the protective mask and clothing or individual weapon for ambulatory patients will be collected and turned in to the battalion S4 for return to the parent unit. The battalion S4 coordinates the return of property and equipment to the casualty's unit.

3. Under combat conditions, protective masks are kept in the immediate proximity of each patient throughout their period of evacuation and stay at any MTF. In other operations, the protective mask policy for patients will be based on the NBC threat and the policy established by higher headquarters.

F. Rules for Employment of Ambulance and Ambulance Personnel.

1. The use of MEDEVAC vehicles will be restricted to—

a. Transportation of sick or injured personnel.

b. Transportation of medical personnel.

c. Transportation of Class VIII supplies/equipment and blood.

2. Medical personnel assigned to the ambulances will—

a. Adhere to the tactical commanders' standards for uniform and camouflage and other requirements identified in the supported unit's TSOP.

b. Participate in the medical training being conducted at the supported medical element.

c. Assist with patient treatment as required.

NOTE

Caution should be exercised by officers-in-charge or noncommissioned officers-in-charge to ensure the ambulance crew has adequate rest in order that they can safely perform their evacuation duties.

d. Perform PMCS on their vehicles.

e. Ensure their vehicle is restocked with required Class VIII supplies and equipment, full of fuel, and ready for the next evacuation mission.

3. Medical personnel assigned to the ambulances that are positioned with the supported medical element will not be required to—

- Perform duties as kitchen police.
- Perform EPW or perimeter guards.
- Perform driver duties for other than their assigned vehicle.

G. Use of Aeromedical Evacuation.

1. Aeromedical evacuation is the preferred method of evacuation and will be used when—

a. Life, limb, or eyesight is in jeopardy (URGENT or URGENT-SURGICAL category).

b. Speed, distance, and time are factors in assuring prompt and adequate treatment.

c. There is a critical need for resupply of Class VIII supplies or whole blood/blood products.

d. There is a critical need for movement of medical personnel and equipment.

2. Helicopter landing zones are established when and where tactical situations permit. A helicopter landing zone should be marked with a letter "H" or a letter "Y," using identification panels or other appropriate marking material. See FMs 8-10-6 and 57-38 for a complete description and guidelines for establishing a helicopter landing zone.

3. Precedence for air ambulance evacuation is provided in FMs 8-10-6 and 8-10-26.

V. DECEASED PERSONNEL

A. Principles Governing Medical Disposition of Deceased Personnel.

1. Deceased personnel are segregated from other casualties.

2. The deceased, as determined by the senior medical authority, are not evacuated with other casualties, nor are they routinely evacuated on medical vehicles. This is especially true if the threat of biological or chemical contamination will render the vehicle unfit for subsequent medical evacuation missions. A FMC should be initiated and attached to the remains, if possible.

3. Medical evacuation resources should not be used to transport deceased personnel.

4. All deceased personnel should have an FMC that is signed by a medical officer prior to their departure from a graves registration collection point operating in forward areas.

B. Coordinate with the Battalion S4 for Transport of Deceased Personnel.

1. The battalion S4 coordinates with mortuary affairs collection point personnel for the evacuation of deceased personnel to the graves registration site (normally near or in the BSA).

2. The battalion S4 obtains and issues pouches for human remains (body bags) to the BAS.

3. Battalion aid station personnel coordinate transport of deceased personnel with the battalion S4.

4. For definitive information on mortuary affairs operations, see FM 10-64.

VI. ENEMY PRISONERS OF WAR

A. All EPW will be provided medical care according to the articles of the GWS, dated 12 August 1949.

B. Enemy prisoner of war patients will be segregated from allied and US personnel.

C. Enemy prisoner of war patients will be reported through normal medical reporting procedures.

D. Enemy medical personnel are considered retained personnel and shall receive the benefits provided by the Geneva Conventions. Retained enemy medical personnel will be used to the maximum extent possible to care and treat EPW patients.

E. Enemy prisoner of war patients will be evacuated through medical channels.

F. Enemy prisoner of war patients will be under armed guard at all times. Guards are the responsibility of the echelon commander. Medical personnel will not be used as guards for EPW according to the Geneva Conventions.

G. Enemy prisoner of war patients will be searched prior to each move in the MEDEVAC system.

H. Information on EPW patients will be coordinated with the prisoner of war information center to maintain accountability of captives in medical channels. See FM 19-4 for additional information on EPW.

VII. CLASS VIII SUPPLY

A. Class VIII Supply Procedures in the Field. In the field, the medical platoon maintains a 2-day (48-hour) stockage of Class VIII supplies within its MES. The following MES are authorized for the medical platoon treatment section:

- Chemical Agent Patient Decontamination, NSN 6545-01-176-4612 (1).
- Chemical Agent Patient Treatment, NSN 6545-01-141-9469 (2).
- Sick Call Field, NSN 6545-01-228-1886 *(2).
- Trauma Field, NSN 6545-01-228-1667 *(2).

* Indicates the numbers of MES authorized for each treatment team.

B. Authorized Stockage and Controls. Medical supply items authorized for use by the medical platoon are normally those items that are identified as part of the MES. Items that are not in the MES must be approved for stockage by the division surgeon. This includes both expendable items and pharmaceuticals. For perishable and dated items that are found in the MES, a DA Form 4998-R, Quality Control and Surveillance Record for TOE Medical Assemblage, is initiated for all expendable and durable items in the medical assemblage. For a medical item that has a shelf life and demands are expected, a DA Form 4996-R will be prepared and maintained in accordance with AR 40-61. These items are referred to as potency and dated items. Controlled pharmaceutical such as R and Q items (see NOTE below) are stored in the most secure container. A record of controlled medical items will be kept on DA Form 3862 (Controlled Substances Stock Record).

NOTE

Controlled substances are drugs so designated by the Drug Enforcement Administration. A list of these drugs and changes are published in the Federal Register and in the SB 8-75 series. Standard controlled substances are identified by Note R and Q in the notes column of the Federal Supply Catalog, DOD Section, Medical Materiel and by controlled inventory items codes R and Q in the Army Master Data File.

QUALITY CONTROL AND SURVEILLANCE RECORD FOR TOE MEDICAL ASSEMBLAGES

For use of this form, see AR 40-61; the proponent agency is OTSG

NO	LOCATION	MANUFACTURER	CONTRACT NO (If available)	LOT/BATCH NUMBER	EXP/MFR DATE (If available)	QTY ON HAND	DATE LAST INSPECTION	DATE NEXT INSPECTION

NSN	DESCRIPTION	UNIT OF ISSUE	NOTES	INSPECTION FREQUENCY	SHELF LIFE/ ESTIMATED SHELF LIFE

SAMPLE

DA FORM 4998-R, APR 1994 REPLACES DA FORM 4998-R, AUG 1981, WHICH IS OBSOLETE USAPA V1.00

Figure F-2. Quality Control and Surveillance Record for TOE Medical Assemblage, DA Form 4998-R.

NSN	DESCRIPTION	INSP FREQ	DATE LAST INSP	DATE NEXT INSP

QUALITY CONTROL CARD

For use of this form, see AR 40-61; the proponent agency is OTSG

NO	MANUFACTURER	LOT NUMBER	EXP DATE	DATE MFG	SHELF LIFE	DATE RECD
1						
2						
3						
4						
5						
6						
7						
8						
9						
10						

NSN	DESCRIPTION	INSP FREQ	DATE LAST INSP	DATE NEXT INSP

SAMPLE

DA FORM 4996-R, APR 1994 REPLACES DA FORM 4996-R, AUG 1981, WHICH IS OBSOLETE USAPA V1.00

Figure F-3. Quality Control Card, DA Form 4996-R.

C. Class VIII Requisitioning Procedures in the Field.

1. Routine Requisitions. Routine requisitions for Class VIII supplies are submitted to the FSMC using a DA Form 2765-1, Request for Issue or Turn-in, for each item in accordance with DA Pam 710-2-1. The FSMC may fill these requests from its MESs or forward them to the DMSO. The DMSO will fill the requests and send items forward to the supporting FSMC via LOGPAC. The FSMC coordinates the delivery of Class VIII supplies to the requesting unit through the FSB support operations section. Requisitions may also be sent via radio or by using a sneaker net method. Prior to deployment, individuals should know the Class VIII resupply requirement and procedures.

2. Emergency Requisitions. Emergency requisitions for Class VIII supplies are sent to the FSMC via radio or telephone. Any item not filled is immediately forwarded to the DMSO. Delivery of emergency requests will be by the most expedient mode of transportation, based on METT-TC factors. Normally, a requisition with a 03 or higher priority will require the approval of the commander.

D. Class VIII Supply Procedures in Garrison.

1. Medical Treatment Facility Operations. Operations of the BAS/MTF at the home station are normally under the control of the medical center (MEDCEN)/MEDDAC commander. For MTF (garrison) operations, Class VIII items may be requisitioned directly from the MEDDAC. Under other situations, the BAS/MTF orders from the DMSO or supporting MEDLOG battalion. The installation medical supply activity is operated by either the MEDDAC or the MEDLOG battalion.

2. Class VIII Requirements. The MEDCEN/MEDDAC commander establishes requirements for MTF operations in garrison. A list of Class VIII items required and authorized at the MTF is identified by hospital SOP or regulation. Medications are identified in the hospital formulary. When operating MTF in a garrison environment, medical platoons are subject to the same standards as any of the MEDCEN/MEDDAC treatment facilities.

3. Requisitioning Information. Requisitioning information includes—

a. Department of Defense activity address code. Example: YNAMTF

b. Serial numbers. The serial number is issued by the document register maintainer and this number is preceded by the Julian date on each request for issue or turn-in. Serial number example: W35MW2.

c. Account processing code. This code is issued by the supporting MEDLOG element and identifies the unit. An example is AWHA.

d. Signature cards will be on file with the MEDCEN/MEDDAC and a copy maintained by the platoon SGT or his designated person for supply actions.

e. The platoon SGT or his designated person for supply activities maintains the document register.

4. Supply Matrix. A supply matrix will be posted in the supply 3-ring binder with the following information:

a. Quantity maintained.

b. Quantity needed to be ordered.

c. Date medication was ordered.

d. Expiration dates of pharmaceuticals (checked each month).

5. Expired Medications. If from the MTF, they are turned in to the MEDCEN/MEDDAC. If from the TOE MES, they are turned in to the DMSO.

6. Multidose Vials. When opened, they are dated and initialed by the user.

7. Medication Refrigerator. Will be cleaned and defrosted monthly.

8. Medication Locker and Safe. Medication lockers are unlocked during sick call but are locked if the care provider leaves the room. The safe with code R and Q items is secured except for removal of medication to fill a prescription. Keys will be maintained by the physician, PA, or platoon SGT. Only the physician, PA, or platoon SGT will have the combination to the safe.

9. Inventories. A 100 percent inventory of all assets belonging to the medical platoon is performed on an annual basis according to DA Pam 710-2-1 and DA Pam 710-2-2.

10. Point of Contact. The telephone number of the MEDCEN/MEDDAC logistics division's is ________________.

VIII. MEDICAL MAINTENANCE

A. Garrison Medical Treatment Facility. Operator maintenance is performed on medical equipment according to MEDCEN/MEDDAC SOP and in accordance with manufacture's instructions. Medical equipment is turned in to the MEDCEN/MEDDAC medical maintenance shop or a work order is phoned in according to MEDCEN/MEDDAC SOP.

B. Medical Maintenance in the Field. Medical maintenance will consist of operator-/user-level maintenance. See FM 4-02.1 for additional information. Medical platoon personnel will exercise their responsibilities by performing operator PMCS. This includes maintaining equipment by performing routine services like cleaning, dusting, washing, and checking for frayed cables, loose hardware, and cracked or rotting seals. In addition, medical platoon personnel will—

- Perform equipment operational testing.

- Replace operator-level spares and repair parts that will not require extensive disassembly of the end item, critical adjustment after the replacement, or extensive use of tools.

IX. MANAGEMENT OF MASS CASUALTIES

A. Mass casualty situations occur when the number of casualties exceed the available medical capability to rapidly treat and evacuate them. The battalion surgeon working with the S4 and the S3 advises the commander on integrating all available resources into an effective mass casualty plan.

B. All companies must have procedures in place to respond effectively to mass casualty situations. The potential of disasters in war and other operations requires that the medical element be prepared to support mass casualty situations. They must be able to receive, triage, treat, and evacuate large numbers of casualties within a short period of time. Contingency plans for supporting mass casualty operations must be developed by all units in coordination with their battalion surgeon and battalion S3. Unit mass casualty plans, as a minimum, will address the following subject areas:

1. Planning and training requirements.

2. Medical duty positions.

3. Nonmedical personnel positions and duties, including litter teams, perimeter guards crowd control, and information personnel.

4. Location of treatment areas, to include triage, immediate care, minimal care, delayed care, and expectant care areas.

5. Support requirements beyond the unit's capability.

6. Medical evacuation.

7. Use of nonmedical transportation assets.

8. Nuclear, biological, and chemical casualties.

9. Return to duty procedures.

10. Medical records and reports.

11. Locating deceased personnel away from and out of sight of all patients.

C. Upon notification of large number of casualties, the medical platoon will request augmentation support from the FSMC. The division or brigade surgeons should be informed of any mass casualty situation by the most expedient means available. As a minimum, information provided should include location, anticipated number of casualties, and additional support required. For additional information, see FMs 8-10, 8-10-1, 8-55.

D. The DSS directs and coordinates CHS requirements for the requesting unit. Supporting corps and division medical units in the chain of evacuation are alerted of the situation.

E. Mass casualty management in garrison is accomplished according to the MEDCEN/MEDDAC SOP.

X. PREVENTIVE MEDICINE

A. The battalion surgeon oversees all PVNTMED activities in the battalion.

B. The division PVNTMED cell located in the MSB (or for Force XXI, the DSB) is responsible for supervising the division's PVNTMED program as described in AR 40-5. This cell ensures PVNTMED measures are implemented to protect division personnel against food-, water-, and arthropodborne diseases, as well as environmental injuries (for example, heat and cold injuries). This cell provides advice and consultation in the areas of environmental sanitation, epidemiology, sanitary engineering, and pest management.

C. Preventive medicine support for the battalion is requested through the FSMC and formal tasking is accomplished through the division or brigade headquarters.

D. All companies in the battalion will establish unit field sanitation teams. Preventive medicine personnel will assist in the training of these teams in the aspects of environmental sanitation and the limited control of animal reservoirs and disease vectors.

E. Company commanders will—

1. Use trained field sanitation team members to assist in preserving the health of the unit and reducing the incidence of DNBI which will hinder mission accomplishment (FM 21-10).

2. Ensure the field sanitation team members take to the field all required field sanitation equipment and supplies to perform their duty (AR 40-5).

3. Enforce food and water safety standards. Unless otherwise stated, water will be treated to at least 5 parts per million chloride residual and will be obtained from approved sources only. Safe handling, storage, and preparation of food will be according to AR 30-21, AR 40-5, FMs 10-23, and 21-10.

4. Plan for the construction of hygienic devices, such as handwashing devices in the unit area. Company commanders will also enforce personal hygiene measures to reduce the threat of disease.

5. Motivate subordinates to execute individual preventive measures (such as carrying an extra pair of dry socks; and/or eating or drinking from approved sources only).

6. Enforce the use of the DOD repellent systems requiring the use of repellents on skin (DEET [75 percent N,N-diethyl-M-tolumide]) and clothing (permethrin).

7. Develop and enforce the unit sleep plan that provides soldiers with a minimum of 4 hours of uninterrupted sleep in a 24-hour period. If sleep is interrupted, then 5 hours should be given. During continuous operations when uninterrupted sleep is not possible, blocks of sleep which add up to 6 hours in a 24-hour period are adequate for most people. Remember, 4 hours each 24-hour period is far from ideal. Do not go with only 4 hours sleep each 24 hours for more than 2 weeks before paying back sleep debt. Recovery time should be approximately 8 to 10 hours sleep each 24 hours over a 5- to 7-day period. If at all possible, give the individuals (such as ambulance drivers) whose key tasks are vulnerable to sleep loss 6 hours of sleep a day.

8. Plan for measures to prevent environmental injuries (such as heat or cold) (see FM 21-10).

9. Obtain and disseminate information on the medical threat so soldiers can reduce their risk of DNBI.

10. Request PVNTMED consultation/assistance. Requests can be submitted to the DSS, the BSS, or any medical company/element in the division.

XI. DENTAL SERVICES

A. Dental treatment facilities are located in each FSMC and in the MSMC (or DSMC for Force XXI). Each medical company establishes dental sick call hours; supported units are notified of the sick call hours.

B. Operational dental care will be provided in the field. This dental care consists of emergency and essential care.

C. Battalion personnel report to the BAS for dental sick call. Dental patients are sent back to the FSMC for dental services.

D. The goal by battalion personnel for dental patients is to relieve pain, to get them back for dental treatment, to get them seen, and to return them to duty.

XII. MENTAL HEALTH/COMBAT STRESS CONTROL

A. Under the guidance of the division psychiatrist and the FSMC MH section, the battalion surgeon is responsible for the battalion's CSC programs.

1. Conduct or request classes on stress control.

2. Identify and eliminate unnecessary stress on unit personnel.

3. Ensure personnel and leaders are familiar with stress-reducing techniques.

4. Ensure all leaders understand the problems associated with sleep deprivation and the consequences of not following the unit sleep plan.

B. The battalion surgeon requests after-action stress debriefings for battalion personnel involved in operations where numerous injuries or loss of life has occurred. He coordinates for critical-events debriefing following a catastrophic incident in the unit. See FMs 8-51 and 22-51 for definitive guidance on stress control.

XIII. EYEWEAR

A. Battalion personnel requiring corrective eyewear will deploy with two pairs of eyewear plus one set of inserts for their protective mask.

B. Personnel authorized to wear contact lenses will deploy with two pair of standard eyewear.

C. The following procedures are followed for repair or replacement of eyewear including—

1. Maintaining the latest prescription on the soldier's field medical record.

2. Sending eyewear that is broken or in need of repair to the division optometry section located in the MSMC (or DSMC for Force XXI).

3. Requesting the replacement of lost eyewear.

XIV. GENEVA CONVENTIONS COMPLIANCE

A. Medical Facilities.

1. All US medical facilities and units, except veterinary, will display the distinctive flag of the Geneva Conventions. This flag consists of a red cross on a white background. It is displayed over the unit or facility and in other places as necessary to adequately identify the unit or facility. Nondisplay of the flag can be ordered by a brigade or higher level commander.

2. Camouflage of the medical facility (medical units, medical vehicle, and medical aircraft on the ground) is authorized when a lack of camouflage might compromise the tactical operation.

3. The order to camouflage can be given by a brigade-level or higher commander.

NOTE

As used in this context, camouflage means to cover up or remove the Geneva Conventions emblem. The black cross on an olive background is not a recognized emblem of the Geneva Conventions.

B. Defense of Medical Units.

1. Medical personnel may carry small arms for personal defense of themselves and defense of their patients. Self-defense of medical personnel or defense by medical personnel of their patients is always permitted. This does not mean that they may resist capture or otherwise fire on the advancing enemy. It means that, if civilian or enemy military personnel are attacking and ignoring the marked medical status of medical personnel, medical transportation, or the medical unit, the medical personnel may provide self-protection. If an enemy military force merely seeks to assume control of a military medical facility or a vehicle for the purpose of inspection and without firing on it, the facility or vehicle may not resist.

2. An overall defense plan may not require medical units to take offensive or defensive actions against enemy troop at any time. If a medical force is part of a defensive area containing nonmedical units, medical personnel may not be responsible for manning part of the overall perimeter. If located in isolation, the medical unit may provide its own local and internal security if other support is not available. However, all soldiers (medical and nonmedical) providing this internal and local security must comply with the requirements in subparagraph 1 above.

XV. MEDICAL REPORTING

Medical reporting requirements for the battalion should be listed in the battalion TSOP. Medical reporting requirements are established by the division and brigade surgeons. Listed below are documents used for reporting or from which information is obtained for reporting purposes.

A. Field Medical Card. A FMC will be initiated for each new patient and for cases required to be carded-for-record only. This will be accomplished according to AR 40-66 and FM 8-10-6. Field Medical Cards will be conspicuously attached to the patient's clothing.

B. Daily Disposition Log. The Daily Disposition Log is maintained by all Echelon I and Echelon II MTFs assigned or attached to the division. Information from this log is extracted, when required, and provided to the S1 or the supported unit requesting the information. The log is also the primary source document for information needed in the preparation of the Patient Summary Report and the Patient Evacuation and Mortality Report.

C. Medical Reports Format. Medical reporting will be accomplished using the FBCB2, FAX, or voice, transmitted via radio/MSE. A manual backup system will be developed. Formats for medical reports are required to maintain consistency and continuity in reporting procedures for information submitted to the BSS and the DSS. Data contained in these reports are required to support the DSS's capability projections and to assist the BSS, HSSO, and FSMC commander in coordinating and planning CHS operations. Data is also extracted for consolidated reporting to higher headquarters. The guidelines presented below should be followed exactly.

1. Each line of information is divided into a number of fields. Each field has a minimum number of alphanumeric characters as indicated in the sample format provided.

2. Each field is separated by a single slash(/).

3. The end of each set of fields is indicated by a double Slash (//).

4. If information from a prior report has not changed, "NC" will be entered in that field (/ NC//).

5. Reports are formatted according to special instructions and reports format.

D. Medical Situational Report, Battalion Aid Station. The MEDSITREP, BAS is a daily patient summary report. This report is used to inform the commander of the battalion's patient, Class VIII, and medical equipment status. This report is submitted daily, covering the events in a 24-hour time period based on time lines provided by the higher headquarters. It is submitted to the supporting medical company. The battalion surgeon (platoon leader) or platoon SGT is responsible for this report. This report could be dispatched via courier, FAX, and/or teletype.

E. Medical Situational Report, Medical Companies. The MEDSITREP, medical companies is a daily patient summary report. This report is submitted daily to the DSS according to time lines provided by higher headquarters. The following information will be included in line six of this report:

1. Status of all assigned and attached ambulances, to include—

a. Total number of ambulances.

b. Number of ambulances that are operational.

c. Number of ambulances that are nonoperational.

2. Status of personnel; identify shortages by AOC or MOS.

3. Treatment of any EPW will be entered in this section.

4. Identify all patients seen during the reporting period with a number and provide the following information in the order provided below:

a. Nationality.

b. Name.

c. Rank.

d. Service number.

e. Unit.

f. Date of birth.

g. Diagnosis.

h. Disposition.

i. Date of disposition.

j. Gaining unit.

5. A hard copy of each aid station's MEDSITREP must accompany the submitting medical company's report.

F. Medical Situational Report, Medical Operations. The MEDSITREP, medical operations is a consolidated patient summary report. This report is consolidated by the DSS and pertains to the previous 24 hours. It is submitted from the DSS daily to the division based on time lines established by the division surgeon.

G. Patient Evacuation and Mortality Report. All Echelons I and II MTFs assigned or attached to the division prepare the Patient Evacuation and Mortality Report. The purpose of this report is to provide a status of patients seen by division MTFs. This is a weekly report compiled as of 2400 each Sunday and distributed each Monday to supported units.

H. Patient Summary Report. The Patient Summary Report provides the status of patients seen by division medical companies and includes their subordinate elements (dental, optometry, MH, or attached units). This report is a weekly report compiled as of 2400 each Sunday. It is prepared by all Echelons I and II MTFs operating in the division AO. It is submitted each Monday to the DSS.

I. Blood Report. The Blood Report is a required report for requesting blood support. Echelon II MTFs will request only Group O Positive and Group O Negative liquid red blood cells.

J. Team Movement Report. The Team Movement Report is used to track the status and location of teams (PVNTMED, CSC, veterinary, ambulance, and treatment teams).

XVI. AUTHORIZED ABBREVIATIONS AND REPORT CODES FOR MEDICAL REPORTS

AUTHORIZED ABBREVIATIONS

ARMS AND SERVICES:

AVN	AVIATION	EMPLOYED MEANS:
ABN	AIRBORNE	MORTARS
AD	AIR DEFENSE	POISON
AMINF	ARMED INFANTRY	ROCKETS
AMPB	AMPHIBIOUS	SABOTAGE
ARMD	ARMORED	TUBEARTY
ARTY	ARTILLERY	MISSILES

AT	ANTITANK
ATAGM	ANTITANK GUIDED MISSILE

COMMAND LEVEL:

AG	ADJUTANT GENERAL
ARMY	ARMY
BDE	BRIGADE
BN	BATTALION
CO	COMPANY
CORPS	CORPS
DIV	DIVISION
GP	GROUP
HQ	HEADQUARTERS
PLT	PLATOON
RGT	REGIMENT

NATIONALITY:

BE	BELGIAN
CA	CANADIAN
GE	GERMAN
NL	NETHERLANDS/HOLLAND
UK	BRITISH
US	AMERICAN

CAUSE OF CASUALTY TO BE USED FOR MASS CASUALTY REPORTING.

ACCIDENT:	AIRCRASH
ACCIDENT:	MARITIME
ACCIDENT:	MOTOR VEHICLE
ACCIDENT:	RAILWAY
ACCIDENT:	FIRE
ACCIDENT:	INDUSTRIAL
ACCIDENT:	POISON
ACCIDENT:	NATURAL DISASTERS
ACCIDENT:	OTHER CAUSES
BATTLE:	CONVENTIONAL
BATTLE:	NUCLEAR
BATTLE:	BIOLOGICAL
BATTLE:	CHEMICAL

GLOSSARY

ABBREVIATIONS AND ACRONYMS

A2C2 Army airspace command and control

ABCA American, British, Canadian, and Australian

ACE analysis and control element

ACM airspace control measures

ACR armored cavalry regiment(s)

ACS armored cavalry squadron

ADA air defense artillery

ADC assistant division commander(s)

ADC-M assistant division commander(s) for maneuver

ADC-OT assistant division commander(s) for operations and training

ADC-S assistant division commander(s) for support

admin administration

AIPD Army Institute for Professional Development

alt alternate

AM amplitude modulated

amb ambulance

AMEDD Army Medical Department

AMEDDC&S Army Medical Department Center and School

AO area of operations

AOC area of concentration

AOE Army of Excellence

AR Army regulation

ARPERCEN Army Reserve Personnel Center

AR–PERSCOM Army Reserve–Personnel Command

ASL authorized stockage level

AT antitank

atk attack

ATM advanced trauma management

attn attention

AXP ambulance exchange point

BAS battalion aid station

BCIS Battlefield Combat Identification System

bde brigade

BF battle fatigue

BICC battlefield information coordination center

BMNT beginning of morning nautical twilight

BMO battalion maintenance officer

bn battalion

BSA brigade support area

BSMC brigade support medical company

BSS brigade surgeon's section

BSU blood supply unit

btry battery

BW biological warfare

C2 command and control

C3 command, control, and communications

C4 command, control, communications, and computers

CA civil affairs

CAC command aviation company

CASEVAC casualty evacuation

cbt combat

CCP casualty collection point

CDR commander

CEMR civilian employee medical record(s)

CHL combat health logistics

CHS combat health support

CI counterintelligence

CIP combat identification panel

cl class

clr clearing

CLS combat lifesaver

cmd command

CNR combat net radio

co company

COA course(s) of action

comm communications

CONUS continental United States

COSCOM corps support command

COTS commercial off-the-shelf

CP command post

CRP common relevant picture

CS combat support

CSC combat stress control

CSM command sergeant major

CSS combat service support

CSSCS Combat Service Support Control System

CTCP combat train command post

CW chemical warfare

DA Department of the Army

DASB division aviation support battalion

DD Department of Defense

DEET N, N-diethyl M tolumide

DISCOM division support command

disp disposition

div division

DIVARTY division artillery

DMC distribution management center

DMLSS Defense Medical Logistics Standard Support

DMOC division medical operations center

DMSO division medical supply office

DNBI disease and nonbattle injury(ies)

DOA dead on arrival

DOD Department of Defense

DODD Department of Defense Directive

DODI Department of Defense Instruction

DOW died of wounds

DS direct support

DSA division support area

DSB division support battalion

DSMC division support medical company

DSS division surgeon's section

DTF dental treatment facility

DVE driver's vision enhancer

EAC echelons above corps

EAD echelons above division

EENT end of evening nautical twilight

emerg emergency

EMT emergency medical treatment

EPLRS Enhanced Position Location and Reporting System

EPW enemy prisoner(s) of war

ETA estimated time of arrival

evac evacuation

EW electronic warfare

FA field artillery

FARE forward air refueling equipment

FAS forward aid station

FAX facsimile

FBCB2 Force XXI Battle Command Brigade and Below System

1SG first sergeant

fld field

FLE forward logistics element

FLOT forward line of own troops

flt flight

FM field manual; frequency modulated

FMC US Field Medical Card

FRAGO fragmentary order

FSB forward support battalion

FSC forward support company

FSE fire support element

FSMC forward support medical company

FSMT forward support MEDEVAC team

FSO fire support officer

FST forward surgical team

G1 Assistant Chief of Staff (Personnel)

G2 Assistant Chief of Staff (Intelligence)

G3 Assistant Chief of Staff (Operations and Plans)

G4 Assistant Chief of Staff (Logistics)

G5 Assistant Chief of Staff (Civil Affairs)

G6 Assistant Chief of Staff (Signal)

GC Geneva Convention Relative to the Protection of Civilian Persons in Time of War

GCSS-A Global Combat Support System–Army

GOTS government off-the-shelf

GP general purpose

GPS Global Positioning System

GPW Geneva Contention Relative to the Treatment of Prisoners of War

GS general support

GSO general supply office

GWS Geneva Convention for the Amelioration of the Conditions of the Wounded and Sick in Armed Forces in the Field

GWS (Sea) Geneva Convention for the Amelioration of the Condition of Wounded, Sick, and Shipwrecked Members of Armed Forces at Sea

HHB headquarters and headquarters battery

HHC headquarters and headquarters company

HHT headquarters and headquarters troop

hlth health

HMMWV high-mobility multipurpose wheeled vehicle

HN host nation

hosp hospital

hq headquarters

HREC health record

HSC headquarters and support company

HSMO health service materiel officer

HSSO health service support office(r)

IAV interim armored vehicle

IBCT interim brigade combat team

IEW intelligence and electronic warfare

IHFR improved high frequency radio

IPB intelligence preparation of the battlefield

IR infrared

JTF joint task force

L light

LAN local area network

LD line of departure

ldr leader

LOA limit of advance

LOC line of communications

LOGPAC logistics package(s)

LOGSITREP Logistical Situational Report

LRSD long-range surveillance detachment

LTC lieutenant colonel

LZ landing zone

maint maintenance

MAJ major

MAS main aid station

MC Medical Corps

MC4 Medical Communications for Combat Casualty Care

MCO movement control office(r)

MDMP military decision-making process

MDT Medical Detachment–Telemedicine

MED medical

MEDCOM medical command

MEDDAC medical department activity

MEDEVAC medical evacuation

MEDLOG medical logistics

MEDSITREP Medical Situational Report

MES medical equipment set(s)

METT-TC mission, enemy, terrain, troops, time available, and civilian considerations

MH mental health

MI military intelligence

MLRS multiple launch rocket system

MMC materiel management center

MMMB medical materiel management branch

MOPP mission-oriented protective posture

MOS military occupational specialty

MP military police

MRI Medical Reengineering Initiative

MRO medical regulating office(r)

MS Medical Service Corps

MSB main support battalion

MSE mobile subscriber equipment

msg message

MSMC main support medical company

MSR main supply route

MTF medical treatment facility(ies)

MTOE modification table(s) of organization and equipment

NATO North Atlantic Treaty Organization

NBC nuclear, biological, and chemical

NCO noncommissioned officer

NCS net control station

NSN national stock number

NVG night vision goggles

O&I operation and intelligence

OAKOC obstacles; avenues of approach; key terrain; observation and fields of fire; and cover and concealment

OBJ objective

OEG operational exposure guide

off officer

OPCON operational control

OPLAN operation plan

OPORD operation order

opr operator

ops operations

OPSEC operations security

OR operating room

PA physician assistant

pam pamphlet

PIC personal information carrier

PL phase line

plt platoon

PMCS preventive maintenance checks and services

pnt patient

POL petroleum, oils and lubricants

PSYOP psychological operations

PVNTMED preventive medicine

PZ pickup zone

QRF quick reactionary force

QSTAG Quadripartite Standardization Agreement

R&S reconnaissance and surveillance

RAS regimental aviation squadron

RECON reconnaissance

regt regiment

rpts reports

RSTA reconnaissance, surveillance, and target acquisition

RT receiver/transmitter

RTD return to duty

S1 Adjutant (US Army)

S2 Intelligence Officer (US Army)

S3 Operations and Training Officer (US Army)

S4 Logistics Officer (US Army)

S6 Communications Staff Officer (US Army)

S&T supply and transportation

sct scout

SEAD suppression of enemy air defense

sec section

2LT second lieutenant

SF standard form

SFC sergeant first class

SGT sergeant

SINCGARS single-channel ground and airborne radio system

sm small

SOI signal operating instructions

SOP standing operating procedure(s)

SP start point

SPC specialist

spt support

SQD squad/squadron

sta station

STANAG Standardization Agreement

TACP tactical air control party

TAMMIS Theater Army Medical Management Information System

TC track commander

TCF tactical combat force

TCON teleconsultation

TDA table(s) of distribution and allowances

TF task force(s)

TMEM telementoring

TMIP Theater Medical Information Program

TOC tactical operations center

TOE table(s) of organization and equipment

trmt treatment

TSOP tactical standing operating procedure(s)

TTP tactics, techniques, and procedures

UMT unit ministry team

US United States

USA United States Army

USAF United States Air Force

USN United States Navy

veh vehicle

WAN wide area network

WARNO warning order

whl wheeled

WIA wounded in action

WIN Warfighter Information Network

XO executive officer

REFERENCES

SOURCES USED

These are the sources quoted or paraphrased in this publication.

NATO STANAGs

These agreements are available on request using DD Form 1425 from Standardization Document Order Desk, 700 Robin Avenue, Building 4, Section D, Philadelphia, Pennsylvania 19111-5094.

2128. *Medical and Dental Supply Procedures. Edition 4.* 21 November 1991.
2454. *Regulations and Procedures for Road Movement and Identification of Movement Control and Traffic Control Personnel and Agencies. Edition 1.* 6 July 1998.
2931. *Orders for the Camouflage of the Red Cross and Red Crescent on Land in Tactical Operations. Edition 2.* 19 January 1998. (Latest Amendment, 3 April 1998.)

ABCA QSTAGs

This agreement is available on request using DD Form 1425 from the Standardization Document Order Desk, 700 Robins Avenue, Building 4, Section D, Philadelphia, Pennsylvania 19111-5094.

435. *Medical Materiel Management During Patient Evacuation. Amendment 1.* 20 June 1984.
436. *Minimum Labelling Requirements for Medical Materiel. Amendment 1.* 8 October 1980.

Joint and Multiservice Publications

Joint Pub 3-07.3. *Joint Tactics, Techniques, and Procedures for Peace Operations.* 12 February 1999.
Joint Pub 4-02. *Doctrine for Health Service Support in Joint Operations.* 26 April 1995.
AR 600-8-14. *Identification Cards for Members of the Uniformed Services, Their Family Members, and Other Eligible Personnel.* AFI 36-3026(I); BUPERS I 1750.10A, Change 2; MCO P5512.11B Change 2; CG M5512.1; Commissioned Corps Personnel Manual 29.2, Instructions 1 and 2; NOAA Corps Regulations, Chapter 1, Part 4. 1 March 1998. (Change 1, 10 August 1999.)
FM 3-3. *Chemical and Biological Contamination Avoidance.* FMFM 11-17. 16 November 1992. (Change 1, 29 September 1994.)
FM 101-5-1. *Operational Terms and Graphics.* MCRP 5-2A. 30 September 1997.
TM 8-227-12. *Armed Services Blood Program Joint Blood Program Handbook.* NAVMED P-6530; AFH 44-152. 21 January 1998.

Department of Defense Directive (DODD)

6490.2. *Joint Medical Surveillance.* 30 August 1997.

Department of Defense Instructions (DODI)

1400.32. *DOD Civilian Work Force Contingency and Emergency Planning Guidelines and Procedures*. 24 April 1995.
3020.37. *Continuation of Essential DOD Contractor Services During Crises*. 6 November 1990. (Administrative Reissuance Incorporating Change 1, January 26, 1996.)
6490.3. *Implementation and Application of Joint Medical Surveillance for Deployments*. 7 August 1997.

Army Regulations (AR)

40-3. *Medical, Dental, and Veterinary Care*. 30 July 1999.
40-5. *Preventive Medicine*. 15 October 1990.
40-61. *Medical Logistics Policies and Procedures*. 25 January 1995.
40-66. *Medical Record Administration and Health Care Documentation*. 3 May 1999.
40-68. *Quality Assurance Administration*. 20 December 1989.
40-501. *Standards of Medical Fitness*. 30 August 1995. (Reprinted with basic including Change 1, 27 February 1998.)
310-25. *Dictionary of United States Army Terms (Short Title: AD)*. 15 October 1983. (Reprinted with basic including Change 1, 21 May 1986.)
350-41. *Training in Units*. 19 March 1993.
351-3. *Professional Education and Training Programs of the Army Medical Department*. 8 February 1988.
750-1. *Army Materiel Maintenance Policy and Retail Maintenance Operations*. 1 August 1994. (Change 1, 1 July 1996.)

Field Manuals (FM)

3-19-1. *Military Police Operations*. 22 March 2001.
4-02.1. *Combat Health Logistics*. Scheduled to be printed FY 2001.
4-02.17 (8-10-17). *Preventive Medicine Services*. 28 August 2000.
4-02.19 (8-10-19). *Dental Service Support in a Theater of Operations*. 1 March 2001.
4-02.21. *Division and Brigade Surgeons' Handbook (Digitized)—Tactics, Techniques, and Procedures*. 15 November 2000.
4-02.24 (8-10-24). *Area Support Medical Battalion—Tactics, Techniques, and Procedures*. 28 August 2000.
8-10-3 (4-02.3). *Division Medical Operations Center—Tactics, Techniques, and Procedures*. 12 November 1996.
8-10-5 (4-02.5). *Brigade and Division Surgeons' Handbook—Tactics, Techniques, and Procedures*. 10 June 1991.
8-10-8 (4-02.8). *Medical Intelligence in a Theater of Operations*. 7 July 1989.
8-55 (4-02.55). *Planning for Health Service Support*. 9 September 1994.
10-64. *Mortuary Affairs Operations*. 16 February 1999.
19-4. *Military Police Battlefield Circulation Control, Area Security, and Enemy Prisoner of War Operations*. 7 May 1993.
21-10-1. *Unit Field Sanitation Team*. 11 October 1989.

24-1. *Signal Support in the AirLand Battle*. 15 October 1990.
27-10. *The Law of Land Warfare*. 18 July 1956. (Reprinted with basic including Change 1, 15 July 1976.)
41-10. *Civil Affairs Operations*. 14 February 2000.
57-38. *Pathfinder Operations*. 9 April 1993.
63-2. *Division Support Command, Armored, Infantry, and Mechanized Infantry Divisions*. 20 May 1991.
63-2-1. *Division Support Command, Light Infantry, Airborne, and Air Assault Divisions*. 16 November 1992. (Reprinted with basic including Change 1, 20 September 1994.)
63-2-2. *Division Support Command (Digitized)*. Scheduled to be printed FY 2001.
63-21. *Main Support Battalion*. 7 August 1990.
63-23-2. *Division Aviation Support Battalion (Digitized)*. Scheduled to be printed FY 2001.
100-5. *Operations*. 14 June 1993.
100-7. *Decisive Force: The Army in Theater Operations*. 31 May 1995.
100-10. *Combat Service Support*. 3 October 1995.
100-14. *Risk Management*. 23 April 1998.

DOCUMENTS NEEDED

These documents must be available to the intended users of this publication.

Joint and Multiservice Publications

FM 4-02.283 (Final Draft). *Treatment of Nuclear and Radiological Casualties*. NTRP 4-02.21; AFMAN 44-161(I); MCRP 4-11.1B. Scheduled to be printed FY 2001.
FM 8-284 (4-02.284). *Treatment of Biological Warfare Agent Casualties*. NAVMED P-5042; AFMAN (I) 44-156; MCRP 4-11.1C. 17 July 2000.
FM 8-285 (4-02.285). *Treatment of Chemical Agent Casualties and Conventional Military Chemical Injuries*. NAVMED P-5041; AFJMAN 44-149; FMFM 11-11. 22 December 1995.
*FM 21-10 (4-25.10). *Field Hygiene and Sanitation*. MCRP 4-11.1D. 21 June 2000.

Field Manuals (FMs)

7-20. *The Infantry Battalion*. 6 April 1992. (Change 1, 29 December 2000.)
7-30. *The Infantry Brigade*. 3 October 1995. (Change 1, 31 October 2000.)
*8-10 (4-02). *Health Service Support in a Theater of Operations*. 1 March 1991.
*8-10-1 (4-02.6). *The Medical Company—Tactics, Techniques, and Procedures*. 29 December 1994.
*8-10-6 (4-02.2). *Medical Evacuation in a Theater of Operations—Tactics, Techniques, and Procedures*. 14 April 2000.
*8-42 (4-02.42). *Combat Health Support in Stability Operations and Support Operations*. 27 October 1997.

*This source was also used to develop this publication.

*8-51 (4-02.51). *Combat Stress Control in a Theater of Operations—Tactics, Techniques, and Procedures*. 29 September 1994. (Change 1, 30 January 1998.)
17-95. *Cavalry Operations*. 24 December 1996.
22-51 (4-02.22). *Leaders' Manual for Combat Stress Control.* 29 September 1994.
*63-20. *Forward Support Battalion*. 26 February 1990.
63-20-1. *Forward Support Battalion (Digitized)*. Scheduled to be printed FY 2001.
71-3. *The Armored and Mechanized Infantry Brigade*. 8 January 1996.

Department of the Army Forms (DA Form)

145. *Army Correspondence Course Enrollment Application*. October 2000.
1155. *Witness Statement on Individual.* 1 June 1966.
1156. *Casualty Feeder Report.* 1 June 1966.
2765-1. *Request for Issue or Turn-In.* April 1976.
3444. *Terminal Digit File for Treatment Record (Orange)*. May 1991.
3862. *Controlled Substances Stock Record.* 1 June 1972.
4996-R. *Quality Control Card (LRA)*. August 1981.
4998-R. *Quality Control and Surveillance Record for TOE Medical Assemblage (LRA)*. August 1981.
8005. *Outpatient Medical Record (OMR) Orange.* November 1991.
8007-R. *Individual Medical History (LRA).* November 1996.

Department of Defense Forms (DD Form)

1380. *US Field Medical Card.* December 1991.
1425. *Specifications and Standards Requisition.* March 1986.
1934. *Geneva Conventions Identity Card for Medical and Religious Personnel Who Serve in or Accompany the Armed Forces*. July 1974.
2697. *Report of Medical Assessment*. February 1993.
2766. *Adult Preventive and Chronic Care Flowsheet*. March 1998.
2766C. *Adult Preventive and Chronic Care Flowsheet (Continuation Sheet)*. March 1998.
2795. *Pre-Deployment Health Assessment Questionnaire*. May 1999.
2796. *Post-Deployment Health Assessment Questionnaire*. May 1999.

Standard Forms (SF)

558. *Medical Record—Emergency Care and Treatment.* September 1996.
600. *Health Record—Chronological Record of Medical Care.* June 1997.
603. *Health Record—Dental.* October 1975.

READINGS RECOMMENDED

These readings contain relevant supplemental information.

Joint and Multiservice Publications

AR 190-8. *Enemy Prisoners of War, Retained Personnel, Civilian Internees, and Other Detainees.* OPNAVINST 3461.6; AFJI 31-304; MCO 3461.1. 1 October 1997.

FM 3-4. *NBC Protection.* FMFM 11-9. 29 May 1992. (Reprinted with basic including Changes 1—2, 21 February 1996.)

FM 3-5. *NBC Decontamination.* MCWP 3-37.3. 28 July 2000.

FM 3-100. *Chemical Operations Principles and Fundamentals.* MCWP 3-3.7.1. 8 May 1996.

FM 8-9 (4-02.11). *NATO Handbook on the Medical Aspect of NBC Defensive Operations AMEDP-6(B), Part I—Nuclear, Part II—Biological, Part III—Chemical.* NAVMED P-5059; AFJMAN 44-151V1V2V3. 1 February 1996.

FM 90-13. *River Crossing Operations.* MCWP 3-17.1. 26 January 1998.

Army Regulations (AR)

30-21. *The Army Field Feeding System.* 24 September 1990.

71-32. *Force Development and Documentation—Consolidated Policies.* 3 March 1997.

700-138. *Army Logistics Readiness and Sustainability.* 16 September 1997.

Department of the Army Pamphlets (DA Pam)

27-1. *Treaties Governing Land Warfare.* 7 December 1956.

350-59. *Army Correspondence Course Program Catalogue.* 1 October 2000.

710-2-1. *Using Unit Supply System (Manual Procedures).* 31 December 1997.

710-2-2. *Supply Support Activity Supply System (Manual Procedures).* 30 September 1998.

Field Manuals (FM)

1-111. *Aviation Brigades.* 27 October 1997.

3-19.1. *Military Police Operations.* 22 March 2001.

3-50. *Smoke Operations.* 4 December 1990. (Change 1, 11 September 1996.)

5-71-100. *Division Engineer Combat Operations,* 22 April 1993.

5-424. *Theater of Operations Electrical Systems.* 25 June 1997.

6-20-2. *Tactics, Techniques, and Procedures for Corps Artillery, Division Artillery, and Field Artillery Brigade Headquarters.* 7 January 1993.

7-92. *The Infantry Reconnaissance Platoon and Squad (Airborne, Air Assault, Light Infantry).* 23 December 1992.

8-10-7 (4-02.7). *Health Service Support in a Nuclear, Biological, and Chemical Environment.* 22 April 1993. (Reprinted with basic including Change 1, 26 November 1996.)

8-10-9 (4-02.9). *Combat Health Logistics in a Theater of Operations—Tactics, Techniques, and Procedures.* 3 October 1995.

8-10-18 (4-20.18). *Veterinary Services—Tactics, Techniques, and Procedures.* 22 August 1997.

8-10-25 (4-02.25). *Employment of Forward Surgical Team—Tactics, Techniques, and Procedures.* 30 September 1997.
8-10-26 (4-02.26). *Employment of the Medical Company (Air Ambulance).* 16 February 1999.
10-23. *Basic Doctrine for Army Field Feeding and Class I Operations Management.* 18 April 1996.
11-43. *The Signal Leader's Guide.* 12 June 1995.
11-50. *Combat Communications Within the Division (Heavy and Light).* 4 April 1991.
11-55. *Mobile Subscriber Equipment (MSE) Operations.* 22 June 1999.
12-6. *Personnel Doctrine.* 9 September 1994.
16-1. *Religious Support.* 26 May 1995.
17-98. *Scout Platoon.* 10 April 1999.
19-40. *Enemy Prisoners of War, Civilian Internees and Detained Persons.* 27 February 1976.
20-3. *Camouflage Concealment, and Decoys.* 30 August 1999.
24-24. *Signal Data References: Signal Equipment.* 29 December 1994.
34-3. *Intelligence Analysis.* 15 March 1990.
34-54. *Technical Intelligence.* 30 January 1998.
44-100. *U.S. Army Air and Missile Defense Operations.* 15 June 2000.
63-21-1. *Division Support Battalion (Digitized).* Scheduled to be printed FY 2001.
63-23-2. *Division Aviation Support Battalion (Digitized).* Scheduled to be printed FY 2001.
63-3. *Corps Support Command.* 30 September 1993.
71-100 (3-71). *Division Operations.* 28 August 1996.
100-103. *Army Airspace Command and Control in a Combat Zone.* 7 October 1987.
101-5. *Staff Organization and Operations.* 31 May 1997.

Supply Bulletins (SB)

8-75-Series. These bulletins provide Army Medical Department supply information.

Technical Bulletin (TB)

38-750-2. *Maintenance Management Procedures for Medical Equipment.* 12 April 1987. (Reprinted with basic including Changes 1—3, 1 November 1989.)

INDEX

References are to paragraph numbers except where specified otherwise.

Made in the USA
Monee, IL
08 November 2026

68028160R00246